Marine Permian of England

THE GEOLOGICAL CONSERVATION REVIEW SERIES

The comparatively small land area of Great Britain contains an unrivalled sequence of rocks, mineral and fossil deposits, and a variety of landforms that span much of the earth's long history. Well-documented ancient volcanic episodes, famous fossil sites and sedimentary rock sections used internationally as comparative standards, have given these islands an importance out of all proportion to their size. The long sequences of strata and their organic and inorganic contents have been studied by generations of leading geologists, thus giving Britain a unique status in the development of the science. Many of the divisions of geological time used throughout the world are named after British sites or areas, for instance, the Cambrian, Ordovician and Devonian systems, the Ludlow Series and the Kimmeridgian and Portlandian stages.

The Geological Conservation Review (GCR) was initiated by the Nature Conservancy Council in 1977 to assess, document and ultimately publish accounts of the most important parts of this rich heritage. Since 1991, the task of publication has been assumed by the Joint Nature Conservation Committee on behalf of the three country agencies, English Nature, Scottish Natural Heritage and the Countryside Council for Wales. The GCR series of volumes will review the current state of knowledge of the key earth-science sites in Great Britain and provide a firm basis on which site conservation can be founded in years to come. Each GCR volume will describe and assess networks of sites of national or international importance in the context of a portion of the geological column, or a geological, palaeontological or mineralogical topic.

Within each individual volume, every GCR locality is described in detail in a self-contained account, consisting of highlights (a précis of the special interest of the site), an introduction (with a concise history of previous work), a description, an interpretation (assessing the fundamentals of the site's scientific interest and importance), and a conclusion (written in simpler terms for the non-specialist). Each site report is a justification of a particular scientific interest at a locality, of its importance in a British or international setting and ultimately of its worthiness for conservation.

The aim of the Geological Conservation Review series is to provide a public record of the features of interest in sites being considered for notification as Sites of Special Scientific Interest (SSSIs). It is written to the highest scientific standards but in such a way that the assessment and conservation value of the site is clear. It is a public statement of the value given to our geological and geomorphological heritage by the earth-science community which has participated in its production, and it will be used by the Joint Nature Conservation Committee, English Nature, the Countryside Council for Wales and Scottish Natural Heritage in carrying out their conservation functions. The three country agencies are also active in helping to establish sites of local and regional importance. Regionally Important Geological/Geomorphological Sites (RIGS) augment the SSSI coverage, with local groups identifying and conserving sites which have educational, historical, research or aesthetic value.

All the sites in this volume have been proposed for notification as SSSIs; the final decision to notify, or renotify, lies with the governing Councils of the appropriate country conservation agency.

Information about the GCR publication programme may be obtained from:

Earth Science Branch,
Joint Nature Conservation Committee,
Monkstone House,
City Road,
Peterborough PE1 1JY.

Titles in the series

1. **Geological Conservation Review**
   An introduction

2. **Quaternary of Wales**
   S. Campbell and D.Q. Bowen

3. **Caledonian Structures in Britain**
   South of the Midland Valley
   Edited by J.E. Treagus

4. **British Tertiary Volcanic Province**
   C.H. Emeleus and M.C. Gyopari

5. **Igneous Rocks of South-West England**
   P.A. Floyd, C.S. Exley and M.T. Styles

6. **Quaternary of Scotland**
   Edited by J.E. Gordon and D.G. Sutherland

7. **Quaternary of the Thames**
   D.R. Bridgland

8. **Marine Permian of England**
   D.B. Smith

9. **Palaeozoic Palaeobotany of Great Britain**
   C.J. Cleal and B.A. Thomas

10. **Fossil Reptiles of Great Britain**
    M. Benton and P.S. Spencer

Marine Permian of England

D.B. Smith

Honorary Senior Research Fellow in Geology,
University of Durham, UK

GCR Editor: L.P. Thomas

JOINT NATURE CONSERVATION COMMITTEE

SPRINGER-SCIENCE+BUSINESS MEDIA, B.V.

First edition 1995

Originally published by Chapman & Hall in 2002

Typeset by Columns Design & Production Services Ltd, Reading

ISBN 978-94-010-4527-8 ISBN 978-94-011-1204-8 (eBook)
DOI 10.1007/978-94-011-1204-8

A catalogue record for this book is available from the British Library

Library of Congress Catalog Card Number: 94-70931

♾ Printed on permanent acid-free text paper, manufactured in accordance with ANSI/NISO Z39.48-1992 and ANSI/NISO Z39.48-1984 (Permanence of Paper).

Contents

Contents

Acknowledgements

The author is pleased to acknowledge the help of Dr A.H. Cooper, Dr M.R. Lee, Dr N.T.J. Hollingworth, Mr J. Pattison and Mr T.H. Pettigrew, each of whom read part or all of the early versions of the text and made many helpful comments and suggestions. Thanks go to Dr W.A. Wimbledon who did some early editing work and to Dr L.P. Thomas who edited the completed volume and drafted the 'conclusions' sections. Thelma Smith cheerfully undertook the task of typing the first draft and most of the subsequent early amendments and additions.

Sincere thanks are also due to the GCR Publication Production Team: Dr D. O'Halloran, Project Manager; Neil Ellis, Publications Manager; Valerie Wyld, GCR Sub-editor and Nicholas D.W. Davey, Scientific Officer (Editorial Assistant). Their efficiency and good humour was a constant source of comfort and encouragement. Diagrams were drafted by Lovell Johns Ltd.

Access to the countryside

This volume is not intended for use as a field guide. The description or mention of any site should not be taken as an indication that access to a site is open or that a right of way exists. Most sites described are in private ownership, and their inclusion herein is solely for the purpose of justifying their conservation. Their description or appearance on a map in this work should in no way be construed as an invitation to visit. Prior consent for visits should always be obtained from the landowner and/or occupier.

Information on conservation matters, including site ownership, relating to Sites of Special Scientific Interest (SSSIs) or National Nature Reserves (NNRs) in particular counties or districts may be obtained from the relevant country conservation agency headquarters listed below:

English Nature,
Northminster House,
Peterborough PE1 1UA.

Scottish Natural Heritage,
12 Hope Terrace,
Edinburgh EH9 2AS.

Countryside Council for Wales,
Plas Penrhos,
Ffordd Penrhos,
Bangor,
Gwynedd LL57 2LQ.

Preface

This book is concerned almost wholly with a diverse suite of carbonate rocks that were formed near the margins of shallow tropical seas during the last 5–7 million years of the Permian period (300–251 Ma). These unique rocks, collectively known as the Magnesian Limestone, have been studied for more than 160 years and the names of some of the early workers - Geinitz, Murchison, Phillips, Sedgwick, Sorby - would grace any geological hall of fame. Despite this formidable assault, and the efforts of a host of later workers, the Magnesian Limestone still retains many of its secrets.

Permian marine rocks crop out on both sides of the Pennines, but those of the Zechstein Sea to the east are by far the thicker and more varied, and in these lie all but one of the sites selected for special protection. Detailed accounts of the rocks in 26 such sites form about half of this book and the normal and special features of these sites are compared, contrasted and placed in their mutual context in the remainder of the book. The sites were selected according to a range of criteria, including uniqueness, representativeness, historical importance and suitability for teaching purposes and research; most are inland quarries but a few are in the unrivalled coastal cliffs of classical County Durham where the main difficulty lies in deciding what not to select. Some sites, especially the coastal cliffs at Blackhalls Rocks, Seaham and between South Shields and Sunderland are worthy aspirants to World Heritage status.

The rocks at the sites selected for protection, in conjunction with those at other exposures and with information from boreholes, reveal much of the dynamic history of the late Permian seas in northern England. They suggest initial creation of the seas by catastrophic flooding of sub-sea-level inland drainage basins (themselves perhaps the product of differential subsidence accompanying post-Variscan crustal cooling and attenuation) and a subsequent complex history of basin filling against a background of ?glacially-triggered sea-level oscillation. Evidence of at least four major sea-level changes is fundamental to the widespread recognition of four main cyclic rock sequences in each basin, the first two of which together filled much of the original basin whereas the others were formed mainly in space created by continuing episodic subsidence. In north-east England, especially, the late Permian rocks of the first and second cycles display clear evidence of formation in a wide variety of nearshore tropical environments including sea-marginal subaqueous slopes, shelves, lagoons and reefs. They have, in addition, been altered both chemically and physically during deep burial and re-emergence, the most spectacular effects being the creation of a bewildering and unique range of calcite concretions that are famous world-wide. Finally, almost all the carbonate rocks at almost all the listed sites bear evidence of the former presence of calcium sulphate crystals and patches, and many of the coastal cliffs vividly demonstrate the disruptive dislocation caused by the dissolution of formerly interbedded thick anhydrite (and probably some halite) deposits.

Chapter 1

The Permian marine rocks of England

GEOLOGICAL SETTING

This volume is concerned with rocks laid down in and near epicontinental tropical seas that covered low-lying tracts of northern Europe perhaps for 5-7 million years during the late part of the Permian period; Britain then lay deep within the Laurasian supercontinent, with a climate that was both hot and dry.

During the early Permian and early late Permian, northern Europe was part of one of the world's great deserts, with a topographic and climatic range matching much of that of the present Sahara; widespread barren uplands created during the Variscan earth movements were gradually ground down as early Permian desert erosion led to extensive peneplanation of vast areas of Carboniferous rocks (especially Westphalian Coal Measures), and post-orogenic and extensional subsidence resulted in the formation of extensive, sub-sea-level, inland drainage basins in areas now occupied by the Irish and North seas and some adjoining land areas (Smith, 1970a, b, 1979; Glennie and Buller, 1983; Glennie, 1984; Smith and Taylor, 1992). It was these basins, surrounded by persistent and inhospitable deserts, that were flooded to form the Bakevellia and Zechstein seas (Figure 1.1).

The inferred mode of creation of the two seas gave rise immediately to classic barred basins, with sills that remained sensitively close to world sea levels. Both seas were particularly prone to relative sea-level changes and the thick sedimentary sequences formed in them display an extraordinary variation in their facies and extent. The sea-level changes stemmed from the complex interplay of several main mechanisms, including glacio-eustatic oscillations, isostatic effects, deepening and shallowing caused by periodic reversals in the relative rates of sedimention and subsidence, and, when the sills became emergent during oceanic low-stands, rapid evaporative downdraw. Cyclicity, caused at least partly by these relative sea-level changes, is on a range of scales and is expressed both by incomplete marginal sequences resulting from repeated transgressions and regressions and in a wide variety of carbonates and evaporites. Some variation of lithology and biota also stemmed from repeated and protracted phases of basinal anoxicity, a natural outcome of restricted circulation in a barred basin; benthic communities at these times were confined to marginal and shoal areas above an oscillating pycnocline.

The marine and associated rocks formed in the late Permian seas of northern Britain have been studied intermittently for more than 150 years, but the recent search for hydrocarbons in the Irish and North seas accelerated the rate of learning and has revealed much new information on the stratigraphy, distribution, depositional environments and diagenesis of the various lithostratigraphical units. The marine and associated Permian strata of the Bakevellia Sea (part of the East Irish Sea) Basin, for example, have been shown to include major deposits of rock-salt, and their thicker counterparts in the English Zechstein Basin have been shown to comprise a complex series of progradational carbonate aand sulphate prisms around the margins, and basin-filling halite in the interior (e.g. Taylor and Colter, 1975; Colter and Reed, 1980; Taylor, 1984).

The age of Zechstein and Bakevellia Sea strata in world chronostratigraphic terms is uncertain because these seas lay in the Boreal Realm and their faunas evolved almost independently of those in the Tethyan Realm upon which late Permian stages are now mainly based; a late Permian age is generally agreed, but it is becoming clear from mainly palynological evidence that the initial marine transgressions were probably appreciably later than the early Permian-late Permian boundary previously accepted, and perhaps as late as mid-Tatarian (see Smith *et al.*, 1974 for discussion). Marine fossils found at the various GCR sites featured in this review are strongly facies-linked and throw little light on the precise age of the rocks.

Although few late Permian shorelines in the British Isles can be recognized with certainty, projection of sedimentological and thickness trends suggests that the Bakevellia and Zechstein seas generally extended no more than a few kilometres beyond the present outcrops of late Permian strata and remained separated by a Proto-Pennine barrier (Figure 1.2) that waned with time but was surmounted by the sea only briefly and locally during the third of four main cycles of the English Zechstein sequence. The Zechstein Sea itself was divided into several sub-basins (Figure 1.1), and early Zechstein Basin marginal depositional environments in north-east England fell into distinct Durham and Yorkshire provinces, separated by the Cleveland High (Figure 1.2). These two provinces form the basis for Chapters 3 and 4.

The low relief and aridity of the Proto-Pennines and most other hinterlands, together with the prevalence of onshore trade winds, ensured that

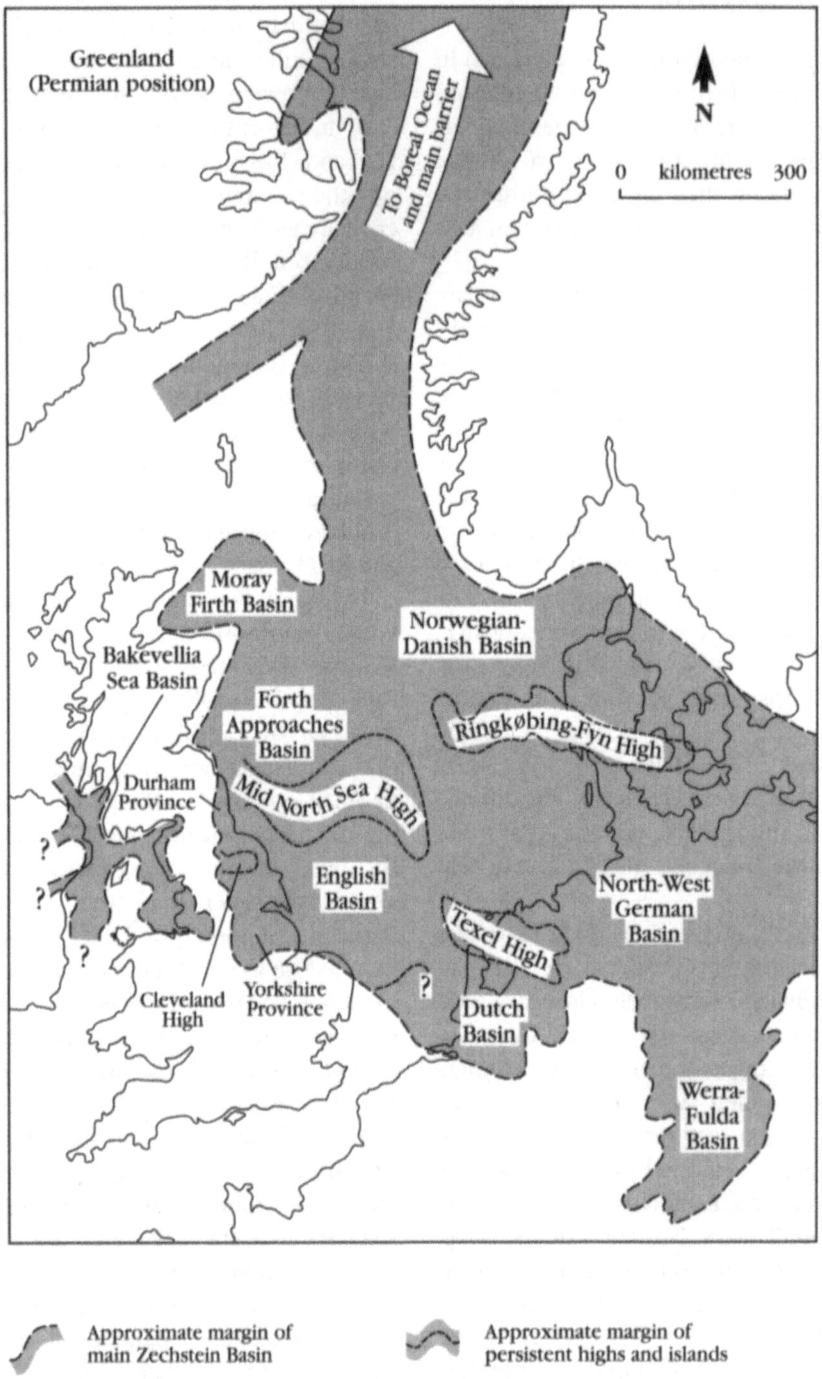

Figure 1.1 The Zechstein Sea and its environs. After Smith (1980a, fig. 1).

the Zechstein Sea in England received relatively little terrigenous sediment and most of the late Permian marine rocks there contain less than 2% of siliciclastic grains (mainly wind-blown); in contrast, the hinterland of much of the Bakevellia Sea was generally more elevated and rugged than that of most of the English Zechstein Sea, and terrigeneous input was correspondingly greater. In both basins, occasional storms resulted in flash floods which swept coarse continental debris far across the coastal plains and led to the formation of marginal breccia lenses and wedges.

The distribution of marine Permian strata in mainland Britain is shown in Figure 1.3, which emphasizes a great disparity between the widely scattered but mainly small occurrences in north-west England and the wide extent and almost continuous outcrop in north-east England. This

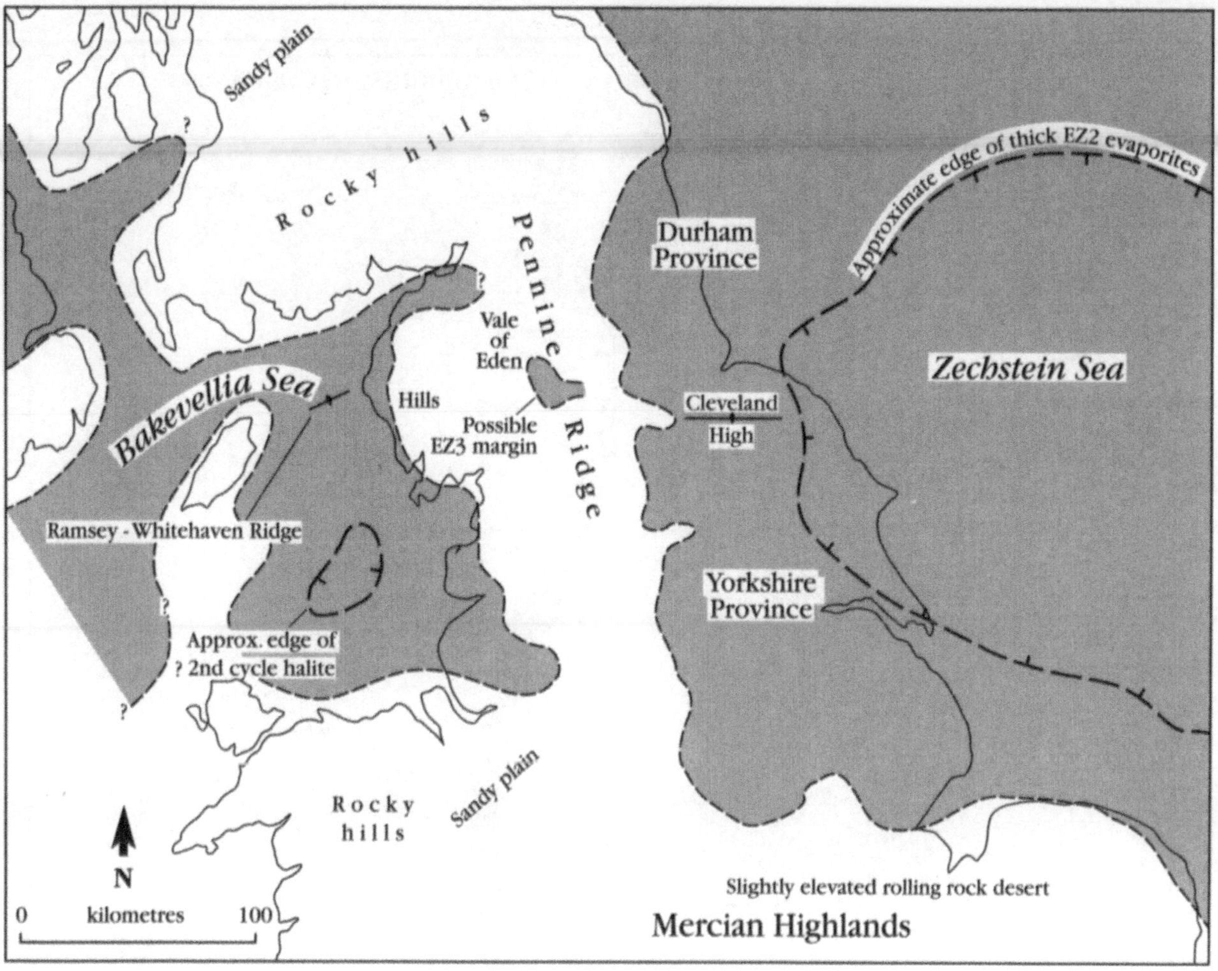

Figure 1.2 Outlines of the late Permian seas of northern England and adjoining areas, showing the persistent Pennine Ridge. After Smith (1992, fig. 9.8).

disparity is reflected in the spread of GCR sites, only one of which is in the north-west and the remainder in the north-east. Altogether, 27 sites have been identified and are described in this volume; almost all meet the exacting requirements laid down by English Nature, the main exception being the site at the former New Edlington Brickclay Pits in Yorkshire, which have recently been filled but are included here pending denotification and the choice of a suitable alternative site demonstrating comparable geological features.

Rationale of marine Permian site selection

The choice of sites described and discussed in this review was governed by a wide range of criteria, including their representativeness, their content of exceptional or unique geological features, their historical and their national and international importance. Together these sites form a dynamic network that covers most of the geology of the late Permian marine sequence in northern England, but some aspects are under-represented pending the choice of appropriately representative additional sites and the network will undoubtedly evolve as sites deteriorate or are superseded. The wealth of suitable high-quality exposures of these rocks in the Durham Province of north-east England is dramatically indicated by the fact that almost all the chosen sites meet at least two of the main selection criteria, and some sites satisfy all the criteria. Several sites, moreover, gain additional importance by forming links in chains of sites that together form a coherent whole but for which, because of the large scale, no single site could give full coverage. In addition to their geological interest, and arising from the specialized flora that characterizes many thin soils

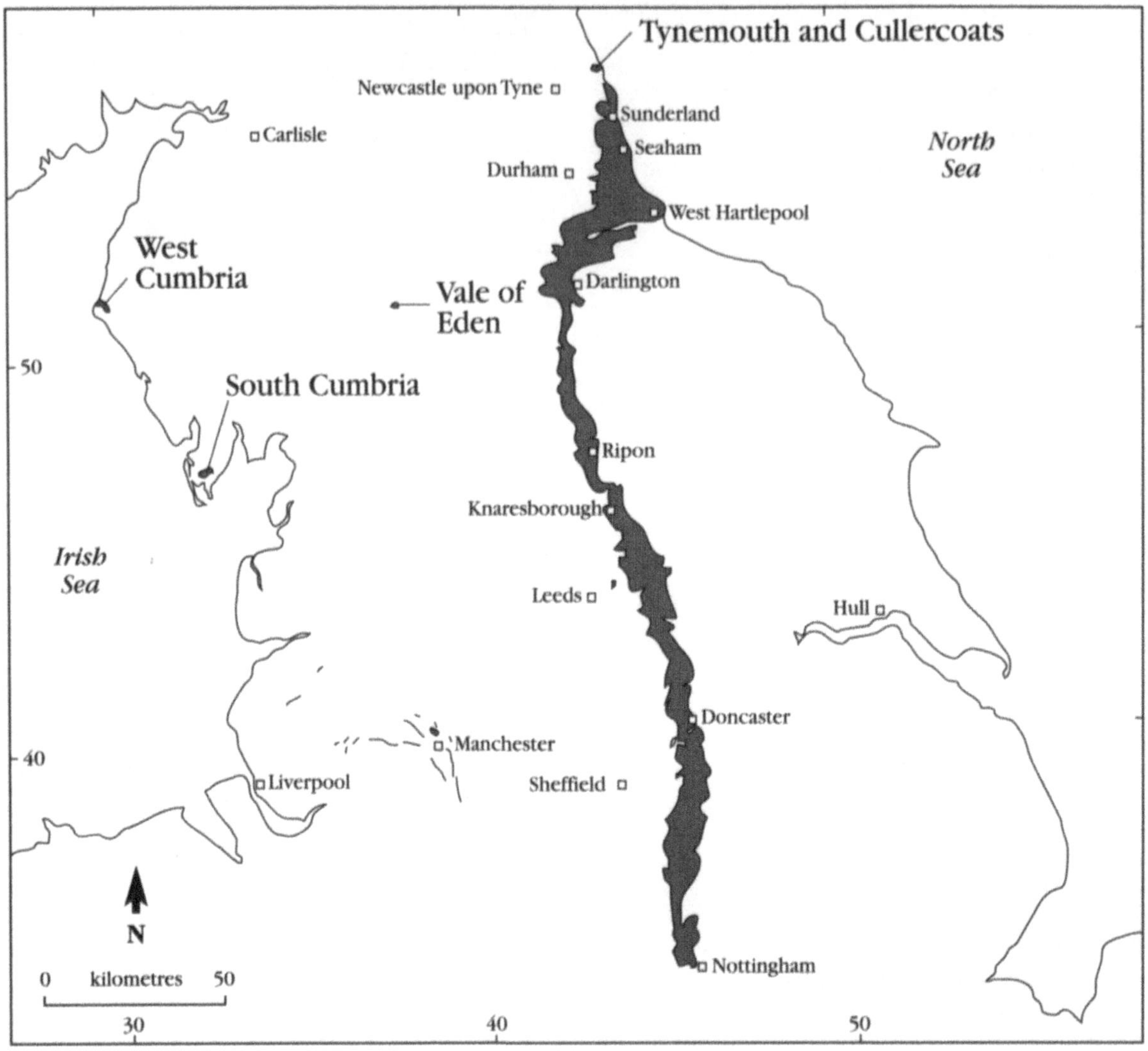

Figure 1.3 Outcrops of marine Permian strata in mainland Britain.

developed on the Magnesian Limestone, several of the GCR sites coincide with (or overlap) sites of special biological interest.

The sites considered here include natural exposures such as river and sea cliffs, but most are quarries (or part of quarries) and cuttings. Several sites include a number of separate exposures and these, and some others, exhibit more than one geological feature worthy of preservation. Outstanding examples of this are the complex of interlinked faces in reef and associated rocks at Ford (Sunderland), that provided the key to the solution of otherwise almost intractable sedimentological and stratigraphical problems, and the spectacular and equally varied coastal cliffs between Trow Point (South Shields) and Sunderland; coastal cliffs at Seaham and Blackhall are other examples. These coastal sites, indeed, are without parallel in Britain and most of Western Europe, and are unrivalled for research and teaching purposes. They are mainly in limestones and dolomites in the upper part of the Permian marine sequence and, in addition to exposing a continuum of rock-types extending basinwards from a marginal barrier system to those deposited near the middle of the basin-margin slope, also furnish magnificent examples of the disruptive effects of evaporite dissolution. Inland, most of the marine Permian GCR sites are in rocks in the lower and middle parts of the sequence, and clearly exhibit the great lateral variation that characterizes carbonate sediments now being formed in tropical nearshore environments; these inland sites include the type sections of three of the main late Permian marine formations.

Late Permian marine and associated strata in north-west England and adjoining areas

Late Permian marine and associated strata are widespread beneath the eastern Irish Sea, but extend far inland only in the Solway Firth and south Lancashire/north Cheshire areas where arms of the Bakevellia Sea extended eastwards. The sequence is up to about 350 m thick in the basin centre, where hydrocarbon exploration has disclosed much new information (Ebburn, 1981; Jackson *et al*., 1987), but thins sharply towards the basin margins where sequences are condensed and incomplete. Exposures on land are concentrated in these basin-marginal areas, especially in west and south Cumbria and around Greater Manchester, and most include intercalations of continental strata that indicate periodic proximity to cyclically migrating shorelines.

Information on Permian strata in Cumbria has been summarized by Arthurton *et al.* (1978) and for the basin as a whole by Smith (1992). Additionally, many new stratigraphical and structural data have been given by Jackson and Mulholland (1993). The nomenclature and classification of these strata in the several outcrop areas have evolved independently and are presented in Table 1.1. Three main cycles and one possible other cycle were recognized in West Cumbria by Arthurton and Hemingway (1972) and four supposedly equivalent cycles (BS1 to 4) were defined by Jackson *et al.* (1987) in the East Irish Sea Basin. Cycles BS1 to BS3 of the Bakevellia Sea sequence probably correlate with Cycles EZ1 (a and b) and EZ2 of the English Zechstein Basin (Smith *et al.*, 1974; Arthurton *et al.*, 1978) and cycles EZ3 and 4 probably equate with red-beds and basin-centre evaporites in the Bakevellia Sea sequences (Jackson *et al.*, 1987; Smith and Taylor, 1992). Attempts to interpret the palaeogeography of the region at various stages during each of the main cycles were given by Smith and Taylor (1992).

The biota of the Bakevellia Sea rocks was summarized by Pattison (1970, 1974) and Pattison *et al.* (1973), and is characterized by its low

Table 1.1 Classification and correlation of Permian marine and associated strata in north-west England and adjoining areas, showing the main depositional cycles. The Manchester Marl of south Lancashire and north Cheshire is a general correlative of the St Bees Evaporites but precise correlation is uncertain. Based on Smith (1992, table 9.2). Cycles after Jackson *et al.* (1987).

Cycles	Manx-Furness Basin (Central area)	South Cumbria	West Cumbria	Vale of Eden
BS4	St Bees Evaporites and Shales: Anhydrite; Anhydrite; Dolomite	St Bees Shales (with Brockram)	St Bees Shales (with Brockram): Blocky Facies	Eden Shales (with Brockram): Blocky Facies; D-Bed; Belah Dolomite
BS3	Halite	Roosecote Anhydrite; Roosecote Dolomite	Fleswick Anhydrite; St Bees Evaporites: Fleswick Dolomite	C-Bed
BS2	Anhydrite; Dolomite	Haverigg Haws Anhydrite	Sandwith Anhydrite; Sandwith Dolomite	B-Bed
BS1	Dolomite, anhydrite, mudstone, etc.	Gleaston Dolomite ('Magnesian Limestone') 'Grey Beds'	Saltom Dolomite; Saltom Siltstone	A-Bed and Hilton Plant Beds

diversity. Pattison (in Pattison *et al.*, 1973, table 3 columns 14 and 15) listed 32 plant and invertebrate species and several additional, but doubtfully recognized, forms; he noted that bivalves are the most widespread and abundant invertebrates and recognized a sequence of six genera that declined to one (*Bakevellia*) and then none in response to inferred, decreasingly favourable, environmental conditions.

Late Permian marine and associated strata in north-east England

Late Permian marine strata crop out almost continuously (partly beneath drift) from Tynemouth southwards to the outskirts of Nottingham and comprise a sequence of mainly dolomitized limestones that individually and together thicken progressively eastwards (Figure 1.4). These carbonate rocks form the basal part of three main evaporite cycles (EZ1 to EZ3), and at depth farther east are separated by evaporites which, in present coastal districts of Yorkshire and much of the North Sea Basin, form most of the sequence. At outcrop the carbonate units are separated by erosion or emersion surfaces, some in combination with siliciclastic beds or evaporite dissolution residues, and some have been affected by large-scale dissolutional foundering. The main Cycle EZ2 carbonate unit does not crop out in Yorkshire but is thick and widespread in County Durham and Tyne and Wear.

The stratigraphical names used in this account are those of Smith *et al.* (1986), which were proposed to eliminate the misleading use of the same names for different rock units in the two provinces and to bring the nomenclature of the late Permian marine strata in north-east England into line with modern practice. The proposed new names are listed in Table 1.2, together with the traditional names used in most of the relevant literature. Type localities for the newly-proposed rock units were nominated by Smith *et al.* (1974) and Smith *et al.* (1986).

Correlation between the late Permian marine sequences in the Durham and Yorkshire provinces (Table 1.2) is firm at the level of the Marl Slate and Seaham/Brotherton formations and reasonably firm between the early EZ1 carbonate formations, and between the EZ2 carbonate formations. There are, however, unresolved doubts on the precise correlation of the later EZ1 carbonate units of the two provinces (the Ford Formation and the Sprotbrough Member of the Cadeby Formation) and on the stratigraphical affinities (EZ1 or EZ2?) of the biostrome member (Figure 1.4) that lies between the EZ1 shelf-edge reef of the Ford Formation and the EZ2 Roker Dolomite Formation in the Durham Province.

An outline of the lithology and distribution of much of the Permian marine sequence at outcrop in north-east England was given by Sedgwick (1829) and details have been added by the authors of Geological Survey Memoirs and many other workers (see individual site accounts for full references); more recent syntheses are by the writer (Smith, 1970a, 1974a, b, 1980a, b, 1992), Pettigrew (1980), Kaldi (1980, 1986a), Harwood (1981, 1989), Aplin (1985), Goodall (1987) and Hollingworth (1987). Data from outcrop and boreholes are now sufficient for a lateral succession of facies belts to be recognized in three of the main carbonate rock units and for palaeogeographical maps of these to be attempted [see Kaldi (1980, 1986a) and Harwood (1981, 1989) for parts of the sequence in the Yorkshire Province and Smith (1989) for the full sequence in the Durham and Yorkshire provinces]. For reasons not yet fully understood, some of the Cycle EZ1 rocks and facies in the Durham Province differ considerably from their counterparts in the Yorkshire Province, but remaining parts of the sequences are broadly comparable. In general, the rocks exposed in the Durham Province lie farther from the basin margin than those at outcrop in the Yorkshire Province, and the two sequences together afford a reasonably complete transect from near the shoreline to below the middle of the basin-margin slope.

The biota of the English Zechstein sequence is generally more abundant and much more diverse than that of the Bakevellia Sea sequence and includes several vertebrate species. Comprehensive lists by Pattison (in Pattison *et al.*, 1973, table 3) include all the species known to those dates, and are subject to minor subsequent amendments and additions; a bivalve-dominated community similar to that in Bakevellia Sea strata is present in nearshore and littoral carbonate rocks, especially in the Yorkshire Province, but is augmented by a wide range of brachiopods and bryozoans slightly farther offshore where environmental conditions are inferred to have been less stressful. The Zechstein fauna reaches peak diversity and abundance in early parts of a massive shelf-edge reef in the Cycle EZ1 Ford Formation of the Durham Province, where a range of complex faunal

Table 1.2 The main stratigraphical units in the marine Permian sequence of the two provinces of north-east England, showing the nomenclature used here and (in brackets) equivalent traditional names. Based on Smith *et al.* (1986, table 1). Note that the Edlington Formation, in parts of the Yorkshire Province where the Kirkham Abbey Formation is absent, extends downwards to the top of the Sprotbrough Member of the Cadeby Formation (see also Table 1.3) and may include an equivalent of the Hayton Anhydrite.

Cycles	Durham Province	Yorkshire Province
	Roxby Formation (Upper Marls)	Roxby Formation (Upper or Saliferous Marls)
EZ4	Sherburn Anhydrite Formation (Upper Anhydrite)	Sneaton Halite Formation (Upper Halite) Sherburn Anhydrite Formation (Upper Anhydrite)
	Rotten Marl Formation (Rotten Marl)	Rotten or Carnallitic Marl Formation (Rotten or Carnallitic Marl)
EZ3	Boulby Halite Formation (Main Salt) Billingham Anhydrite Formation (Billingham Main Anhydrite) Seaham Formation (part of Upper Magnesian Limestone)	Boulby Halite Formation (Middle Halite) Billingham Anhydrite Formation (Billingham Main Anhydrite) Brotherton Formation (Upper Magnesian Limestone)
EZ2	Edlington Formation (Middle Marls or Lower Evaporites) Roker Dolomite Formation (Hartlepool and Roker Dolomite) Concretionary Limestone Formation (Concretionary Limestone) } Part of Upper Mag. Lst.	Edlington Fm in West : Fordon Evaporite Fm in East (Middle Marls) (Lower Evaporites) Kirkham Abbey Formation (Unnamed, ?absent at crop)
EZ1	Hartlepool Anhydrite Formation (Hartlepool Anhydrite) b Ford Formation (Middle Magnesian Limestone) a Raisby Formation (Lower Magnesian Limestone) Marl Slate Formation (Marl Slate)	Hayton Anhydrite Formation (Unnamed) Sprotbrough Member ('upper subdivision') Wetherby Member ('lower subdivision') } Cadeby Formation (Lower Magnesian Limestone) Marl Slate Formation (Marl Slate)

communities has been recognized by Hollingworth (1987). By contrast, most basinal Zechstein rocks contain no benthic fauna and generally sparse remains of a nektonic fauna.

The Zechstein biota in north-east England, like that of equivalent strata in Germany, has been studied by many workers. Early observations by Winch (1817) and Sedgwick (1829) were followed by more comprehensive works by Howse (1848, 1858), King (1848, 1850) and Geinitz (1861). King's 1850 work was an outstanding monograph that remains a standard work. Lesser, but nevertheless important, studies were contributed by Kirkby between 1857 and 1866, and, considerably later, by Trechmann (1945). More recently, the Durham fauna (especially the bivalves) were studied by Logan (1962, 1967), the Durham reef bryozoans by Southwood (1985), the palaeoecology of the Durham reef by Hollingworth (1987) and the gastropods by Hollingworth and Barker (1991). Work on the microfauna includes important contributions from Robinson (1978) and Pettigrew *et al.* (in press) on the ostracods, Swift and Aldridge (1982) and Swift (1986) on the conodonts (mainly in the Yorkshire Province) and Pattison (1981) on the foraminifera. Several of the early workers touched on plant fossils of the English Zechstein, and the macroflora, mainly of algae and of land plants washed into the sea, was studied by Stoneley (1958) and

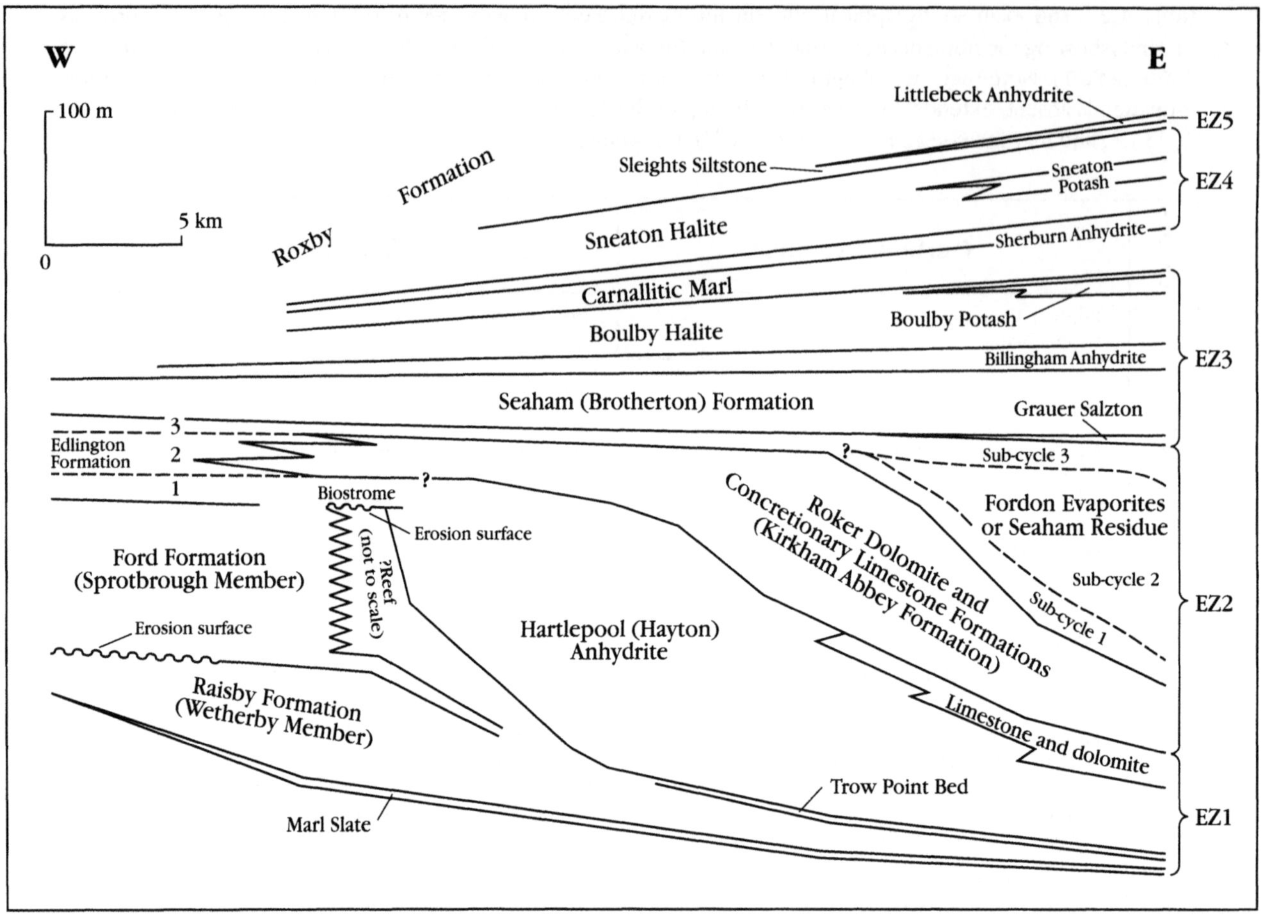

Figure 1.4 Late Permian (Zechstein) lithostratigraphical units in north-east England; names as in the Durham Province with Yorkshire Province names (where different) in brackets. In Yorkshire, the Wetherby Member and Sprotbrough Member together comprise the Cadeby Formation. The erosion surface shown between the Wetherby and Sprotbrough members is the Hampole Discontinuity which lies up to 3 m below the top of the Wetherby Member; it has not been recorded in the Durham Province. Slightly modified from Smith (1989, fig. 1).

Schweitzer (1986). The microflora was investigated by Clarke (1965) and summarized by Warrington (in Pattison *et al.,* 1973).

Despite many variations caused by local factors, chiefly around the margins of the basin, the main structural controls and environmental influences on late Permian marine sedimentation in north-west Europe were shared throughout the Zechstein Sea. For this reason, the patterns of sedimentation and the biotas of the several sub-basins have much in common with each other, so that the results of research in one area are commonly applicable to other areas. This research, initially in England and Germany but later spreading to Denmark, Holland and Poland, has shown that the similarities of geometry, lithology, sedimentology and biota of the various sequences permit reasonably firm correlation between the carbonate units of the first three cycles in England with those of northern Germany and Holland (Table 1.3) and thence basinwide. Problems remaining include uncertainty over the number (two or three?) and precise equivalence in England of the subcycles of the Cycle 1 carbonate unit in Germany and Poland and the widespread doubts on the position of the Cycle 1/Cycle 2 contact in places where Cycle 1 evaporites are absent.

Table 1.3 Classification and correlation of Permian marine and associated strata in north-east England, showing the main depositional cycles and a representative sequence from Holland and northern Germany. After Smith (1989, table 1).

Groups	**Cycles**	**Yorkshire Province (Outcrop area)**	**Durham Province (County Durham, east Tyne and Wear, County Cleveland)**	**Yorkshire Province (East and North Yorks and Humberside)**	**North Germany and Holland**	**Cycles**
Eskdale Group	EZ5	Roxby Formation	Roxby Formation	Littlebeck Anhydrite Formation Sleights Siltstone Formation	Ohre Anhydrit Unterer Ohre Ton	Z5
Staintondale Group	EZ4	Sherburn Anhydrite Fm Rotten Marl Formation	Sherburn Anhydrite Formation Rotten Marl Formation	Sneaton Halite Formation Sherburn Anhydrite Formation Upgang Formation Carnallitic Marl Formation	Aller Salze Pegmatitanhydrit Thin unnamed carbonate Roter Salzton	Z4
Teesside Group	EZ3	Billingham Anhydrite Fm Brotherton Formation ? ? ?	Boulby Halite Formation Billingham Anhydrite Formation Seaham Formation	Boulby Halite Formation Billingham Anhydrite Fm Brotherton Formation Grauer Salzton	Leine Salze Hauptanhydrit Plattendolomit Grauer Salzton	Z3
Aislaby Group	EZ2	Edlington Formation	Seaham Residue Roker Dolomite and Concretionary Limestone Formation	Fordon Evaporite Formation Kirkham Abbey Formation	Stassfurt Salze and Basalanhydrit Hauptdolomit and equivalents	Z2
Don Group	EZ1 b a	Cadeby Fm: b Sprotbrough Member a Wetherby Member	Hartlepool Anhydrite Formation Ford Formation Raisby Formation Marl Slate Formation	Hayton Anhydrite Formation Cadeby Formation Marl Slate Formation	Werraanhydrit Zechsteinkalk Kupferschiefer	Z1
		Basal Permian (Yellow) Sands and Breccias	Yellow (Basal Permian) Sands and Breccias	Basal Permian Sands and Breccias	Rotliegendes	

Chapter 2

North-west England

INTRODUCTION

Marine Permian strata in north-west England are reasonably well known from widely scattered borehole data and from temporary exposures, but good permanent surface exposures are extremely rare and are mainly of strata low in the Permian marine sequence. Only one of these exposures - the sea cliff backing Barrowmouth Beach at Saltom Bay, West Cumbria - has been selected as a GCR site and the documentation and discussion of this imposing exposure therefore forms almost all of this brief chapter. The general geological background is discussed in Chapter 1.

Evidence relating the deposits of the Bakevellia Sea to those of the Zechstein Sea is scanty, but the faunal assemblage of the Saltom Dolomite is nevertheless strikingly similar to that in comparable marginal parts of the Cadeby Formation in the Yorkshire Province and approximate correlation is indicated; amongst GCR sites in the Yorkshire Province, that at Ashfield Brick-clay Pits (Conisbrough, Chapter 4) is probably the closest faunal and sedimentological match with the Saltom Dolomite of Saltom Bay.

BARROWMOUTH BEACH SECTION, SALTOM BAY (NX 9573 1572–9594 1603)

Highlights

This coastal section is the best exposure of late Permian marine strata in Cumbria and is also one of the best exposures of early Permian continental breccias (brockram) and of the underlying Carboniferous-Permian unconformity; the strata exposed lie at the base of the local Permian sequence, higher parts of which are not exposed, but are known from many local exploratory boreholes.

The marine strata are represented by about 4.6 m of varied shallow-water shelly dolomite that was formed near the eastern margin of the Bakevellia Sea, and the underlying breccia is thought to have been a water-laid desert sheet gravel and which may have become stabilized so as to form a desert pavement.

Introduction

The classic Barrowmouth Beach section lies at the southern end of Saltom Bay, Whitehaven, Cumbria and comprises a rugged marine rock platform up to about 25 m wide backed by a near-vertical sea-cliff a few metres high. About 7.0 m of Permian strata dip south-westwards at about 5 degrees and are exposed for about 150 m.

The section was first mentioned by Sedgwick (1836), and was later discussed by Binney (1855), Murchison and Harkness (1864), Goodchild (1893) and Eastwood *et al.* (1931); more recently it was recorded by Arthurton and Hemingway (1972), who nominated it as the type locality of the Saltom Dolomite, and by Macchi (1990) who presented a graphic section. All these authors agree on the Permian age and basin-marginal significance of the 'Magnesian Limestone' that forms the main part of the exposure, but they differ in detail in their recording of the section.

Description

The position of the site and its boundaries is shown on Figure 2.1. The Permian strata display appreciable lateral variation which probably

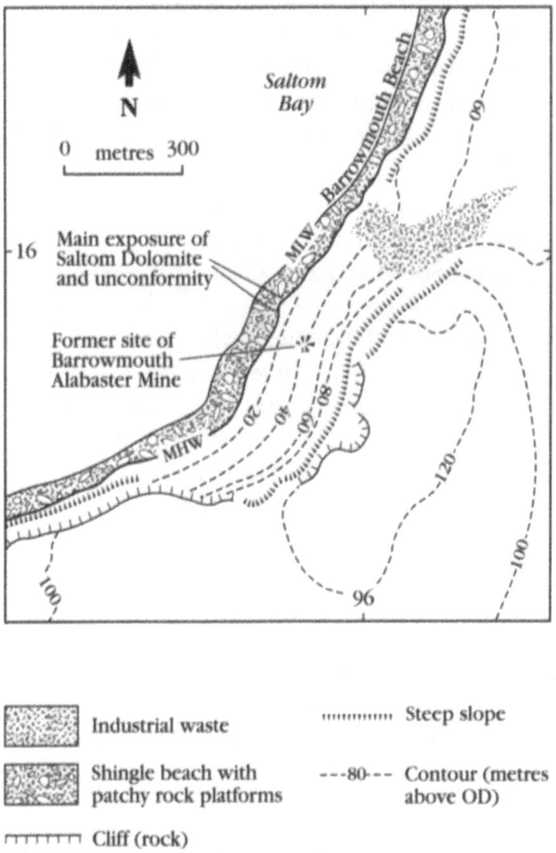

Figure 2.1 The Barrowmouth Beach GCR site, Saltom Bay, Whitehaven, showing the position of the main features of geological interest.

accounts for some of the differences in recording; the following section is based on observations by the writer in 1983 and 1991.

	Average thickness (m)
7 Clay/mudstone, ?dolomitic, variegated in shades of yellow, buff, grey and pale purple, flaky	0.15+
6 Dolomite, calcitic, or dolomitic limestone, mainly olive-buff and grey-buff, finely crystalline, in four to six uneven thin beds; some thin ?argillaceous layers; 0–40 mm nodular bed at top	*c.* 0.52
5 Dolomite, argillaceous or dolomitic mudstone, buff and buff-grey, soft, shaly to flasery, with scattered poorly-preserved plant remains; 40–70 mm bed of buff and grey, microporous, finely crystalline dolomitic limestone, 0.40–0.50 m above base	*c.* 0.75
4 Dolomite, dark buff, calcitic, finely saccharoidal,in uneven thin beds with wavy partings of shaly dolomite; slightly uneven top with thin discontinuous layer of blue-black hematitic dolomite	*c.* 0.50
3 Dolomite, buff, calcitic, finely saccharoidal, unevenly thin- to medium-bedded, partly plane-laminated	*c.* 0.45
2 Dolomite, buff, calcitic, finely saccharoidal, thin- to thick-bedded with flasery olive partings towards top; partly cavernous, with few to abundant bioclast moulds; many purple-stained angular rock fragments in lowest 0.25–0.30 m; slight onlap at sharp base (relief *c.* 0.15 m)	*c.* 3.10
1 Breccio-conglomerate of purple and red-stained, angular to subrounded pebbles, cobbles and small boulders; mainly crudely-bedded; tough matrix of dolomite-cemented sandstone	*c.* 1.60

Unconformity, sharp, local relief 0.3 m, with polygonal neptunian sandstone dykes extending up to 0.9 m into underlying reddened (purple) Whitehaven Sandstone (Upper Carboniferous).

In this section, bed 1 is the Brockram of B. Smith (1924), Lower Brockram of Hollingworth (1942) and Basal Breccia of Arthurton and Hemingway (1972); in parts of the designated area it thins to as little as 0.2 m. Details of this continental deposit are beyond the scope of this volume but were given by, amongst others, B. Smith (1924), Eastwood *et al.* (1931), Arthurton and Hemingway (1972) and Macchi (1990). Higher beds in the section (Figures 2.2 and 2.3) all form part of the St Bees Evaporites as defined by Arthurton and Hemingway and beds 2 to 6 inclusive comprise their Saltom Dolomite (formerly the Magnesian Limestone).

All the carbonate rocks in the Saltom Bay section contain scattered, mainly small cavities (now calcite-lined) after former sulphates, and in some beds small concordantly-elongated cavities are so numerous as to give the rock a 'birdseye' (fenestral)-like fabric. No true fenestral fabric was seen by the writer, however, who was also unable to identify with certainty the algal lamination recorded high in the section by some authors. Dedolomitization of parts of the carbonate rocks was inferred by Arthurton and Hemingway (1972), and presumably accompanied the dissolution of the former sulphate during Tertiary or later uplift. Although a number of the thin argillaceous layers in the section resemble dissolution residues from bedded evaporites, none is associated with evidence of collapse-brecciation and an origin as residues after substantial thicknesses of evaporite is therefore unlikely. The dark grey to black layer at the top of bed 4 may be a mineralized crust or hardground.

The abundant fossils in bed 2 belong to a few species of bivalves and gastropods and were listed by Eastwood *et al.* (1931, p. 217) and Pattison (1970). Pattison noted that the bivalves *Bakevellia (B.) binneyi* and *Permophorus costatus* are the commonest fossils in this bed here, at nearby surface exposures and in borehole cores from the vicinity; he figured *Permophorus* from Saltom Bay (1970, plate 21, fig. 9). The foraminifer *Agathammina pusilla* is also present (Pattison, 1969).

The steep unstable slope backing the designated area is overgrown and has a confused hummocky surface caused by tipping, hill-creep, landslips and mudflows. Few exposures are of rocks *in situ*, but earlier records (e.g. Sedgwick, 1836; Binney, 1855; Eastwood *et al.*, 1931) show that the slope is composed mainly of thinly interbedded and interlaminated dark brick-red

Figure 2.2 Lower beds of the Permian sequence at the south-west end of the Barrowmouth Beach (Saltom Bay) section, showing Saltom Dolomite resting on the slightly uneven surface of the Brockram or Basal Breccia. Hammer: 0.33 m. (Photo: D.B. Smith.)

mudstone and siltstone of the St Bees Shales (?60 m) which overlie the St Bees Evaporites (*c.* 20 m) of which the Saltom Dolomite is here the basal unit. A 7.6 m gypsum/anhydrite noted low in the sequence by Binney (1855) was formerly worked in galleries up to 9 m high at the nearby Barrowmouth Alabaster Mine which closed in 1908 when the workings encountered an unprofitably high proportion of anhydrite; Binney's view on the low position of the gypsum/anhydrite bed was disputed by B. Smith (in Eastwood *et al.*, 1931) who suggested that it was separated from the limestone by 'a considerable thickness of shale'. It is likely to be either the Sandwith Anhydrite of Arthurton and Hemingway (1972) or a combination of the Sandwith Anhydrite and the Fleswick Anhydrite.

Figure 2.3 Higher beds of the Permian sequence in Barrowmouth Bay, a few metres southwest of the view in Figure 2.2. The line of small cavities in dolomite near the base of this photograph is also visible near the top of the section in Figure 2.2. Most or all of the beds here form part of the Saltom Dolomite, but it is possible that bed 6 may be part or all of the Fleswick Dolomite. Hammer: 0.33 m. (Photo: D.B. Smith.)

Interpretation

The low cliff overlooking Barrowmouth Beach is by far the largest and best exposure of the 'Magnesian Limestone' of West Cumbria and the wave-swept rock platform affords an unrivalled view of the underlying 'Brockram' and of the unconformity between Permian strata and the eroded and deeply reddened Upper Carboniferous Whitehaven Sandstone; the hiatus represented by the unconformity probably spans at least 60 million years. The section is doubly important in that it is also the best exposure of marginal strata of the Bakevellia Sea Basin; it lies close to, or on, the Ramsey-Whitehaven Ridge and thus throws light on Permian sedimentation near the margins of both the Solway Firth and East Irish Sea sub-basins.

Permian continental breccias and 'Magnesian Limestone' are, or have been, exposed in a number of small quarries and natural exposures in the Whitehaven area and were recorded by Eastwood *et al.* (1931); some of the quarries have now been filled and most of the other sections are overgrown. The surface exposures, however, afford only a small part of the available information on the local Permian sequence, most of which comes from numerous coal, iron ore and gypsum/anhydrite exploration boreholes (Eastwood *et al.*, 1931; Taylor, 1961; Meyer, 1965; Arthurton and Hemingway, 1972). Analysis of the borehole information by Arthurton and Hemingway revealed that the basal marine units are sharply varied in thickness, up to a maximum of about 15 m, and comprise a diverse mosaic of facies types indicative of a range of nearshore depositional environments; the lower part of the basal carbonate rocks passes south-eastwards into dolomitic siliciclastic siltstone, the Saltom Siltstone, which is thought to be absent at the Barrowmouth section unless it is represented by bed 5.

If Arthurton and Hemingway were correct in assigning beds 2–6 inclusive to the Saltom Dolomite, the carbonate rocks at Barrowmouth would be the whole of the carbonate member of the incomplete first or Saltom Cycle (= Cycle BS1 of Jackson *et al.*, 1987) (Table 1.1). It is possible, however, that the dark grey to black hematitic layer at the top of bed 4 is correlative with the zone of 'intense haematitic staining' at the top of the Saltom Dolomite in Borehole S9 (about 1.5 km SSW of the Barrowmouth Beach section) and if this were so, then the section would include parts or all of the carbonate members of both the Saltom and Sandwith (BS2) cycles.

Future research

Full understanding of the significance of the Barrowmouth Beach section is severely hampered by an almost complete lack of detailed knowledge of the petrography, geochemistry and sedimentology of the rocks exposed. These aspects ought to be addressed in the light of further detailed analysis of cores from nearby boreholes.

Conclusions

The cliff section backing Barrowmouth Beach is the only GCR site in marine Permian strata near the eastern margin of the Bakevellia Sea basin. It is the best coastal exposure of Permian sedimentary rocks and of the underlying Carboniferous-Permian unconformity in Cumbria. The cliff exhibits a sequence of early Permian continental breccias (brockram), the products of erosion of the early Permian uplands, overlain by late Permian shallow marine dolomitized limestones that contain fossil-rich layers. The site has considerable potential for future study and research, particularly into the petrological, geochemical and sedimentological aspects of the sequence, together with the relationships of this site to sequences revealed by detailed analysis of cores from nearby boreholes.

Chapter 3

North-east England (Durham Province)

INTRODUCTION

Most of the GCR sites documented and discussed in this chapter lie within perhaps 30 km of the original western shoreline of the late Permian Zechstein Sea and north of the Cleveland High (Figure 1.2). There is no evidence of a connection with the Bakevellia Sea at any time during the first two main sedimentary cycles (EZ1 and EZ2) and only limited evidence of a brief connection with the Vale of Eden inland sedimentary basin during Cycle EZ3 (but see Holliday, 1993 and Smith, in press).

The Cleveland High (Figure 1.2) projected eastwards into the basin a few kilometres south of Darlington and strongly influenced sedimentation during Cycle EZ1 and part of Cycle EZ2. It was a broad, gentle, topographical feature where subsidence may have been relatively slightly slower than that to the north and south, and it remained emergent until eventually buried by onlapping sediments of the Edlington Formation. Following its burial, the Cleveland High appears to have exerted little or no effect on sedimentation and Cycle EZ3, and later English Zechstein strata in the Durham and Yorkshire provinces are similar to each other.

Rather more than half of the designated marine Permian sites are in the Durham Province and several of these are large complex coastal or inland exposures of international importance; amongst these outstanding sites are the Blackhalls Rocks coast section, the Claxheugh Rock - Ford Quarry section, Fulwell Hills Quarries and the unrivalled coastal cliffs between (and including) Trow Point (South Shields) and Whitburn.

The location of all the GCR sites in the Durham Province is shown in Figure 3.1.

Together the sites in the Durham Province span the whole of the local marine sequence, almost all the major formations and their varied facies being represented at one or more sites. No major carbonate rock unit is unrepresented, but the thick evaporites known in the subsurface farther east and south have been dissolved at outcrop where their place is taken by dissolution residues; this dissolution had the effects firstly of delaying understanding of the stratigraphical and sedimentological relationships of the younger members of the sequence, especially of the Cycle EZ2 carbonate rocks, and secondly of furnishing a wide and instructive range of subsidence features ranging from regional foundering by more than 100 m, to spectacular collapse-breccias and late-stage breccia-gashes. The complexity of the rocks at several of the Durham sites is daunting, and many problems remain to be solved. The sites nevertheless have outstanding qualities as outdoor classrooms for the demonstration of the effects of geological processes and afford abundant material for future research.

Consideration of the early Permian Basal (Yellow) Sands in the Durham Province is inappropriate here, and full discussion will appear in the companion Review volume on the continental Permo-Triassic Red Beds of Britain. Nevertheless, the top of the Formation is exposed at Raisby Quarries, in Frenchman's Bay (South Shields) and at Claxheugh Rock, and is described briefly in the site accounts here for the sake of completeness. The involvement of the top of the desert Yellow Sands at Claxheugh Rock in massive submarine sliding, indeed qualifies the formation there for inclusion in this volume. A product of such a seemingly improbable combination is to be seen at Tynemouth Cliff, 12 km to the north, where a marine debris flow at the top of the Raisby Formation comprises pebbles and cobbles of shelly dolomite in a matrix rich in aeolian sand grains (Smith, 1970c).

No localities specifically listed for their exposures of the Marl Slate have been included in the Marine Permian Review, but are expected to feature in the volumes on palaeobotany and Palaeozoic fish. Nevertheless, normal Marl Slate is exposed at Claxheugh Rock, Frenchman's Bay and Raisby Quarries, and is described in the appropriate accounts. At Frenchman's Bay the top of the Marl Slate has been removed by end-Raisby Formation submarine slumping and sliding, and these processes have removed the whole of the Marl Slate for more than 150 m at the north-eastern end of the Claxheugh Rock section.

Carbonate rocks of the Raisby Formation are the sole subject at the Dawson's Plantation (Penshaw) and High Moorsley sites and are the main subject at the type locality at Raisby Quarries; they also feature at Claxheugh Rock and from Trow Point to Whitburn and are considered briefly in the accounts of those sites. The listing of the Dawson's Plantation and High Moorsley Quarry sites is founded on the evidence of downslope sediment slumping and sliding exposed there superbly, including crumpled strata and atypically fossiliferous debris flows up to 1 m thick; the presence of these features is a link in the chain of evidence favouring a slope origin for these strata in this northern part of the province, though such slumping was not endemic and an external initiating

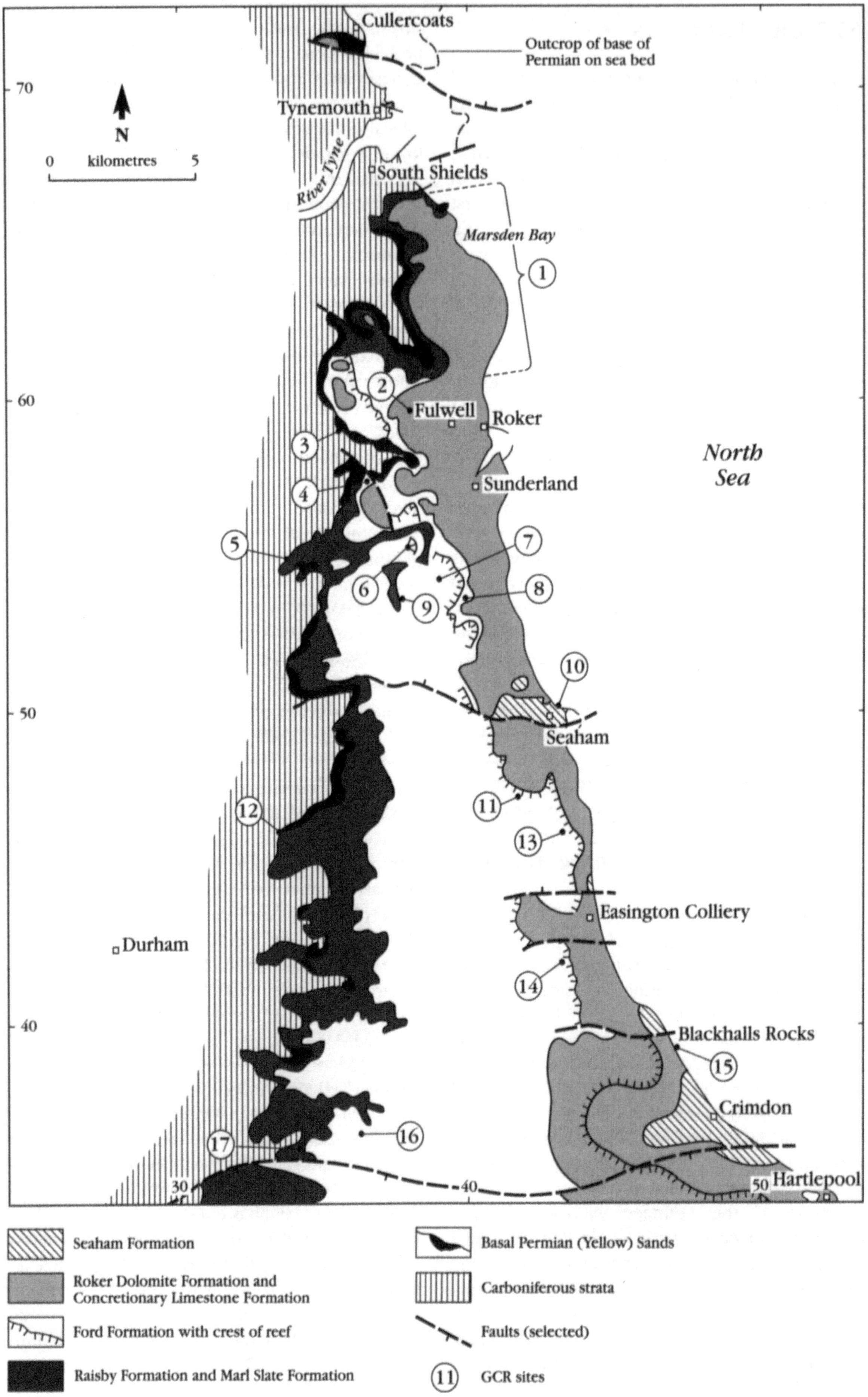
Cullercoats
Outcrop of base of Permian on sea bed
70
N
0 kilometres 5
Tynemouth
River Tyne
South Shields
Marsden Bay
1
2
Fulwell
Roker
60
3
North Sea
4
Sunderland
5
7
6
8
9
10
50
Seaham
11
12
13
Easington Colliery
Durham
14
40
Blackhalls Rocks
15
Crimdon
16
17
30
40
50 Hartlepool
Seaham Formation
Roker Dolomite Formation and Concretionary Limestone Formation
Ford Formation with crest of reef
Raisby Formation and Marl Slate Formation
Basal Permian (Yellow) Sands
Carboniferous strata
Faults (selected)
11 GCR sites

stimulus such as an earth tremor has been inferred (Smith, 1970c). Raisby Quarries afford a complete eponymous sequence through the formation, and have been freely cited in the literature since 1914; they are famous for the unusual presence of thick primary limestone and for an extensive suite of secondary minerals. Fossils (especially some large brachiopods such as *Horridonia*) are locally well-preserved in Raisby Quarries, but evidence of the widespread slumping low and high in the formation has not been recognized here. The thin Raisby Formation at Trow Point is unusual for its evidence of prevalent bioturbation and for the presence, at its top, of a spectacularly complex mass of slide-blocks (olistoliths) of Raisby Formation strata; at Frenchman's Bay most of the formation is inferred to have been removed by the same episode of massive submarine sliding (the Downhill Slide, named after Downhill Quarry, West Boldon) and the largest of the slide-blocks is about 72 m long and 7.5 m thick.

Rocks of the Ford Formation are amongst the most varied of the Magnesian Limestone sequence and feature in more scheduled sites than those of any other late Permian rock-unit in the Durham Province; this is mainly because of the renowned fossil content of the shelf-edge reef which extends from Sunderland to Hartlepool (Figure 3.1) and has attracted attention for almost two centuries. Each of the main sub-facies of the reef is represented at one or more GCR sites and together the several reef sites provide a cross-section through most of the reef and afford full opportunities for further research. The scheduled exposures are reasonably representative of the reef as a whole, but reefs are notoriously variable and surviving unscheduled reef exposures such as those at Dalton-le-dale (NZ 408476), Easington Colliery (NZ 434438), Beacon Hill (Hawthorn) (NZ 442453) and Castle Eden Dene (NZ 4440) reveal other aspects of the reef not clearly seen at the scheduled sites. Details of the lithology, biota and structure of the reef are given in the individual site accounts, which draw extensively on historical records and on more recent work by Aplin (1985), Hollingworth (1987) and the author (Smith, 1981a; 1994).

Landward (west) of the reef, dolomitized carbonate rocks of the Ford Formation are exposed in three scheduled sites, including Ford Quarry where the reef/backreef contact is uniquely clearly exposed, and Gilleylaw Plantation Quarry (Silksworth) where a lagoonal patch-reef is overlain by shallow-water oncoidal dolomite of a type known at only two other localities in the Magnesian Limestone. The third site, Trimdon Grange Quarry, exposes diagenetically altered backreef or lagoonal oolite several kilometres west of the reef, but this widespread facies is under-represented in the Durham site network and Trimdon Grange Quarry has to be viewed in conjunction with other exposures such as those in the local nature reserves at Bishop Middleham (NZ 3332) and Wingate (NZ 3737) quarries.

Seaward of the reef, the youngest and in some respects the most enigmatic accepted member of the Ford Formation is exposed at and between Trow Point and Frenchman's Bay, South Shields. This, the thin but distinctive Trow Point Bed, is present here in both its peloid/oncoid and columnar-stromatolitic modes, which are also known from cored boreholes immediately offshore, as well as in many North Sea hydrocarbon boreholes and, far to the east, in surface exposures and boreholes in Germany and Poland.

The youngest carbonate rock-unit doubtfully assigned to Cycle EZ1, and therefore to the Ford Formation, is the Hesleden Dene Stromatolite Biostrome, superbly exposed at Blackhalls Rocks and less well seen at Hawthorn Quarry. At the latter the biostrome is seen to rest on an erosion surface cut onto the shelf-edge reef of the Ford Formation and is overlain by supposed Roker Dolomite. The doubts about the age of the biostrome stem from the local presence of fragments of it in collapse-breccias thought to be related to the dissolution of the Cycle EZ1 Hartlepool Anhydrite, and therefore younger than at least part of the latter. However, the approximate aerial coincidence of the reef and biostrome, plus their faunal affinities, slightly favour a Cycle EZ1 age rather than a Cycle EZ2 age for the biostrome, unless its fauna was derived. The exposures of

Figure 3.1 The distribution of Permian marine rocks in the Durham Province, showing the location of Permian marine GCR sites: 1, Trow Point to Whitburn Bay; 2, Fulwell Hills Quarries; 3, Hylton Castle Cutting; 4, Claxheugh Rock, Cutting and Ford Quarry; 5, Dawson's Plantation Quarry, Penshaw; 6, Humbledon Hill Quarry; 7, Tunstall Hills (north); 8, Tunstall Hills (south) and Ryhope Cutting; 9, Gilleylaw Plantation Quarry; 10, Seaham; 11, Stony Cut, Cold Hesledon; 12, High Moorsley Quarry; 13, Hawthorn Quarry; 14, Horden Quarry; 15, Blackhalls Rocks; 16, Trimdon Grange Quarry; 17, Raisby Quarries. The map is based on Smith (1980b, fig. 9).

dolomitized algal stromatolites and cobble-boulder conglomerate at Blackhalls Rocks are particularly impressive and throw much light on contemporary geography and processes.

The carbonate rocks of Cycle EZ2 form almost all the coastal cliffs in the Durham Province and are the main interest in the outstanding site stretching south-eastwards from Trow Point, South Shields to Whitburn Bay; they comprise the predominantly shallow-water shelf and uppermost slope carbonates of the Roker Dolomite Formation and the roughly synchronous slope dolomites and limestones of the Concretionary Limestone Formation. Equivalent strata are unknown in the Yorkshire Province except in boreholes, so that the northern Durham cliffs and quarries are the only large-scale surface exposures available for detailed study of the complex sedimentology and diagenesis of these Cycle EZ2 strata. They are also the only places where the spectacular effects of large-scale foundering may be clearly related to the dissolution of former thick evaporites.

There are no GCR sites in which the Roker Dolomite Formation is the main interest, but it is present incidentally and with its normal lithology and fauna, at Whitburn, Seaham, Hawthorn Quarry, Blackhalls Rocks and in the Ryhope Cutting (as collapse-breccia). At Blackhalls Rocks and Hawthorn Quarry the formation displays no evidence of having foundered onto the underlying biostrome, implying that the Cycle EZ1 anhydrite did not overlap the reef here, but the formation has foundered by perhaps 50–100 m at the Ryhope and Seaham exposures. At Ryhope the collapse-brecciated Roker Dolomite is within 200 m of the reef-front, supporting the evidence from other localities such as West Boldon (NZ 3561), Easington Colliery and Horden that the Hartlepool Anhydrite once lay against the reef at Ryhope even if it did not overlap it. The Seaham exposure is of interest in that the uppermost few metres of the formation have been much fractured and dedolomitized, probably by the formation and dissolution of a complex network of evaporite veins related to the formerly overlying Fordon Evaporites. Large exposures of gently foundered Roker Dolomite extend northwards from the Seaham site and form the coastal cliffs at Roker (NZ 4059) and Whitburn (NZ 4161); here it overlies the well-known Cannon-ball Limestone near the contact with the Concretionary Limestone Formation. The Roker Dolomite lies in its normal stratigraphical position at Hartlepool, where thick Cycle 1 anhydrite has resisted dissolution.

Carbonate rocks of the Concretionary Limestone Formation are the sole interest at the Fulwell Hills Quarries and in the Marsden Bay to Whitburn area of the Trow Point to Whitburn GCR site. Faces preserved in the formerly vast complex of quarries at Fulwell Hills are remarkable for the great range of calcite concretions for which this formation is justly famous, and incidentally display sedimentological evidence favouring a submarine slope origin for these strata; they also include fish-bearing calcite-laminites, a characteristic shared with equivalent beds at Marsden Bay from which fossil fish were first collected from this formation. The Fulwell Hills sections display ample evidence of foundering through dissolution of the former underlying Hartlepool Anhydrite, but the evidence of foundering is especially dramatic and inescapable at Trow Point and Frenchman's Bay where almost completely brecciated dedolomitized Concretionary Limestone overlies the thin dissolution residue of the Hartlepool Anhydrite. Farther south, the effects of the major foundering are spectacularly displayed in Marsden Bay, where late-stage collapse of large dissolution-induced cavities is believed to have been the cause of a number of massive subvertical 'breccia-gashes' or 'breccia-pipes'. The main phase of foundering was Palaeocene or earlier, judging from the mutual relationship of foundered ?Roker Dolomite and the *c.* 58 million-year old Hebburn or Monkton Dyke, the crop of which was discovered recently on the coast at Whitburn by Mr G. Fenwick.

The Marsden Bay section, though extensively calcitized in addition to the brecciation, is still mainly of dolomite and comprises a mid-slope complex of interbedded sapropelic organic-rich fine laminates, graded turbidites and slumped beds (including oolite grainstones translated from the Roker Dolomite shelf). These strata and their associated secondary effects (calcitization and foundering) continue in the coastal cliff south of Lizard Point where they merge gradually with shelly carbonate rocks inferred to have been formed in oxic conditions in the upper part of the basin-margin slope.

The Concretionary Limestone rocks exposed at Fulwell Hills and in the Trow Point to Whitburn GCR site are reasonably representative of this most variable of formations, and include the foraminifer-gastropod-bivalve-ostracod-rich rocks that abound in high-slope facies such as those in coastal cliffs a few hundred metres south of Lizard Point. They are, however, relatively poor in calcite spherulites such as typify rocks of this formation in most inland exposures in the South Shields to Whitburn area. They nevertheless afford unrivalled opportunities for the study of the sedimentology and complex diagenesis of the formation and for observing the profound effects of the dissolution of thick underlying evaporites.

The insoluble remains of the youngest Cycle EZ2 strata - the Fordon Evaporites - comprise the striking Seaham Residue and are the main feature of the northern end of the Seaham site; they also crop out at the south end of the Blackhalls Rocks site. Both exposures illustrate graphically the effects of evaporite dissolution on underlying and overlying carbonate rocks, though the Fordon Evaporites were probably thinner there than the Hartlepool Anhydrite at Trow Point and Frenchman's Bay and the foundering was correspondingly less disruptive. The residue at Seaham is many times thicker than that of the Hartlepool Anhydrite, implying that the Fordon Evaporites contained a much higher proportion of insolubles, and it has been strongly contorted by plastic flow (perhaps whilst the evaporites, probably including salt, were still present).

Light is thrown here on the depositional environment of the Fordon Evaporites by the sedimentary features of an ooidal limestone in the Seaham Residue at Seaham, which was probably formed at or very near contemporary sea level. Undissolved Fordon Evaporites (mainly salt) have been recorded in a borehole 12 km ENE of Sunderland (Smith and Taylor, 1989) and are about 15-30 m thick in northern County Cleveland, approximately 25 km along strike from Seaham.

Carbonate rocks of Cycle EZ3 crop out in only limited areas of the Durham coast, mainly in synclines at Seaham and north and south of Blackhalls Rocks; all are estimated to have foundered by at least 120 m as a combined result of the dissolution of the Hartlepool Anhydrite and the Fordon Evaporites. At its type locality in the walls of Seaham Harbour, the Seaham Formation is mainly of thin-bedded limestone with a restricted biota and a range of shelf-type sedimentary structures, but both there and to the south of Blackhalls Rocks it also features bizarre calcite concretions similar to some of those in the Concretionary Limestone Formation farther north. Foundering is expressed by medium-scale tilting and dislocation of blocks of the Seaham Formation at its main exposures, and at Seaham was accompanied or followed by the creation of breccia gashes that contain fragments of strata (including the Rotten Marl) that are now otherwise eroded from the area.

The main features of the GCR Marine Permian sites in the Durham Province are summarized in Table 3.1, and the approximate stratigraphical positions of most of them are shown in Figure 3.2.

TROW POINT (SOUTH SHIELDS) TO WHITBURN BAY (NZ 388383–410612)

Highlights

The sea cliffs of this classic site (box 1 in Figure 3.2) provide the key to understanding much of the Magnesian Limestone sequence. In the north, from Trow Point to Frenchman's Bay, lowest beds exposed include the Yellow Sands, Marl Slate and Raisby Formation, and these are overlain, in turn, by (1) the unique algal Trow Point Bed, (2) the dissolution residue of the Hartlepool Anhydrite, (3) collapsed and brecciated Concretionary Limestone strata and (4) possible lower beds of the Roker Dolomite Formation; the upper part of the Raisby Formation was affected by massive submarine slumping (the 'Downhill Slide') and at both Trow Point and Frenchman's Bay contains piles (olistostromes) of large slumped masses (olistoliths). Slightly higher strata exposed between Frenchman's Bay and Lizard Point are almost all of the Concretionary Limestone and feature both spectacular evidence of foundering and brecciation and also primary sedimentary lamination, turbidites and submarine slumps. Strata from Lizard Point southwards are mainly less-obviously affected by foundering and brecciation but feature abundant evidence of sedimentation higher on an unstable submarine slope and contain an important but restricted range of shelly fossils.

Introduction

The bewilderingly varied Permian sedimentary rocks exposed in the sea cliffs between Trow Point and Whitburn Bay, Tyne and Wear, are mainly of the Concretionary Limestone Formation but also include glimpses of the Yellow Sands (1.2 m+) and Marl Slate (0.1-1.5 m) in Frenchman's Bay, extensive exposures of the Raisby Formation (up to 13 m) between Trow Point and Frenchman's Bay and intermittent views of possible Roker Dolomite in cliffs and rock platforms from Whitburn southwards. In northern parts of the site, the Raisby Formation is seen to be overlain by the unusual

Table 3.1 Main geological features of the marine Permian GCR sites in the Durham Province of the English Zechstein

DURHAM PROVINCE		
	Site	**Interest**
Cycle 3 Seaham Formation	Seaham	Type section; complex calcite concretions; *Calcinema;* crinkled algal stromatolites; foundered strata
	Blackhalls Rocks	Calcite concretions; foundered, partly collapse-brecciated
Cycle 2 Seaham Residue (of Fordon Evaporites)	Seaham	Type section; distinctive lithology; plastic deformation; dedolomites
	Blackhalls Rocks	Incidental occurrence
Roker Dolomite Formation	Seaham	Typical lithology passing up to dedolomitized brecciated rock at top
	Blackhalls Rocks	Typical lithology
	Ryhope Cutting (part of Tunstall Hills south)	Partly dedolomitized collapse-breccia with infiltrated cavity-fill
	Hawthorn Quarry	Slightly atypical lithology, partly dedolomitized; collapse-brecciated in east
Concretionary Limestone Formation	Fulwell Hills quarries	Bizarre calcite concretions; Fulwell Fish-bed and other laminites; foundered strata
	Trow Point to north end of Marsden Bay, South Shields	Dedolomitized collapse-breccias with infiltrated cavity-fill
	Marsden Bay, South Shields	Interbedded laminated and turbiditic dolomitized slope carbonate mudstones to grainstones; calcite concretions; dedolomites; foundered strata and breccia-gashes
Cycle 1 Residue of Hartlepool Anhydrite	Trow Point to Frenchman's Bay, South Shields	Typical evaporite-dissolution residue underlying collapse-breccias
	Ryhope Cutting (part of Tunstall Hills south)	Near-reef evaporite-dissolution residue; evidence of past plastic flow
?Ford Formation, Heselden Dene Stromatolite Biostrome	Blackhalls Rocks, Hawthorn Quarry	Coarse conglomerate of rolled blocks of dolomitized reef boundstone overlain by dolomitized algal laminites with spectacularly large domes
Ford Formation, Trow Point Bed	Trow Point	Type section of Trow Point Bed; a distinctive thin unit of marine oncoids, peloids and columnar stromatolites, partly dedolomitized
Ford Formation, shelf-edge reef facies	Claxheugh Rock, Cutting and Ford Quarry, Hawthorn Quarry, Humbledon Hill Quarry, Hylton Castle Cutting, Stony Cut (Cold Hesledon), Tunstall Hills (N and S), Horden Quarry	Massive mainly dolomitized fossiliferous reef boundstone, comprising several sub-facies: reef-base at Claxheugh Rock and Humbledon Hill; basal coquina at Tunstall Hills (N); reef-core at Claxheugh Rock, Cutting and Ford Quarry, Hylton Castle, Humbledon Hill and Tunstall Hills (N and S); reef-backreef contact at Ford Quarry; reef-flat at Hawthorn Quarry and Stony Cut; reef talus at Tunstall Hills (S); reef fissures at Tunstall Hills (N); reef crest at Ford Quarry, Horden Quarry and Stony Cut; reef-top erosion surface at Hawthorn Quarry. Humbledon Hill Quarry and Tunstall Hills are renowned historical faunal sites

Table 3.1 *(continued)*

DURHAM PROVINCE		
	Site	**Interest**
Ford Formation, backreef facies	Claxheugh (Ford) Cutting and Ford Quarry	Reef-backreef contact; sparingly fossiliferous dolomitized mudstone/wackestone with allochthonous slide-blocks or olistoliths (best seen in cutting)
	Gilleylaw Plantation Quarry, Silksworth	Dolomitized ooid grainstones overlain by shelly algal-bryozoan patch-reef; coarse oncoids and lamellar stromatolites at top
	Trimdon Grange Quarry, Trimdon	Typical cross-laminated shallow-water ooid grainstones, extensively replaced by calcite after secondary ?anhydrite; bioturbated
Raisby Formation	Raisby Quarries	Type locality; thick primary limestones; diagenetic breccia; mineralized
	Dawson's Plantation Quarry	Debris flow near base of formation; typical lithology; spatulate listric joints and fractures
	High Moorsley Quarry	Typical lithology with thin debris flow and evidence of large-scale downslope sediment sliding; mineralized; cambered (Quaternary feature)
	Trow Point	Typical lithology; much evidence of bioturbation; major submarine slide plane overlain by debris flow with exceptionally large slide-blocks (olistoliths)
Marl Slate	Claxheugh Rock, Frenchman's Bay	Typical lithology; was locally fluidized and injected downwards into fissures; partly removed by submarine sliding
	Raisby Quarries	Typical lithology; thins against crest of ridge in Basal Permian Sands
Basal Permian Sands (mainly pre-Cycle 1)	Claxheugh Rock, Frenchman's Bay, Raisby Quarries	Typical lithology; top involved in submarine slide-breccia at Claxheugh Rock; remains of fluidized Marl Slate in fissures at Claxheugh Rock; forms ridge in floor of Raisby Quarry and at head of Frenchman's Bay

and exceptionally persistent Trow Point Bed (0–0.60 m), the sole representative of the Ford Formation which was probably more than 100 m thick only 6 km to the west, and this bed is succeeded by the thin (0–0.15 m) dissolution residue of the Hartlepool Anhydrite.

All beds of the Concretionary Limestone have foundered by the former thickness of the dissolved anhydrite (?100 m+ at Marsden) and have responded in a number of ways ranging from barely disturbed (especially in higher parts of the formation) to completely brecciated and dedolomitized; the limestone collapse-breccias at the base of the formation are particularly resistant and are mainly responsible for the ruggedness of the coast between Trow Point and the northern end of Marsden Bay, whereas the less resistant overlying dolomite has been differentially eroded to form Marsden Bay and its neighbour to the south. Farther south, varied secondary limestones, though less resistant than the massive collapse-breccias, have given rise to a variety of lower sub-vertical cliffs and minor bays, and, by their recession, to exceptionally wide rock platforms off Whitburn (=White Burn, an allusion to the white-capped breakers that occur here during certain combinations of tide and weather).

The cliffs, especially those between Trow Point and Lizard Point and around Byer's Hole and Byer's Quarry some distance farther south, have featured freely in the literature. Early mentions were by Winch (1817), who recorded the discovery by

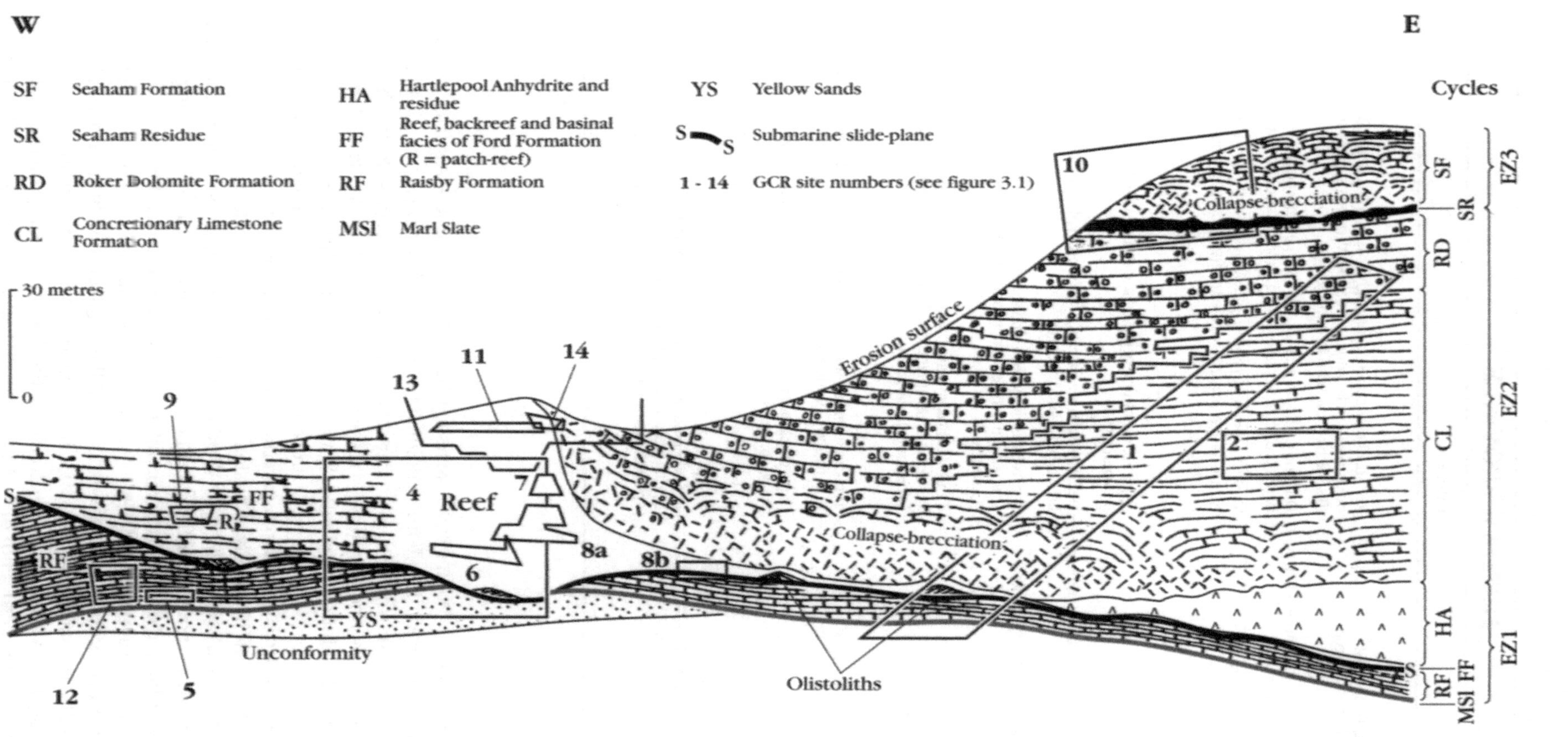

Length of section is approximately 7 kilometres; some features exaggerated

Figure 3.2 Approximate stratigraphical position of GCR marine Permian sites in the northern part of the Durham Province of north-east England (diagrammatic). Some sites in the southern part of the Durham Province cannot be accommodated on this line of section and have been omitted. The Hartlepool Anhydrite would not normally be present so close to the present coastline but is included for the sake of completeness.

Nichol of flexibility in dolomite laminites in Marsden Bay, and by Sedgwick (1829) who also noted the flexibility and graphically described the rocks there, concentrating on the disturbance and brecciation. Bivalves from Byer's Quarry were figured and/or cited by Howse (1848), King (1850) and Logan (1967) and fish remains were found in Marsden Bay in 1836 or 1837 by Miss Green (Kirkby, 1864; Howse, 1891); Howse briefly described the whole section. Clapham (1863) published analyses of three varied samples from Trow Point, Browell and Kirkby (1866) analysed limestone from Byer's Quarry, and Trechmann (1914) gave analyses of specimens from both these locations. Lebour (1884) reviewed the 'gash-breccias' and Card (1892) investigated the flexibility of dolomite laminites from Hendon and Marsden. The sections from Trow Point to Marsden Bay were then exceptionally fully described and illustrated by Woolacott (1909, 1912), who claimed to recognize evidence of low-angle thrusting, a theme returned to by Trechmann (1954), but this interpretation was not generally accepted and the evidence has been reinterpreted by Smith (1970a, c, 1985a) as more consistent with large-scale submarine slumping and collapse-brecciation. Burton (1911) published details of cavity-fill and chert in the breccias at Trow Point and elsewhere.

More recent works include brief reviews of the collapse-breccias by Hickling and Holmes (1931) and Smith (1972), complete geological map coverage on a scale of 1:10560 (Smith, 1975a, b; Land and Smith, 1981, based on fuller notes and scale drawings of all the cliffs and lodged in the fieldnote files of the British Geological Survey), several illustrations and interpretative drawings of strata at Trow Point and in Marsden Bay by Pettigrew (1980) and detailed analyses of a number of rocks from Trow Point and near the Grotto in Marsden Bay by Al-Rekabi (1982). Lastly, Braithwaite (1988) published photomicrographs of samples from near the Grotto and from nearby Marsden Hall Quarry, and discussed the origin of many of the secondary (diagenetic) features in the Concretionary Limestone exposed there. All the northern exposures have also been visited repeatedly by geological excursion parties and numerous guides and excursion reports have been published by local and national geological societies (e.g. Smith, 1973a).

Description

The scheduled site comprises the steep sea cliffs and rock-shore platforms extending uninterruptedly for about 6.5 km between the north side of Trow Point (NZ 667384), South Shields and The Bents (NZ 409613) at Whitburn (Figure 3.3). For the purposes of description it is convenient to divide the site into several sectors, which are described from north to south and the rocks in the order in which they are encountered; summaries of the geology of key parts of these sectors were given by Smith (1975a, b).

Sector 1: Trow Point to Frenchman's Bay, inclusive (Figure 3.4)

Despite great lateral variation, these cliffs display a broadly uniform sequence that has been described partly or wholly by Woolacott (1909), Trechmann (1954), Smith (1970a, c, 1973a) and Land and Smith (1981); the sequence is shown below:

	Thickness (m)
Drift deposits, including boulder clay	up to 6.00
---- unconformity ----	
Concretionary Limestone Formation, mainly brecciated, with much internal sediment (cavity-fill)	up to 11.00
Hartlepool Anhydrite Formation (dissolution residue of)	0–0.15
Ford Formation, Trow Point Bed; peloidal dolomite and dedolomite with oncoids and columnar stromatolites	0–0.60
Raisby Formation, low-slope facies, with disturbed (slumped) beds (0.75–8 m) overlying a discordant slide plane cut into undisturbed beds	0–15.00
Marl Slate (Frenchman's Bay only)	0.10–1.50
Yellow Sands (Frenchman's Bay only)	1.20+

The relationships of the several stratigraphical units are shown diagrammatically in Figure 3.5.

The Concretionary Limestone Formation here mainly comprises a massive, resistant breccia of angular fragments of thinly interbedded laminated and unlaminated calcite mudstone (dedolomite) in

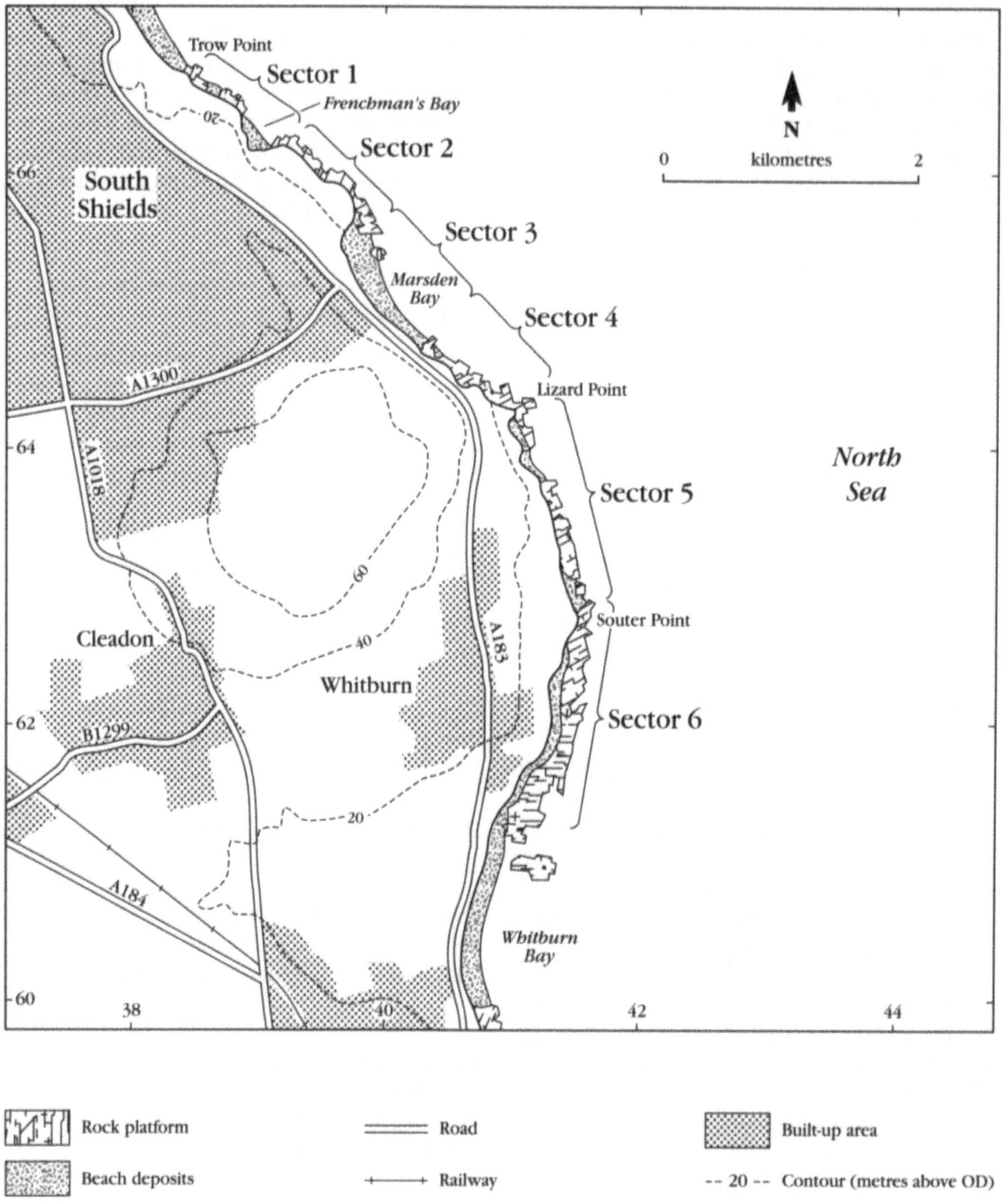

Figure 3.3 Location of the Trow Point to Whitburn Bay GCR site, showing the sectors described in the text.

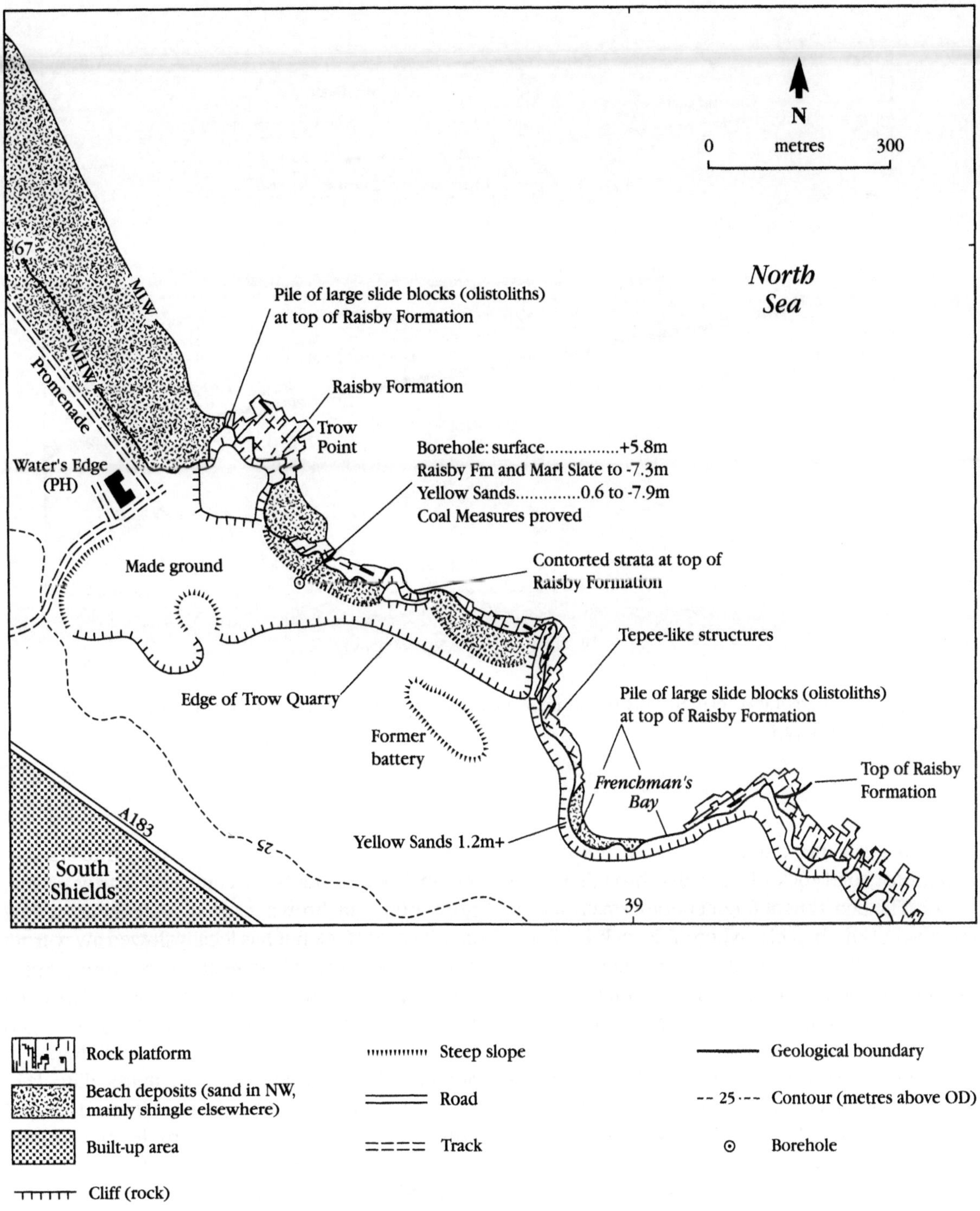

Figure 3.4 The Trow Point to Frenchman's Bay sector, showing the main features of geological interest. In general, strata above high-tide level are collapse-brecciated rocks of the Concretionary Limestone Formation and those below are of the Raisby Formation.

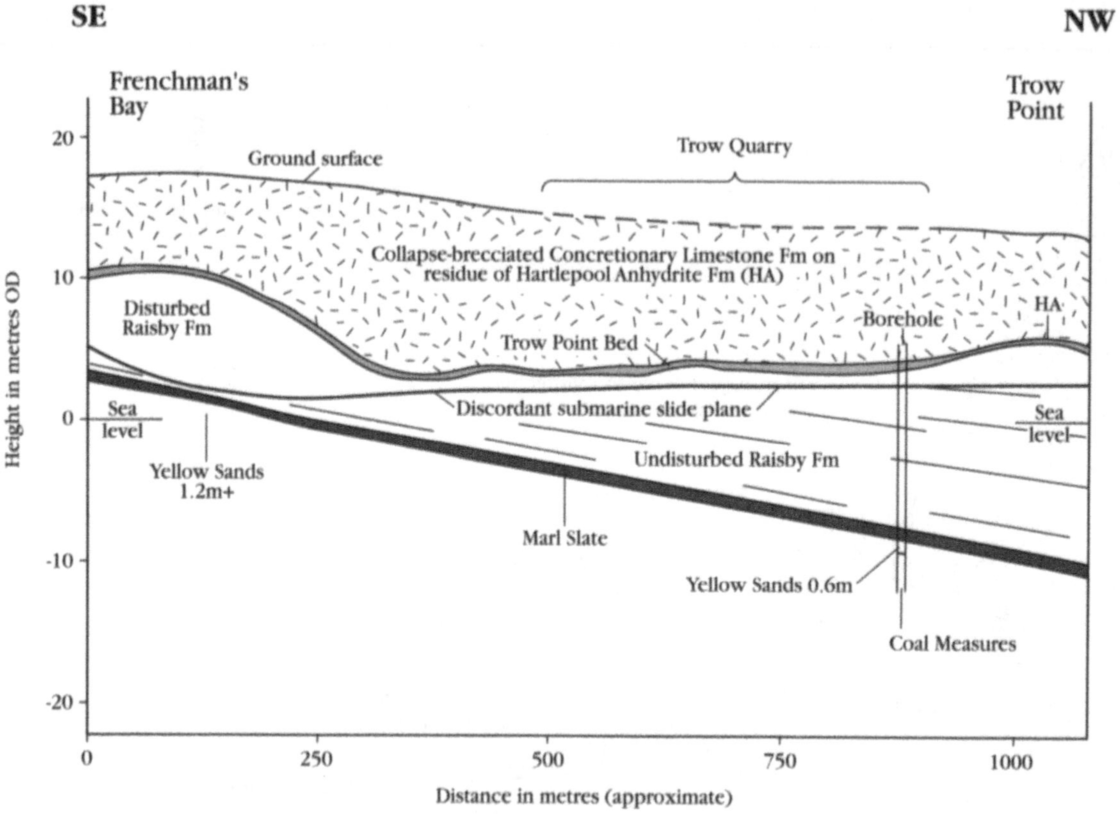

Figure 3.5 Stratigraphical relationships of Permian rock units in the Trow Point to Frenchman's Bay sector, as seen from the north-east.

a microcrystalline calcite matrix; it forms the uppermost solid rock of the cliffs throughout this sector. The constituent fragments are smallest near the base of the breccia, where the rock is almost entirely calcitic (dedolomite), with blocks of disarticulated beds recognizable in higher parts where some dolomite remains; there is much evidence of repeated fracturing and re-cementation, and 'cellular breccias' (Sedgwick, 1829) or 'negative breccias' (Lebour, 1884) in which the clasts having proved to be less resistant to weathering than the matrix are present on the north-east side of Trow Point (Figure 3.6). Silt-grade infiltrated laminar calcite cavity-fill is widespread, especially near the base of the breccia, and contains clasts of detached roof-rocks; the fill bears abundant evidence of intermittent accumulation punctuated by episodes of contortion, tilting and brecciation. The well-documented report of grains of the Yellow Sands in cavity-fill at Trow Point (Burton, 1911) cannot now be verified but is puzzling in view of the known depth of at least 13 m to the top of the 0.6 m Yellow Sands Formation there.

The residue of the Hartlepool Anhydrite is a thin variable bed of unevenly laminated, partly plastic, grey, buff and brown clay; it is generally a few centimetres thick but has locally flowed away from eminences in the substrate and is correspondingly thicker and contorted nearby. A sample of this bed from the south side of Trow Point (NZ 3841 6660) was found by R.K. Harrison and K.S. Siddiqui (in Smith, 1972, p. 260) to comprise micas, illite, kaolinite, gypsum and subordinate calcite, together with detrital quartz, apatite, rutile and zircon; Clapham (1863) reported 10% of silica and 35% of magnesia in a sample apparently from this bed.

The remarkable Trow Point Bed at its type locality has been described in detail by the writer (Smith, 1986). It comprises up to 0.6 m of buff dolomite with subordinate grey limestone (dedolomite), and drapes the underlying hummocky substrate (Figure 3.7) with primary dips of up to 40° and a local relief at Trow Point of about 3 m; in the sector as a whole, its relief relative to the base of the Raisby Formation is at least 14 m. The deposit commonly comprises two beds and is thickest in

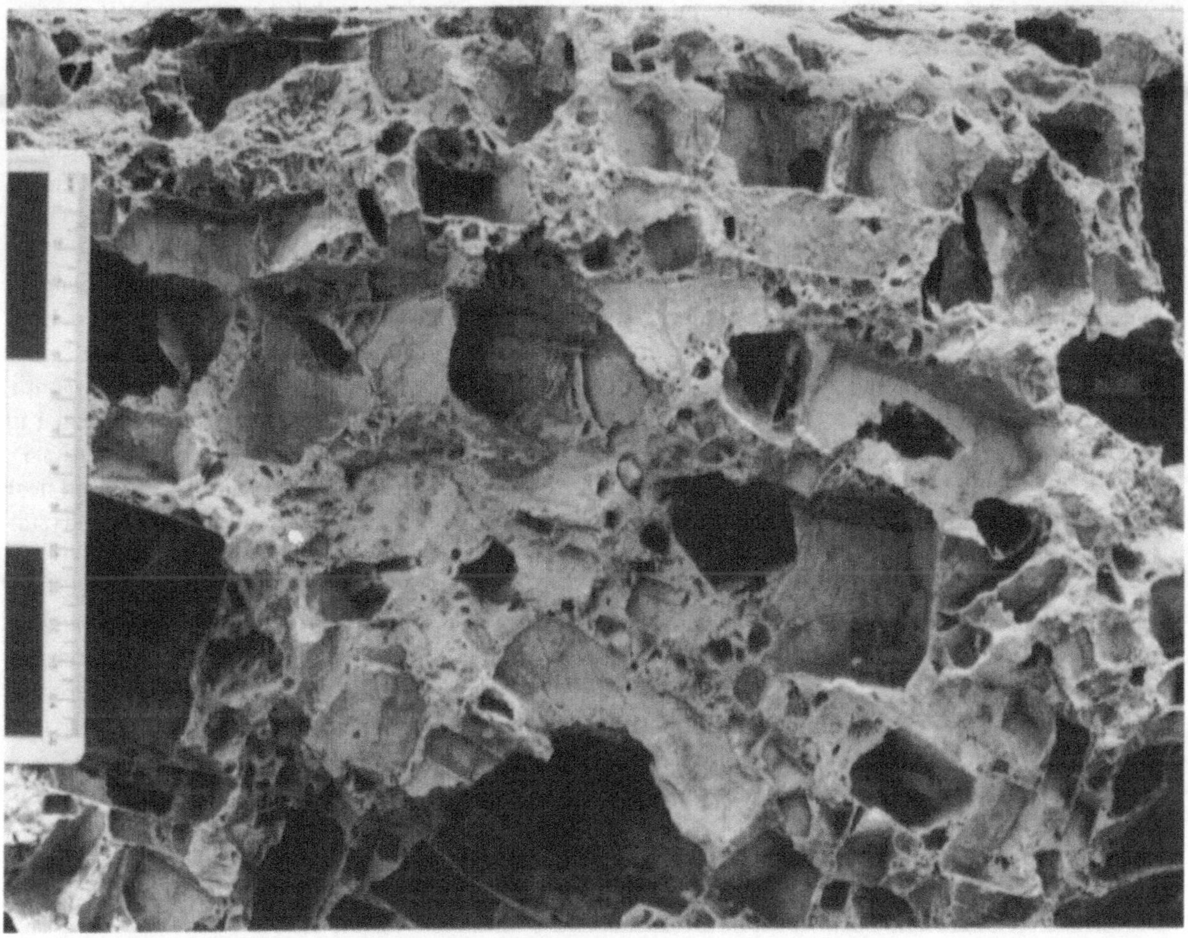

Figure 3.6 'Negative breccia' in collapse-brecciated strata of the Concretionary Limestone Formation at the north-east corner of Trow Point. The clasts (?dolomite) have been removed by weathering so as to leave the more resistant network of calcite veins and matrix. Bar: 0.16 m. (Photo: D.B. Smith.)

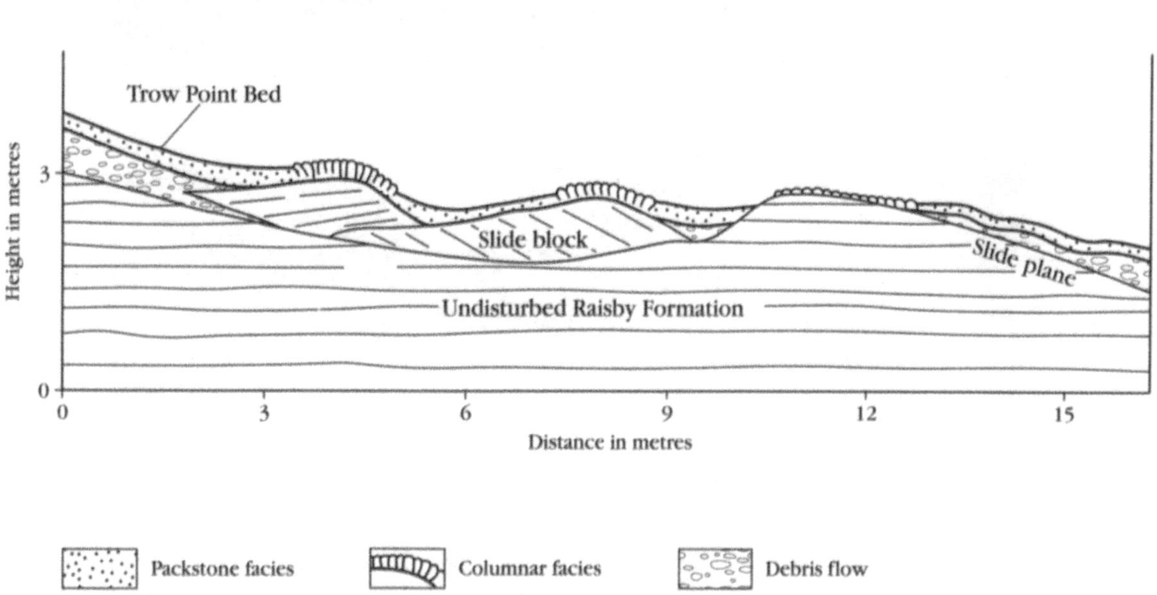

Figure 3.7 Mutual relationships of facies of the Trow Point Bed at its type locality. After Smith (1986, fig. 3).

the hollows where it is mainly an unsorted or poorly-sorted peloid-oncoid packstone; intervening eminences bear one or, more commonly, two layers of radial arrays of narrow (0.01-0.05 m) columnar stromatolites individually up to 0.15 m tall. The packstones contain up to 2% of quartz silt and fine sand and a restricted marine assemblage of foraminifera and ostracods.

The Raisby Formation comprises a disturbed sequence up to about 8 m thick and an underlying undisturbed sequence up to about 11 m thick. The disturbed sequence is extremely varied in thickness, and, where thickest (as at Trow Point and along the south side and head of Frenchman's Bay) is composed of large slide-blocks (many contorted) of buff thin-bedded dolomite lying on a discordant surface interpreted as a major synsedimentary submarine slide plane (Smith, 1970c) (Figure 3.8). Pockets between, beneath and above the slide-blocks are filled with an unsorted mixture of dolomite clasts in a vuggy (i.e. containing many small cavities) dolomite matrix, and similar rock, with scattered slide-blocks or olistoliths, forms a single continuous bed commonly 0.6-1.5 m thick in the cliffs between Trow Point and Frenchman's Bay (Smith, 1970c, plate 2, fig. 2; Pettigrew, 1982, plate 7). The largest slide-block seen by the writer is at the head of Frenchman's Bay and measures about 72 m long and 7.5 m thick; it was also noted by Woolacott (1909, figs 2, 12) and Trechmann (1954, fig. 5), who both interpreted it as a thrust mass. Other slide-blocks, including that illustrated by Pettigrew (1982, fig. 12), contain up to 6.5 m of deformed Raisby Formation strata.

Undisturbed Raisby Formation strata beneath the disturbed sequence have been proved by a borehole (NZ 3847 6652) to be thickest near Trow Point, but they are cut out progressively south-eastwards by the slide plane and are less than 1 m thick at the head of Frenchman's Bay (Figure 3.5). Upper beds, about 5 m of which are exposed near Trow Point, are cream and buff, very finely crystalline dolomites in slightly uneven beds 0.05-0.2 m thick; many feature abundant evidence of bioturbation. The rocks contain small bioclasts, including foraminifera, bivalves, crinoid columnals and obscure plant remains, and also scattered to abundant oval to irregular calcite-lined cavities after former anhydrite; Lee (1990) noted narrow calcitized zones around these cavities. Trechmann

Figure 3.8 Large slide-block (olistolith) of thin-bedded dolomite mudstone/wackestone of the Raisby Formation, resting on a slightly discordant major submarine slide-plane cut onto undisturbed dolomite near the base of the Raisby Formation. The block moved from left to right (i.e. north-eastwards). Coastal cliffs at the north-west side of Frenchman's Bay, South Shields. Bar: 1 m. (Photo: D.B. Smith.)

(1914, p. 245) analysed a sample of the Raisby Formation from Trow Point and reported a dolomite content of 97.19%, and Lee (1990) determined the isotopic composition of secondary limestone from near the top of the formation and found it to be closely comparable with that in the overlying collapse-breccias. Lower beds of the Raisby Formation are exposed progressively south-eastwards, where there is less evidence of bioturbation, fewer cavities, and local traces of graded bedding. Unusual features include tepee-like structures up to 0.5 m high on the shore platform (NZ 3888 6631) about 140 m north of Frenchman's Bay (Figure 3.9) and a complex of low-angle intersecting minor movement planes in the cliff (NZ 3886 6629) slightly farther south (Smith, 1994, plate 4); these terminate sharply upwards at the base of the disturbed beds, here only a few metres above the base of the formation.

Marl Slate is exposed periodically around much of Frenchman's Bay, depending on beach accumulations and rockfalls; it is up to 1.5 m thick on the south side of the bay but thins to 0.1-0.3 m at the head of the bay where onlap against a ridge of Yellow Sands is apparent, and is up to 0.8 m thick in the north of the bay. It is a dark grey pyritic finely laminated argillaceous dolomite, with a thin dolomite bed near the top, and abundant fish scales.

Figure 3.9 'Tepee'-like structures in thin-bedded dolomite mudstone of the Raisby Formation on the shore platform about 140 m north of Frenchman's Bay, South Shields. Hammer: 0.33 m. (Photo: D.B. Smith.)

The oldest Permian rocks of this sector are the early Permian Yellow Sands, which lie at the foot of the cliffs at the head of Frenchman's Bay and form a ridge up to 1.2 m high; their base is not exposed. The sands are of normal lithology for this formation and cross-bedding is inclined mainly northwards; only the uppermost 5-10 cm of the formation is cemented.

Sector 2: Frenchman's Bay to Velvet Beds (Figure 3.10)

The general rock sequence in this 800 m sector is similar to that in Frenchman's Bay, but the Yellow Sands and Marl Slate lie below beach level and the uppermost few metres of the Raisby Formation are exposed only in the north and in a small sharp anticline 200-300 m farther south-east (Land and Smith, 1981). Collapse-brecciated, largely dedolomitized Concretionary Limestone up to 10.5 m thick makes up most of the cliffs in the sector, and has been sculpted into a range of rugged shapes including natural arches; breccias similarly make up the low promontory of Velvet Beds, named after the fine quality of grass formerly present on the thin drift capping.

The most complete sequence in this sector lies in the small anticline, where the dissolution residue of the Hartlepool Anhydrite lies immediately beneath the collapse-breccia and is up to 15 cm thick. The Trow Point Bed, here 0.05-0.60 m thick, is mainly oncoidal but includes ooidal dolomite and unusually narrow columnar stromatolites; in the northern limb of the anticline it cloaks the hummocky upper surface (relief 1.5 m) of the Raisby Formation that here comprises a somewhat enigmatic disturbed sequence (1-4 m) and an underlying undisturbed sequence of thin-bedded finely crystalline dolomites (5 m). Features of unusual interest in the southern limb of the anticline are widespread replacive patches and thin veins of pink and white baryte and some chert in all beds below the residue, and the presence of a well-marked, SSE-facing, 4 m high, steep step in the Trow Point Bed that may be a margin of a minor slump canyon of late Raisby Formation age. The base of the disturbed sequence north of the step is unusually discordant, cutting across the truncated edges of more than 2 m of undisturbed Raisby Formation strata in a horizontal distance of only 10 m. The rising baryte-depositing brines may have been trapped in the anticline by the former anhydrite seal.

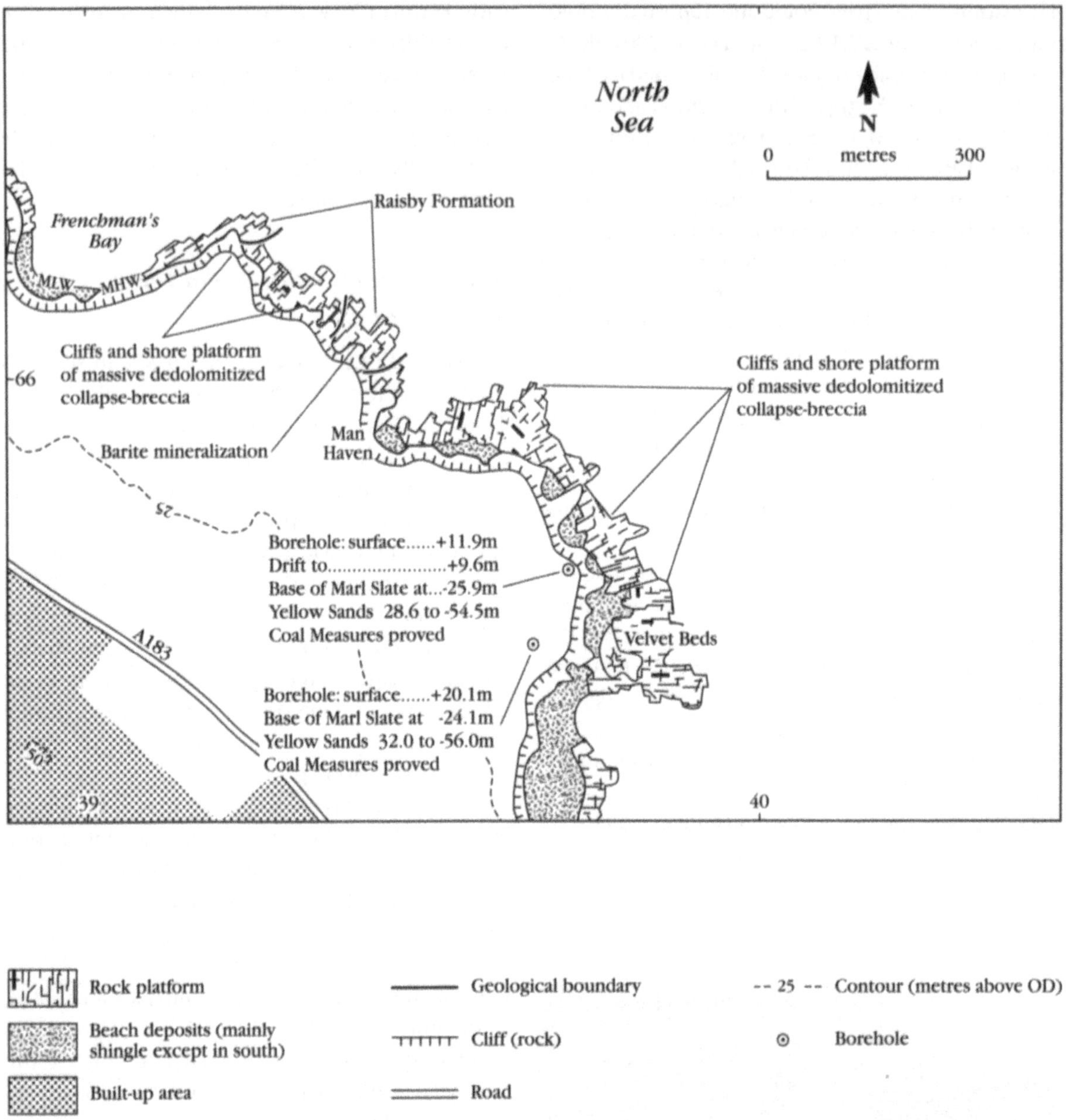

Figure 3.10 The Frenchman's Bay to Velvet Beds sector, showing the main features of geological interest. Except in Frenchman's Bay and in an anticline *c.* 100–200 m north of Man Haven, all the strata are collapse-brecciated rocks of the Concretionary Limestone Formation.

Sector 3: Marsden Bay (Velvet Beds to Marsden Rock) (Figure 3.11)

The broad sweeping curve of Marsden Bay is backed by 15–30 m subvertical cliffs cut in foundered lower strata of the Concretionary Limestone Formation. The response to foundering was extremely varied, with severely collapse-brecciated rocks dominating the northern and southern flanks of the bay and with less dislocated and less altered rocks in the middle. Strata below the so-called 'Flexible Limestone', including the collapse-breccias, were traditionally classified as 'Post-reef Middle Magnesian Limestone' (i.e. Ford Formation) but were reclassified as part of the Concretionary Limestone Formation following detailed mapping and the discovery of a typical Cycle EZ2 fauna in turbidite lags well below the 'Flexible Limestone' (Smith, 1971a).

The disposition of strata in Marsden Bay was dramatically illustrated by Woolacott (1909, plate 2, fig. 8), partly reproduced here as Figure 3.12; drift,

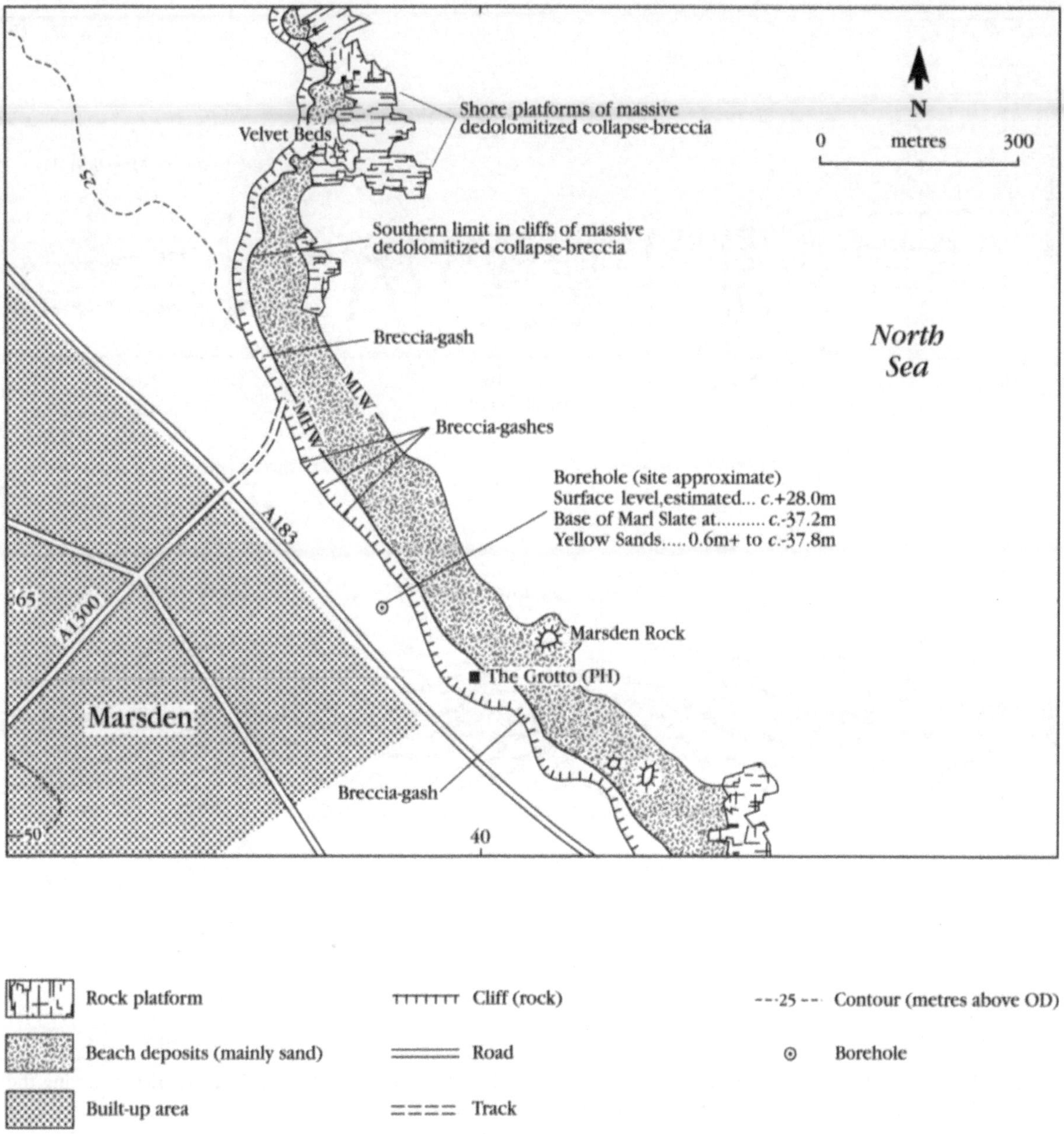

Figure 3.11 The Velvet Beds to Marsden Rock sector, showing the main features of geological interest. All exposed solid strata are of the Concretionary Limestone Formation.

not shown by Woolacott, is generally less than 2 m thick and is a sparingly pebbly silty deposit known locally as the Pelaw Clay.

Massive, dedolomitized collapse-breccias like those between Trow Point and Velvet Beds continue southwards into the north flank of the bay, with the top of the Raisby Formation lying an estimated 5–15 m below beach level at Velvet Beds and the top of the Yellow Sands lying at 24.1 m below Ordnance Datum at the site of the nearby Harton Borehole (NZ 3966 6564). Strata on average dip gently south-eastwards and the sharp and jagged top of the collapse-breccias gradually declines below beach level; the breccias are overlain by about 54 m of cream, buff and grey dolomites and limestones that have been variably dislocated and contain many late-stage 'breccia-gashes' (Figure 3.13) filled with fragments of host rocks and of strata now otherwise eroded off. The most spectacular breccia-gashes are a few metres across and have vertical sides, but some are much larger and have been involved in complex multi-stage foundering and brecciation; the gashes are circular, linear or cruciform in plan.

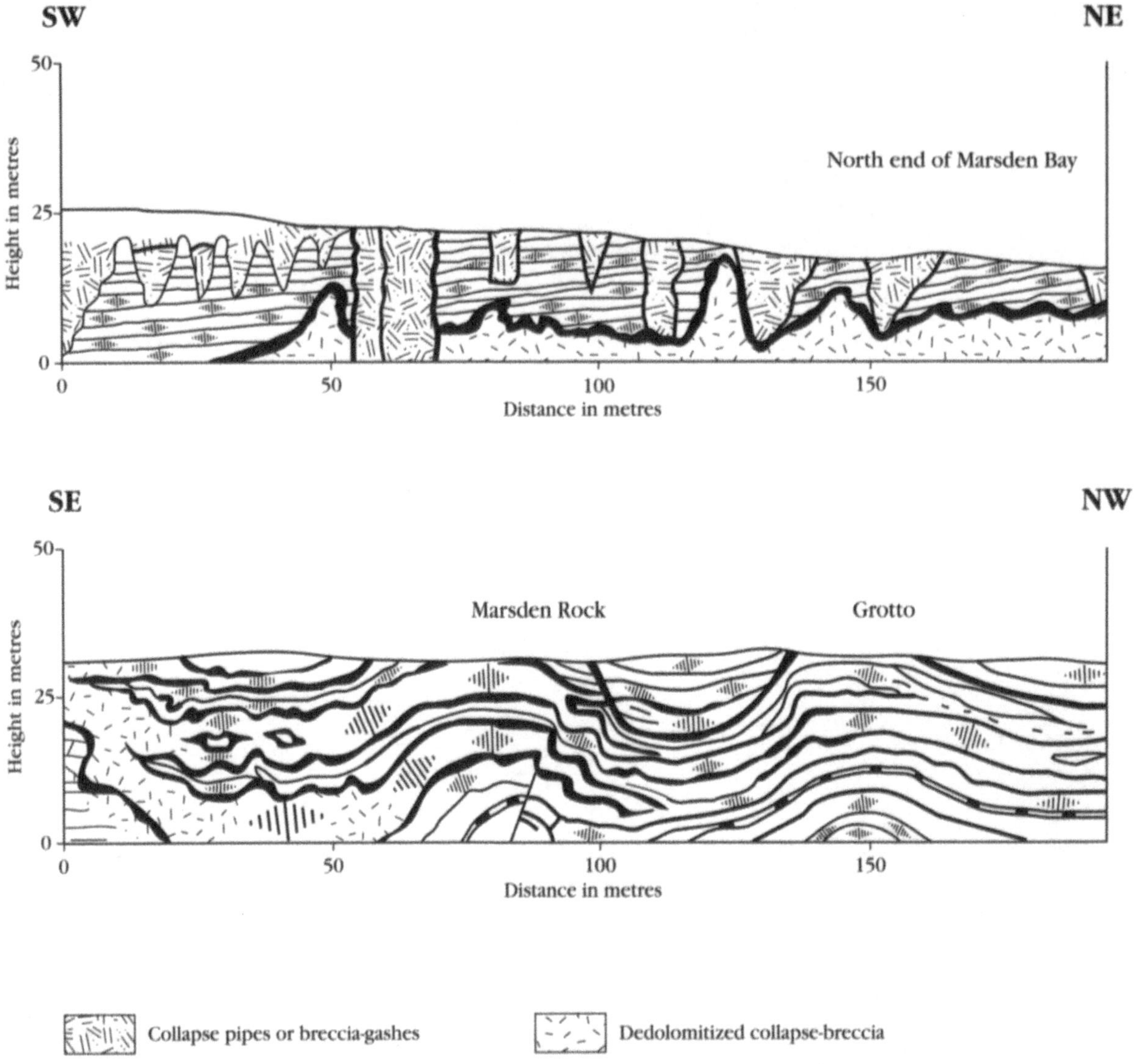

Figure 3.12 Sketch of the cliffs at the northern (above) and southern (below) ends of Marsden Bay, showing the main collapse-related features. All the strata depicted are of the Concretionary Limestone Formation and, where least altered, comprise an interbedded mid-slope sequence of slightly bituminous, finely laminated dolomite mudstones, and sparingly fossiliferous turbiditic and/or slumped dolomite packstones and grainstones. Where severely altered, much of the rock is a hard crystalline secondary limestone (dedolomite). Sketch after Woolacott (1909, plate 2). See Figure 3.13 for detailed distribution of rock types near Velvet Beds.

Least altered and least-brecciated rocks in Marsden Bay comprise a thinly interbedded sequence of plane-laminated dolomite mudstones and dolomite wackestones, packstones and grainstones. The laminites comprise couplets (commonly 15–25 per centimetre), each composed of a carbonaceous film and a thicker dolomite mudstone layer; they contain no shelly fauna but fish remains were recorded (Kirkby, 1864; Howse, 1891) from the 'Flexible Limestone' here, a particularly finely laminated 3–4 m bed that first appears high in the cliff in the northern part of the bay and dips below the beach south of the Grotto (Figure 3.12). Most of the unlaminated beds are a few millimetres to a few centimetres thick and are of silt-grade dolomite; some are graded or reverse-graded. Other unlaminated beds are up to 4 m thick, and comprise dolomite packstones and grainstones (at least some ooidal) that feature a wide range of overfolds, contortions and shear-planes and commonly overlie a slightly discordant erosion surface or slide-plane; some of these thicker units have a basal lag concentrate of gastropods, bivalves and ostracods. Lenses of chert are not

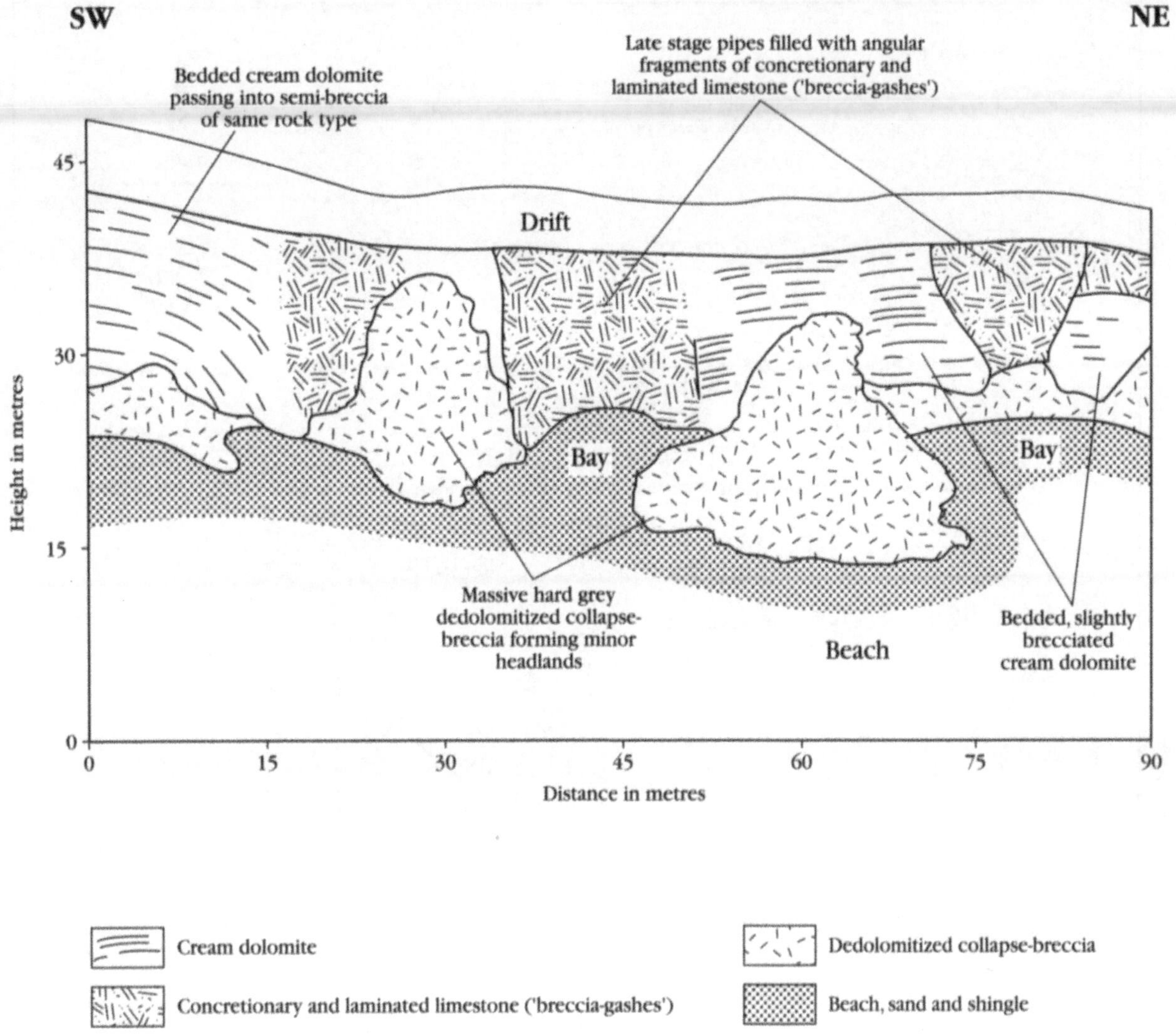

Figure 3.13 Foundered strata of the lower part of the Concretionary Limestone Formation, showing massive dedolomitized collapse-breccias sharply overlain by slightly to severely collapse-brecciated dolomite and limestone; late-stage breccia-gashes (or collapse-pipes) cut the latter. The residue of the Hartlepool Anhydrite probably lies 2-5 m below the lowest rocks shown. Cliffs at Velvet Beds, north end of Marsden Bay, South Shields. The field of view lies near the northern end of the cliffs shown in Figure 3.12 (upper section). After Smith (1994).

uncommon, and irregular to ovoid cavities after former replacive and displacive anhydrite are widespread and locally abundant.

In addition to passing laterally into collapse-breccias, all the dolomite rocks locally pass laterally into secondary, grey or brown, crystalline limestone, which also forms a thick concretion-rich bed high in the cliffs in the northern part of the bay and forms much of the cliff and parts of the stacks near the Grotto.

Sector 4: Marsden Rock to Lizard Point (Figure 3.14)

The geology of this 1.1 km sector of the GCR site is poorly documented in the literature, but the north-western end was included in Woolacott's (1909) drawing of strata in Marsden Bay and the whole sector was summarized by the writer (Smith, 1975a) in notes 1-5 on Geological Survey 1:10,560 Sheet NZ 46 SW; drawings and more detailed descriptions are in the fieldnote files of the British Geological Survey. The subvertical cliffs are generally 15-25 m high but locally approach 30 m, and drift (mainly Pelaw Clay) is generally 0.5-1.5 m thick.

The sequence in the north-western part of the sector is a continuation of that in Marsden Bay, with partly to severely dislocated collapse-brecciated laminated and unlaminated cream dolomites and

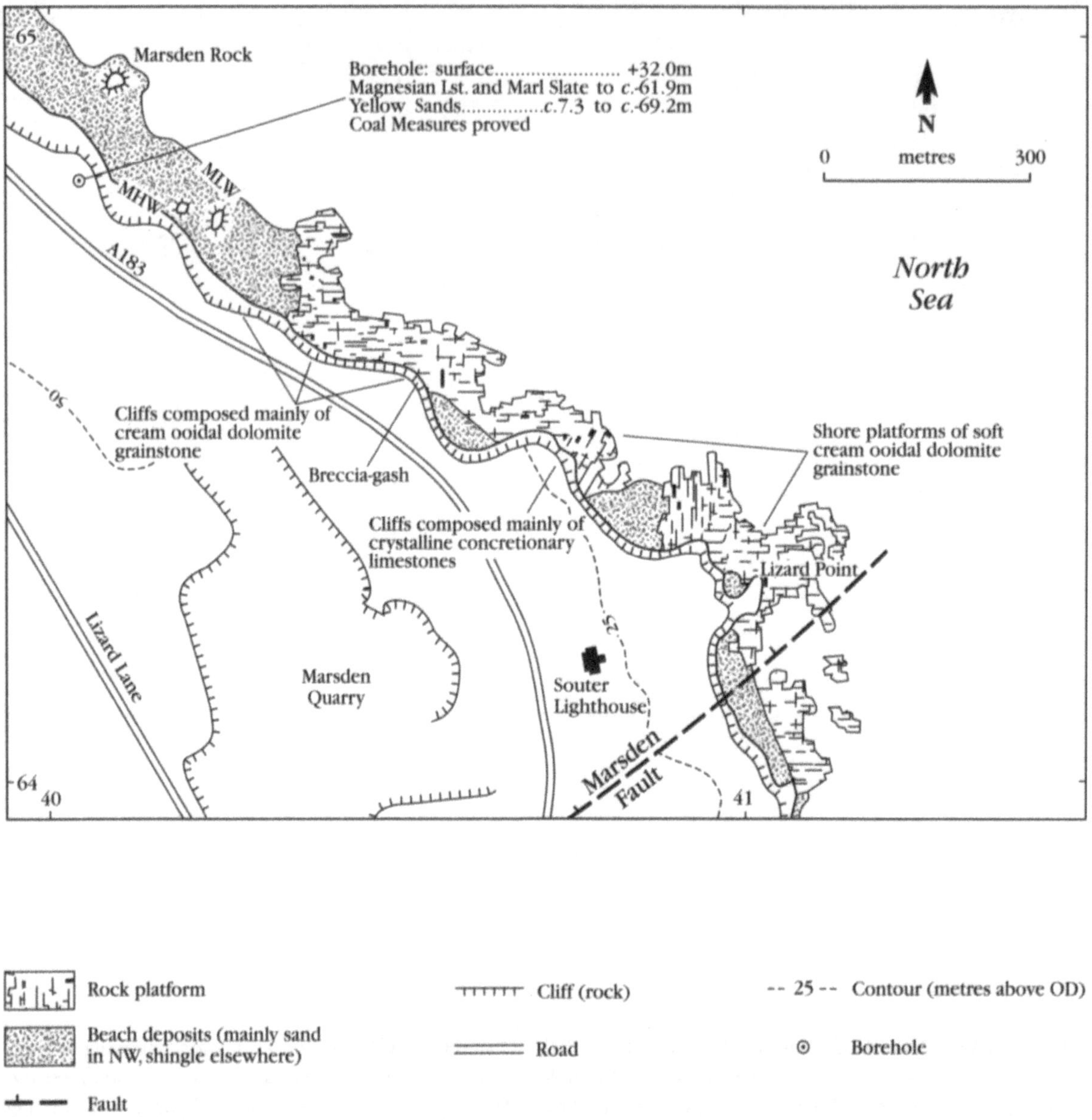

Figure 3.14 The Marsden Rock to Lizard Point sector, showing the main features of geological interest. All exposed solid strata are of the Concretionary Limestone Formation. For further details of strata see British Geological Survey 1:10,560 Sheet NZ 46 SW.

grey to brown secondary limestones forming most of the cliffs; sparingly shelly, vaguely bedded, dolomite ooid packstones/grainstones, however, form the basal few metres of a relatively undisturbed sequence from about 130 to 200 m south-east of the Grotto.

Rocks forming the cliffs in most of the central and southern parts of the sector mainly comprise up to 20 m of vaguely-bedded to massive, altered dolomite ooid packstones/grainstones, but these are overlain by concretionary limestones from about 600 to 725 m north-west of Lizard Point. These limestones reappear at the cliff top about 400 m north-west of Lizard Point and dip gently south-eastwards so as gradually to form the whole cliff; they are at least 14 m thick. The packstones and grainstones are divisible into a variable lower unit in which they are unevenly coarsely interbedded with discontinuous sheets and lenses (?rafts) of laminated dolomite mudstone (some contorted and sheared), and a more uniform 9 m upper unit. Both units contain scattered to abundant bivalves, gastropods and ostracods which, in the lower unit, are concentrated near the base of the thicker ooid beds and lenses. The contact between the ooidal dolomite and the overlying concretionary limestones is marked by an almost continuous, thin, brecciated layer rich in cannon-ball concretions and a similar layer lies 3.5-4 m higher; the concretionary beds themselves are mainly thin- to thick-bedded (locally massive) crystalline limestones in which spherulites are generally abundant and in places form most of the rock. In a few places the concretion-bearing limestones pass laterally into thin-bedded finely-laminated dolomite with only scattered mainly small incipient calcite concretions. Some spherulites in the concretionary limestones are nucleated onto tumid well-preserved bivalves.

Sector 5: Lizard Point to Souter Point (Figure 3.15)

The cliffs in this 1.6 km sector are highest - commonly exceeding 10 m - in the north, but gradually decrease in height from Byer's Hole southwards and are only a few metres high between Wheatall Way and Souter Point (for locations see Figure 3.15); drift (mainly Pelaw Clay) is generally less than 2 m thick except near Whitburn Colliery village where it reaches 5 m for a short distance.

Magnesian Limestone strata in the cliffs and shore platforms here all belong to the middle and upper parts of the Concretionary Limestone Formation and, despite having foundered by at least 100 m through the dissolution of the Hartlepool Anhydrite, are mainly structurally simple and only locally collapse-brecciated; north of the Lizards Fault they comprise a gently rolling strike sequence totalling perhaps 20-25 m thick but an additional 10-15 m of strata may be present to the south of the fault. Cliffs in the most northerly 150 m of the sector, at and immediately south of Lizard Point, are almost entirely of thin-bedded to massive crystalline limestone (mainly spherulitic), and thick-bedded to massive spherulitic limestone forms the cliffs at Souter Point and for about 100 m to the north. Between these stretches the cliffs are mainly composed of a laterally variable interbedded sequence of unlaminated and laminated mainly thin-bedded grey limestone, grey and brown (locally red and black) spherulitic limestone and lenticular to relatively persistent thick beds of cream finely crystalline to powdery dolomite (some possibly of altered oolite); most of the latter, and many of the thick unlaminated limestone beds, have been weakly to strongly contorted and locally brecciated by contemporaneous downslope movement (Figure 3.16).

The remarkable lateral variability of these strata is expressed in several ways, including the proportion of calcite spherulites and thin calcite lenses present and in changes of bed thickness; thus, for example, there are several places where substantial units of well-bedded laminated or unlaminated dolomite- or calcite-mudstone pass abruptly or at a stepped contact into coarsely crystalline spherulitic or (uncommonly) reticulate limestone, and other places where thin-bedded limestones or dolomites pass laterally into thick units with only vague bedding traces. Elsewhere there is convincing evidence of the partial dissolution of carbonate beds, leading to the collapse and brecciation of immediately overlying strata (Smith, 1994, plate 30). Idiomorphic calcite scalenohedra up to 0.05 m long, though also present elsewhere in the district, are a feature of patches of powdery dolomite between concretions in this sector. The sequences in individual parts of the cliffs are summarized in notes 5-12 on British Geological Survey 1:10,560 Sheet NZ 46 SW and NW (Smith, 1975b) and detailed scale drawings of all the cliffs are lodged in the Geological Survey fieldnote files.

The Concretionary Limestone at Byer's Hole and in the adjoining Byer's Quarry (now filled) is well known for its foraminifera, annelid, gastropod, bivalve and ostracod fauna (Figure 3.17) which comprises abundant individuals of a restricted range of species; well-preserved remains of plants have also been reported (Trechmann, 1914). The fauna was noted and listed by Howse (1848, 1858), King (1850), Kirkby (1858), Trechmann (in Woolacott, 1912), Logan (1967), Pattison (Geological Survey internal reports 1967; 1977) and Pettigrew (1980); both King and Logan figured several specimens from here, including some designated as types or syntypes. Additionally, King (followed by Logan who used many of King's specimens) cited 'Souter Point, Marsden' as a bivalve source locality, though it is possible that he meant Lizard Point. Howse (1848) noted the exceptional preservation of bivalve shells at Byer's Quarry, where the original shell has been replaced by crystalline calcite, in contrast to most Magnesian Limestone fossils which are known only from casts; Kirkby (1858) noted that the ostracods, too, are locally exceptionally well-preserved and abundant (see also Pettigrew, 1980, plate 13).

Analyses of grey limestones from Byer's Quarry (Browell and Kirkby, 1866; Woolacott, 1912;

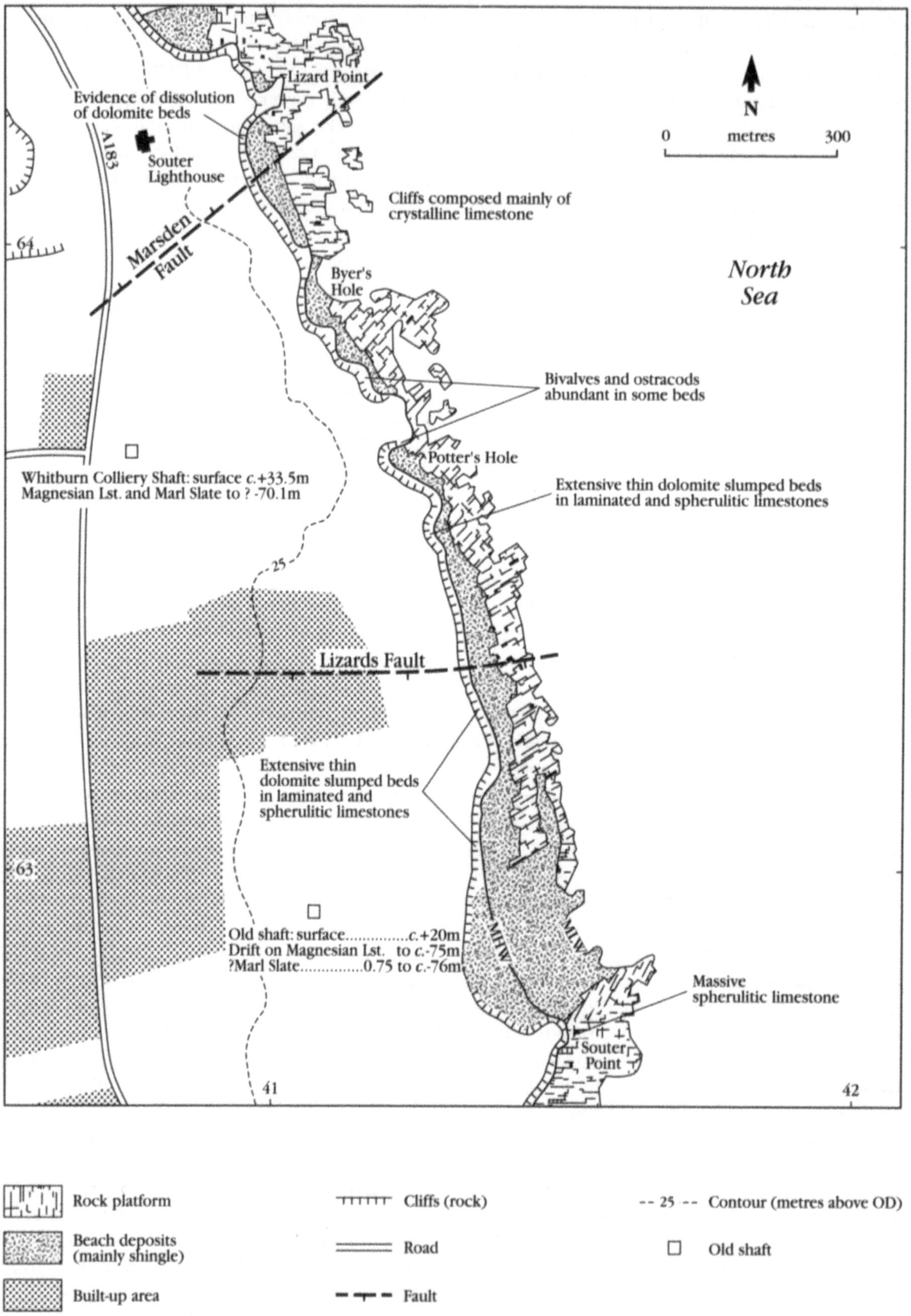

Figure 3.15 The Lizard Point to Souter Point sector, showing the main features of geological interest. All exposed solid strata are of the Concretionary Limestone Formation. For further details of strata see British Geological Survey 1:10,560 Sheet NZ 46 SW.

Figure 3.16 Tight slump folds in high-slope thin-bedded calcite mudstones of the Concretionary Limestone Formation. Coastal cliffs *c.* 500 m south of Potter's Hole, Whitburn Colliery. Hammer: 0.33 m. Reproduced by permission of the Director, British Geological Survey: NERC copyright reserved (NL 138).

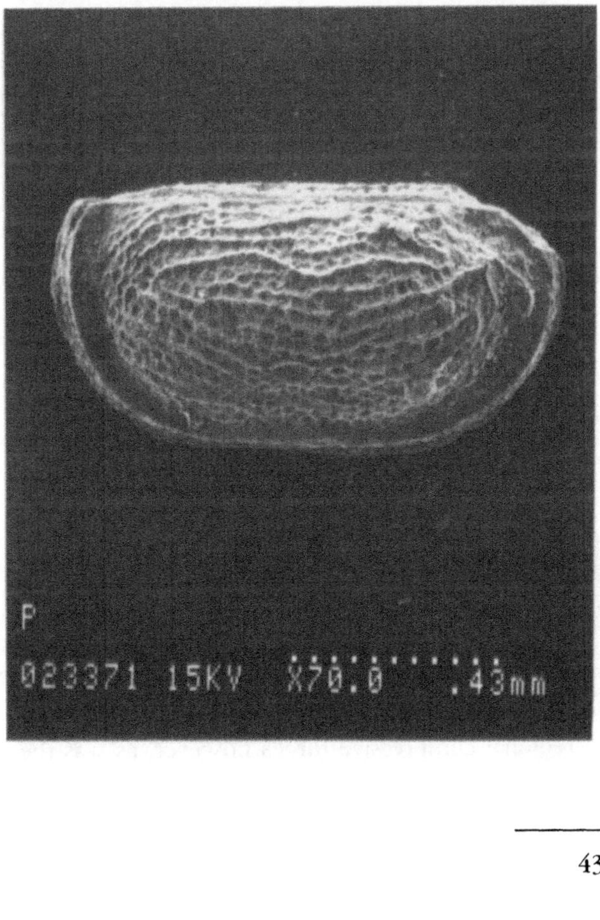

Figure 3.17 *Kirkbya permiana* (Jones), a typical ostracod from high-slope calcite mudstones of the Concretionary Limestone Formation. Top of coastal cliffs on the south side of Byer's Hole, Whitburn Colliery. Bar: 0.43 mm. (Photo: Sunderland Museum TWCMS: P1004.)

Trechmann, 1914) show that they are amongst the purest limestones in the area, with calcium carbonate contents ranging from 96.94 to 98.04% (three analyses); an interbedded brown friable bed analysed by Trechmann had a calculated composition of 93.2% of dolomite and 5.8% of calcite.

Sector 6: Souter Point to Whitburn Bay (Figure 3.18)

The cliffs in this most southerly sector (1.6 km) of the Trow Point to Whitburn Bay site are generally 6–10 m high, of which drift forms the uppermost 2–3 m from Souter Point to about 150 m north of White Steel; the drift then thickens gradually southwards so as to form the whole of the cliff from a point about 250 m north-east of The Bents. The thin drift in most northern parts of the sector is mainly of the sparingly stony Pelaw Clay but that in the south also includes Durham Lower Boulder Clay and interbedded laminated clay and sand and features widespread and locally intense contortion and involution (Smith, 1981c, fig. 6).

Foundered Magnesian Limestone strata in this sector are structurally and lithologically more varied than in the sector to the north but are similarly gently rolling; dip is generally eastwards at perhaps 3–5°, with a strike section between Souter Point and Rackley Way Goit passing southwards into a broad shallow apparent syncline with its axis about 50 m south of White Steel. Breccia-gashes occur at intervals throughout the sector and, together with several minor faults, are particularly well-exposed in the cliffs and wide shore platforms in the south of the sector (see Geological Survey 1:10,560 Sheet NZ 46 SW and notes and scale drawings in British Geological Survey fieldnote files).

Strata in the strike section north of Rackley Way Goit are probably about 12–15 m thick and belong to the upper part of the Concretionary Limestone Formation; they are lithologically and faunally similar to those exposed in the sector to the north and display comparably extreme lateral and vertical variation. Limestones in the cliffs in the most northerly 450 m of the sector (and locally elsewhere) have been partly to severely brecciated to depths of as much as 4 m below rockhead, probably by Devensian periglacial cryoturbation; drift erratics are mixed with angular limestone debris in the uppermost metre or so of this brecciated sequence.

Strata in the apparent syncline south of Rackley Way Goit total perhaps 20 m and may belong partly or wholly to the Roker Dolomite Formation. They are mainly of thin- to thick-bedded porous, cream, saccharoidal dolomite (probably mainly altered oolite) and include a spectacular basal 5 m bed packed with mutually-interfering 0.05–0.25 m calcite spheroids; this bed, which could be assigned either to the Concretionary Limestone or the Roker Dolomite, may equate with the famous Cannon-ball Rocks' at Roker, 2.5 km farther south. Highest strata in the sector occupy a breccia-gash at the eastern tip of White Steel (NZ 4133 6192) and include dolomite laminites (?algal stromatolites) and remarkable cellular massive limestones interpreted by Dr G.M. Harwood (pers. comm. 1988) as compressed bivalve coquinas.

The subvertical Hebburn (or Monkton) tholeiite dyke, previously not known to crop out in the coastal area, was discovered in 1993 in the cliffs (NZ 4108 6156) of this sector by Mr G. Fenwick of Sunderland University. The dyke and its effects on the host-brecciated limestone are now being investigated.

Interpretation

The Magnesian Limestone rocks of the Trow Point to Whitburn Bay site together constitute a unique assemblage of exposures of truly international significance; the display of submarine slump products in the most northerly sector ranks high in such features anywhere in Britain, the overlying Trow Point Bed is unique in its extent and its range of mixed coated grains and sessile stromatolites, and the effects of evaporite dissolution on overlying Cycle EZ2 carbonate strata are spectacularly exposed between Trow Point and Lizard Point; where least affected by collapse brecciation, the Cycle EZ2 rocks bear striking evidence of carbonate deposition on an unstable submarine slope. The Yellow Sands, Marl Slate and undisturbed lower beds of the Raisby Formation are normal for those formations and require no special comment.

Disturbed beds of the Raisby Formation

The compelling evidence of lateral movement of large masses of bedded dolomite at the top of the Raisby Formation at Trow Point and in Frenchman's Bay was recognized and illustrated by Woolacott (1909, 1912) and Trechmann (1954), who attributed it to tectonic thrusting and interpreted the underlying plane of discontinuity as a thrust plane. The inferred directions of movement of the displaced masses were inconsistent with regional compressive forces however, as was the

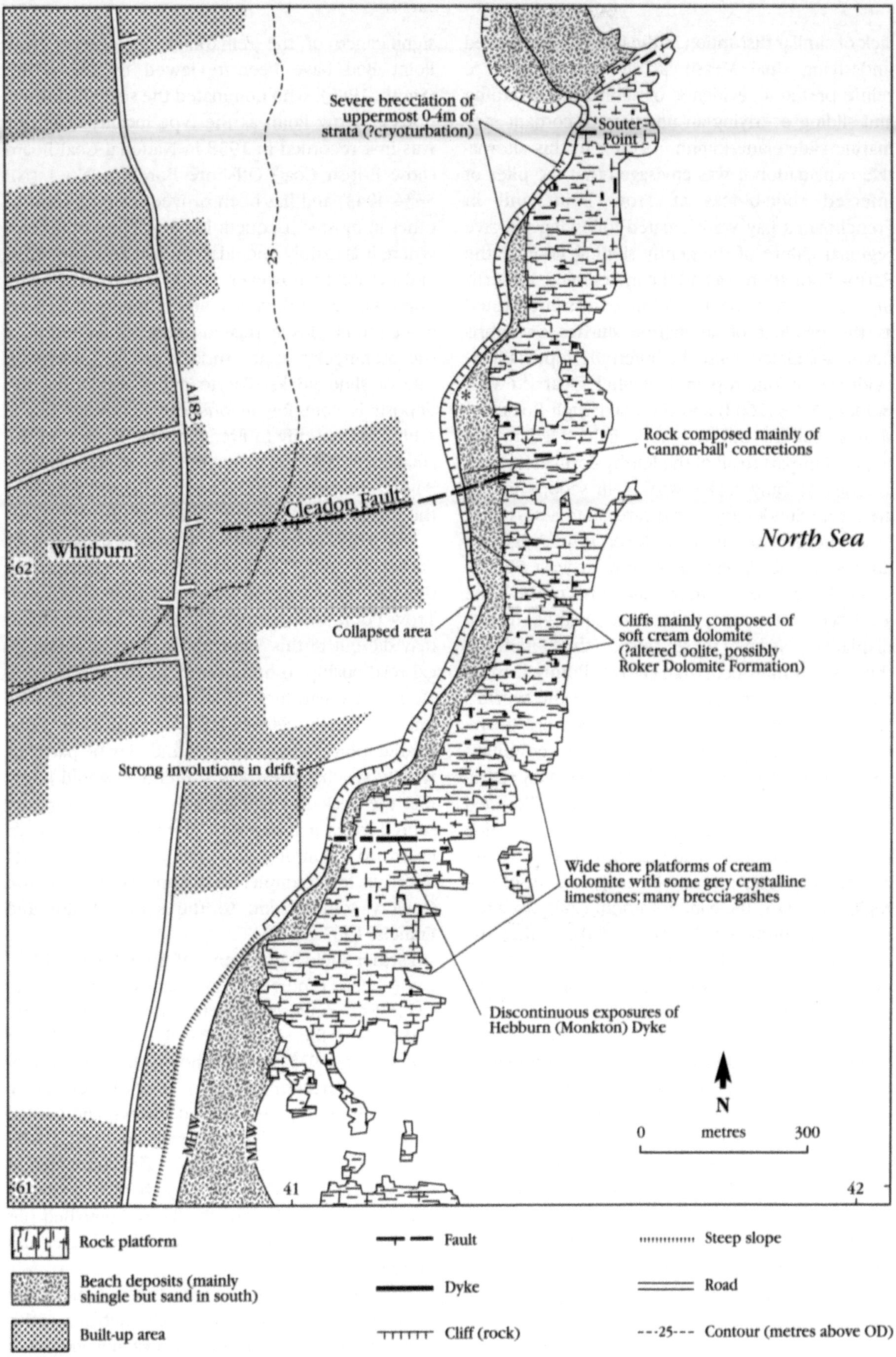

Figure 3.18 The Souter Point to Whitburn Bay sector, showing the main features of geological interest. All exposed strata north of Rackley Way Goit (*) are of the Concretionary Limestone Formation but those to the south may include lower beds of the Roker Dolomite Formation.

lack of similar disruption in the intensively worked underlying Coal Measures, and the data were reinterpreted as evidence of submarine slumping and sliding overlying an undulate discordant submarine slide-plane (Smith, 1970c). In this alternative explanation it was envisaged that the piles of inferred slide-blocks at Trow Point and in Frenchman's Bay were created following massive regional failure of the gently sloping floor of the Raisby Formation sea and the intervening, underlying and overlying pebbly dolomite was interpreted as the product of submarine slurries or debris flows associated with the inferred slope failure. Evidence at outcrop in Downhill Quarry (West Boldon, NZ 347602) and the Claxheugh Rock site shows that the inferred slope failure led to the removal of part to all of the Raisby Formation from a series of canyon-like WSW/ENE scoops in the area from Sunderland northwards, and assessment of all the data south of the River Tyne suggested downslope displacement there of at least 50 million cubic metres of strata; data north of the Tyne are fewer but suggest at least an equal volume of displaced strata there also, thus ranking the Downhill-Claxheugh-Trow Point-Frenchman's Bay submarine slides amongst the world's largest. Contemporary earth tremors were postulated as the cause of the instability and are thought also to have been responsible for abundant spatulate minor movement-planes and clay (liquified Marl Slate) intrusions in strata below the slide plane; such tremors could have originated through movement along one of the major faults of the region, perhaps the Ninety Fathom Fault. An alternative interpretation, by Lee (1990), is that the slope-failure may have resulted from a major decline of sea level following deposition of the Raisby and Ford formations.

Regardless of the cause of the Downhill Slide, it might be supposed that the downslope submarine movement of so vast a volume of sediment would generate a substantial tsunami, though it is not clear if or how such an event would be recorded in the rock record or if it could be distinguished from the record of the event itself. A thin unit of sandstone or siltstone has been reported at the appropriate stratigraphical level in a number of coal exploration boreholes off the Durham coast, and the possibility that this may be the product of a tsunami cannot be excluded.

Trow Point Bed

The distribution, character and environmental significance of the remarkably persistent Trow Point Bed have been reviewed by the writer (Smith, 1986), who nominated the several cliff sections at Trow Point as the type locality. The bed was first recorded in 1958 in National Coal Board (now British Coal) Offshore Borehole No 1 (NZ 5334 4043) and has been proved in some form or other in most subsequent local offshore boreholes where it is mainly oncoidal; the key to its great lateral variability, however, is only apparent at Trow Point where the thickness and lithology of the bed is seen to be closely related to the configuration of the hummocky upper surface of the underlying pile of slide-blocks. Baryte mineralization of the deposit is common in offshore boreholes but is seen onshore only in Frenchman's Bay and in the small bay 200–300 m south-east of Frenchman's Bay where the disturbed and undisturbed beds at the top of the Raisby Formation are also affected.

The Trow Point Bed is the youngest Cycle EZ1 carbonate unit of the Ford Formation basinward of the reef, but it has not been identified between Trow Point and the toe of the reef foreslope and may die out in this 5 m-wide belt. Its stratigraphical relationship to both the reef and the Hesleden Dene Stromatolite Biostrome are therefore unknown, although it clearly occupies the same stratigraphical slot as the reef and may be partly or wholly synchronous; a post-reef age would imply that the period of reef growth is wholly unrepresented by basinal deposits, emphasizing the sharp eastward thinning of reef-equivalent strata inferred, for example, from tunnels at Easington Colliery some 22 km to the south (Smith and Francis, 1967, fig. 21).

Farther afield, equivalents of the Trow Point Bed have been reported immediately beneath the Cycle EZ1 anhydrite at depth in North Yorkshire and parts of the Southern North Sea Basin (Taylor and Colter, 1975), in boreholes and surface outcrops in northern Germany (Füchtbauer, 1968; Richter-Bernburg, 1982) and in widely spaced boreholes in Poland (e.g. Peryt and Peryt, 1975; Peryt and Piatkowski, 1976). The provings in Germany suggest that the bed there is probably closely comparable with the Trow Point Bed but the Polish occurrences are somewhat thicker and more varied.

The environmental interpretation of the Trow Point Bed is a matter of lively debate. Smith (1970a), influenced by then prevailing views on the uniquely peritidal growth of columnar algal stromatolites and of net-fabric sulphate rocks, initially inferred that the bed formed near contemporary sea level, which implied marine drawdown equivalent to the height (100 m+) of the reef foreslope. This view was subsequently modified when more recent work showed (a) that many modern stromatolites are formed subtidally (e.g. Monty, 1973) and (b) that the net-fabric of the Hartlepool Anhydrite may be secondary and therefore does not necessarily indicate intertidal sabkha accumulation. The Trow Point Bed is now interpreted as a basin-floor deposit, that probably accumulated slowly in somewhat unusual marine conditions at a depth of perhaps 25–100 m; deep drawdown is not necessitated (but is not excluded) on the evidence in north-east England, though Peryt and Piatkowski (1976) deduced such drawdown on the evidence of inferred pedogenic features in the supposedly equivalent beds in Poland.

The Hartlepool Anhydrite Residue and overlying collapse breccias

Though previously described as 'a sort of mylonite' by Woolacott (1909), who regarded it as part of his evidence for regional thrusting, the thin mixed layer between the Trow Point Bed and the Cycle EZ2 breccias is now accepted as the dissolution residue of the Hartlepool Anhydrite; in coal exploration boreholes offshore, the same stratigraphical interval is occupied by up to 150 m of massive anhydrite, which is directly underlain by the Trow Point Bed (Magraw *et al.*, 1963; Smith and Francis, 1967, plate 13A; Smith, 1986, fig. 10). The anhydrite is almost devoid of siliciclastic impurities (Trechmann, 1913, p. 243), which accounts for the remarkable thinness of the residue at most localities.

The varied and spectacular breccias of the cliffs between Trow Point and Lizard Point have been noted and described by Winch (1817), Sedgwick (1829), Howse and Kirkby (1863), Lebour (1884), Woolacott (1909, 1912, 1919a), Hickling and Holmes (1931), Trechmann (1954) and Smith (1972, 1985a, 1994). The brecciation and associated mineralogical changes are greatest in the lowest 10–30 m of the formation and diminish unevenly upwards, but there are places where large blocks of strata have foundered with relatively little brecciation or alteration, and other places where severe brecciation and diagenetic changes extend well up into the formation and even into the overlying Roker Dolomite Formation. Sedgwick (1829) was the first to deduce that fracturing and cementation of many of the breccias had taken place repeatedly, and Howse and Kirkby (1863) were the first to suggest that the late-stage breccia-gashes (their 'breccia-dykes') were formed by the collapse of the roofs of large cavities. Early suggestions that the more extensive brecciation of rock in the coastal cliffs might have accompanied or followed the dissolution of interbedded evaporites were strongly supported by Trechmann (1913) in view of the known presence of thick anhydrite beneath the Roker Dolomite at Hartlepool, but the confirmation of the precise stratigraphical position of the anhydrite awaited the drilling of cored coal-exploration bores offshore (Magraw *et al.*, 1963). The calcitization ('dedolomitization') of the breccia clasts was investigated by Woolacott (1919a) who concluded that it resulted from the reaction between dolomite and calcium sulphate solution (Von Morlot's reaction) and Gillian Tester (pers. comm., 1988) records clear evidence that patchy chert and chalcedony nodules in the breccias have replaced both gypsum and anhydrite. Also at Trow Point, Al-Rekabi (1982, p. 106) reported fibrous chalcedony after calcite.

The discovery near Whitburn of a surface exposure of the Hebburn or Monkton Dyke by Mr G. Fenwick is important partly because of its bearing on the time when the Hartlepool Anhydrite was dissolved. The dyke is one of a swarm with a radiometric age of about 58 million years (Evans *et al.*, 1973; Mussett *et al.*, 1988, fig. 2) and, judging from its partly dendritic shape at outcrop, almost certainly intruded country rock that had already been brecciated. The brecciation, and thus the dissolution of the anhydrite here, is therefore probably Paleocene or older.

Cavity-fill in the breccias was first mentioned by Burton (1911) and has since been found to be extensive, and Hickling and Holmes (1931) recorded stalactitic cavity lining. Smith (1972) has drawn attention to the critical influence on the shape and size of breccia clasts played by the creation in basal post-evaporite carbonate rocks of a dense rectilinear network of sulphate veins, itself possibly related to high-pressure fluid injection following burial-related expulsion of formation brines or dehydration of primary gypsum.

The evaporite-dissolution collapse-breccias of north-east England all lie east of the shelf-edge reef of the Ford Formation, and occupy a NNW/SSE belt that extends for 2–5 km beneath the North Sea. Farther to the east and deeper, increasing thicknesses of anhydrite remain undissolved and overlying Concretionary Limestone rocks are progressively less brecciated. West of the present coastline the Concretionary Limestone in northern

Durham passes into the Roker Dolomite and the character of the breccia clasts, as seen in the Ryhope Cutting GCR site, changes accordingly. The collapse-breccias between Trow Point and Lizard Point are amongst the most convincing of their type anywhere in Britain; other excellent (but generally less accessible) exposures of such breccias are in coastal cliffs between Ryhope and Horden (Smith, 1972) and in the Wear Gorge at Sunderland.

Although foundering and collapse following evaporite dissolution are now regarded as the main cause of brecciation in the Concretionary Limestone Formation, there are a number of places in the coastal cliffs (as, for example, just south of Lizard Point) where collapse-brecciation has resulted from the dissolution of carbonate beds (Smith, 1973a, 1994) and many places where partial to complete brecciation has been caused by interstratal carbonate dissolution and stylolite formation during late diagenesis (Braithwaite, 1988). Finally, as in the sector south of Souter Point, local severe brecciation of beds near rockhead appears to have been caused by intense periglacial cryoturbation.

Sedimentology and diagenesis of the Concretionary Limestone Formation

Some aspects of the sedimentology of the Concretionary Limestone are touched on in the account of the Fulwell Hills Quarries site, but most of the critical evidence on which current interpretations are founded is superbly exposed in the cliffs between Velvet Beds and Souter Point. Here, as the sector accounts and literature (Smith, 1970a, 1971a, 1980a, b, 1985a; Smith and Taylor, 1989) show, least-altered strata in the north comprise interbedded finely laminated and unlaminated graded carbonate mudstones in which grainstones and packstones with an exogenous fauna locally form discordant sheets and lenses, whilst strata farther south contain fewer laminites but many disturbed beds with an abundant benthic fauna. These features, coupled with others seen in quarries and borehole cores, have led to interpretation of the Concretionary Limestone as a submarine slope deposit, with strata exposed north of Lizard Point being formed mainly in anoxic or semi-oxic conditions on middle parts of the slope, below an oscillating pycnocline, and those to the south being formed in oxic conditions (i.e. above the pycnocline) higher on the slope and perhaps towards its top (Smith, 1994). In this interpretation, the laminites are envisaged as quiet-water deposits, perhaps as annual couplets (summer sapropel, winter carbonate mud), the graded unlaminated beds are seen as distal turbidites and the disturbed beds are viewed as proximal to medial submarine slumps that may pass downslope into the turbidites. The overall picture is of a gentle subaqueous slope several kilometres long on which differentially high carbonate mud productivity and sedimentation on the upper part resulted in inherent oversteepening and endemic sediment instability. The mud may have been derived by winnowing of the grainstone shoals and back-barrier lagoon of the equivalent shelf facies (i.e. the Roker Dolomite) and the shoals presumably were also the source, through shelf-edge and high-slope failure, of the grainstone sheets and lenses in the mid-slope domain in Marsden Bay as far south as Lizard Point.

Diagenetic changes in the Concretionary Limestone are discussed in the account of the Fulwell Hills site, and most of the secondary features seen in the Fulwell exposures are seen also in the cliffs between Velvet Beds and Souter Point; they were considered in detail by Al-Rekabi (1982) who illustrated and analysed rocks from Marsden Bay and by Braithwaite (1988) who deduced a long and complex diagenetic history. Calcite concretions in the Whitburn to Marsden area are predominantly spherulitic, lacking, however, some of the great range of concretionary patterns seen at the Fulwell Hills site and in coastal cliffs at Hendon (Sunderland). From this point of view, therefore, the Whitburn–Marsden cliffs are perhaps not the best place for the study of these enigmatic structures.

Future research

Although most major aspects of the geology of this remarkable stretch of coastal cliffs and shore platforms have been investigated during the last few years, and several aspects have been researched in detail, many parts of it remain poorly understood and much remains to be discovered. In particular, the detailed sedimentology and local stratigraphy and variation of the Concretionary Limestone are worthy of further detailed research, as are the nature and ecology of the indigenous fauna of the Concretionary Limestone and the diagenetic history of the collapse-breccias.

Conclusions

This very extensive GCR site is of international importance because it constitutes a unique set of exposures which display firstly, a whole range of marine Permian depositional features which characterize the western margin of the Zechstein Sea in north-eastern England, and secondly, the post-depositional effects of evaporite dissolution and associated foundering.

Notable are the Trow Point Bed, which can be traced eastwards across the Zechstein Sea into Germany and Poland, the Hartlepool Anhydrite dissolution residue, and the foundered and brecciated Concretionary Limestone beds. Within these strata are found an important but restricted shelly fauna, much of it transported from more congenial environments nearer to the land.

The section has long been studied, and much of it has been well documented in the literature. However, many parts still remain to be studied and understood, so that there is a need for future research, particularly on the sedimentology of the Concretionary Limestone, its associated fauna and the diagenetic history of the foundered strata.

FULWELL HILLS QUARRIES (MAINLY SOUTHWICK QUARRY) (NZ 3859)

Highlights

The several preserved quarry faces in the Concretionary Limestone of Fulwell Hills, Sunderland (box 2 in Figure 3.2), are representative of more than 40 former exposures. They contain a unique range of complex and spectacular calcite concretions, including some claimed to simulate organic structures such as those of some corals and blue-green algae, and have yielded many fish remains from a thin bed near the base of the exposed sequence; the fish include *Acentrophorus varians* (Kirkby), this being the type locality. The rocks are unevenly gently folded and fractured, probably mainly by differential foundering caused by dissolution of the formerly underlying thick Hartlepool Anhydrite.

Introduction

The vast complex of quarries in the Concretionary Limestone of Fulwell Hills, in the north-western outer suburbs of Sunderland, has long been justly famous for its bewildering array of bizarre calcite concretions. Quarrying started before 1746 and ceased in 1957. In that time almost all the concretion-bearing beds were removed from an area exceeding a 0.5 km^2, largely for lime burning and building purposes; much of the output was transported by wagonways to ships on the River Wear, 2 km to the south.

Only a few of the many faces once worked have been preserved, but records of 36 faces examined in 1954 are lodged in the fieldnote files of the British Geological Survey. About 26 m of strata are now visible, out of a former total of about 35.7 m, and are thought to lie in about the middle of the Concretionary Limestone Formation. Concretions from Fulwell are to be found in many museums, with substantial collections at the Hancock Museum (Newcastle upon Tyne), Sunderland Museum, Nottingham University and the British Museum of Natural History; Abbott (1914) also cites collections at Oxford and Aberdeen University museums and museums at Haslemere and Copenhagen. They feature strikingly in many local walls and buildings, and in hundreds of private and public gardens in the South Shields and Sunderland areas.

The calcite concretions of Fulwell Hills were first noted by Winch (1817) and described in more detail by Sedgwick (1829); they were further described, classified and freely illustrated by Abbott (1907, 1914), Holtedahl (1921) and Tarr (1933). Briefer descriptions and attempts (so far not wholly successful) to explain the genesis of the concretions in the formation as a whole have been made by Garwood (1891), Woolacott (1912, 1919a, b), and Holmes (1931), and a detailed study on calcitization and compaction in rocks of the formation was made by Braithwaite (1988). Browell and Kirkby (1866) and Trechmann (1914) published much-quoted analyses of a selection of rocks from Fulwell Quarries and the site is also the type locality of the Fulwell Fish-Bed from which Kirkby (1863, 1864, 1867) recorded two species of fish. Finally, Fulwell Quarries and other local localities in these strata have been mentioned and/or illustrated in a number of regional accounts (Trechmann, 1925; Smith, 1970a, 1980b, 1994; Pettigrew, 1980) and for many years they have been a favourite venue for excursion parties whose visits have been recorded in the proceedings of various learned societies. Drift deposits formerly present at about +45 m O.D. on the north side of the hills contained a gravel lens interpreted by Howse (1864) and Woolacott (1897, 1900a, b) as a Quaternary raised beach, and Kirkby (1860) describes sand-filled pipes up to 3.5 m deep in the underlying limestone here.

Tarr (1933, p. 268) waxed lyrical over the limestones of Fulwell Quarries and wrote 'This exposure should be permanently preserved as one of the most outstanding examples of nature's ability to build artistically in stone'. The faces to which he referred have, unfortunately, long since disappeared, but those now preserved are reasonably representative of the many formerly available for study.

Description

Most of the several quarries on Fulwell Hills (including Fulwell Quarry itself) have been filled and partly or wholly landscaped, though a number of small faces have survived in addition to those scheduled for preservation. The quarries were worked under a number of names, but the main preserved faces are in the former Southwick Quarry; the scheduled areas are shown on Figure 3.19.

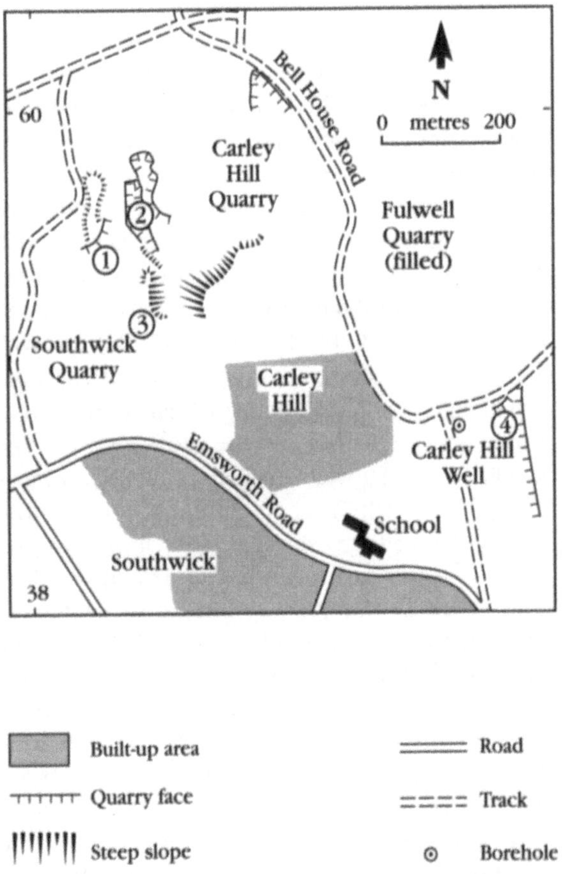

Figure 3.19 Preserved faces within the complex of former limestone quarries on Fulwell Hills, Sunderland; numbers refer to quarry faces described in the text.

The Concretionary Limestone strata of the Fulwell Hills Quarries totalled about 35.7 m in thickness, and comprise four main lithological units; the general sequence formerly visible is shown below.

	Thickness (m)
Drift, mainly red-brown, silty, stony clay, but with patches, sheets and lenses of limestone brash, gravel, sand and laminated clay	0–6.0
---- unconformity ----	
Limestone, grey and brown, very finely crystalline, mainly finely and evenly laminated, but with many thin unlaminated beds (some graded), interbedded with and passing into subordinate very finely saccharoidal, cream and buff dolomite; the laminated limestones contain abundant, but patchy, coarse, radially crystalline, brown calcite, including coarse spherulites, and many displacive lenses and tongues of white calcite; the unlaminated limestones contain patchy, radial/concentric calcite concretions and the dolomites contain scattered to abundant subspherical to lobate calcite concretions. Marked lateral variation	*c.* 21.0
Limestone, grey and brown, very finely to very coarsely crystalline, mainly finely and evenly laminated, but with many thin unlaminated beds (some graded) and with a widespread 0.5 m bed of cream and buff, very finely saccharoidal dolomite 2.4–2.9 m from top. Most of the limestone is massive and comprises a wide range of spectacular reticulate, lobate and spherulitic calcite concretions, with some lateral passage into finely saccharoidal dolomite containing scattered to abundant subspherical and/or lobate calcite concretions; some beds deformed by submarine slumping	*c.* 7.6
Limestone, grey and buff, very finely crystalline, mainly very finely and evenly laminated, but with some thin	

unlaminated beds (some graded); scattered, coarse, radial calcite crystals in the laminated beds, and small subspherical calcite concretions in the unlaminated beds	1.8–2.0
Dolomite, cream-grey, very finely saccharoidal, soft	5.5+

Of these units, the uppermost is by far the most laterally variable; even when many faces were fully exposed, correlation between sections was uncertain unless the faces were either very close or in contact; most parts of this unit are now obscured. In contrast, the 7.6 m unit comprises relatively laterally extensive thick and massive beds that could be traced readily with the aid of the recessive 0.5 m dolomite bed; this 7.6 m unit was the main quarrying target and the floor of the quarry widely followed its base. The 1.8–2 m bed near the base of the sequence is the 'Flexible Limestone' of the literature, although only the thinnest laminae here are noticeably flexible; it includes, near the top, the slightly bituminous Fulwell Fish-Bed discovered and recorded by Kirkby (1863, 1864, 1867) but which is only very sparingly fossiliferous in the solitary section now exposed. The lowest unit is the Great Marl Bed of the quarrymen (Woolacott, 1912) although Kirkby (1863, 1864, 1867) and Browell and Kirkby (1866) used the term in a different (perhaps the original and more correct) sense for a bed at the base of the 21 m unit.

Except in the fish-bed, fossils are very uncommon at Fulwell but a few were recorded by Kirkby from near the base and top of the Flexible Limestone, from four levels low in the 7.6 m unit and from a single level high in this unit; they comprise the conifer *Ullmannia frumentaria*, obscure plant remains, and the nektonic fish *Acentrophorus varians* (Kirkby) and *Acrolepis* (rare); Pettigrew (1980) records that whole *Acentrophorus* (0.04 m), for which this is the type locality, have been found within the visceral cavity of the much larger (0.25–0.35 m) predatory *Acrolepis*. Kirkby (1864) commented that only the fish-bed was worth examining for fossils and that fish discoveries in other beds were almost invariably accidental and made in the course of quarrying. Although bivalves commonly form the nucleus of (or lie within) concretions in the Marsden and Roker areas, none have been found in the Fulwell Hills Quarries (Woolacott, 1912).

Almost all the strata in the Fulwell Hills Quarries (including those in faces now covered) are gently folded and fractured, with dips of up to 40°, but generally 5 to 10°; a few small almost-completely brecciated areas also occur, mainly in faces now covered. The pattern of folds and fractures appears to be random, and is unrelated to well-documented structures in underlying coal workings.

The calcite concretions in the rocks at Fulwell Hills Quarries are bewilderingly varied, but there is a general tendency for the most complex types to be concentrated in the 7.6 m unit where they characterize beds traceable for some hundreds of metres. The most spectacular concretions are in the laminated limestones and comprise three-dimensional reticulate combinations of rhythmic bands and radial calcite crystals on scales ranging from millimetres to several decimetres (depending on the spacing of nucleation centres); spaces between the concretionary calcite are either empty or partly to wholly occupied by powdery cream dolomite which, in places, contains calcite scalenohedra up to 0.05 m long. Many of the concretions in the laminated limestone are clearly spatially and genetically related to joints, cracks and bedding planes, but others appear to be unrelated to such features; as Sedgwick (1829, p. 95) noted, the lamination commonly passes uninterruptedly through the concretions, although slight disruption is common. Concretions also persist through many of the thin unlaminated beds, whereas the thicker unlaminated beds are mainly of dolomite and are either concretion-free or patchily contain rod-like, lobate, or subspherical ('cannon-ball') types (Figure 3.20); some of the last exceed 0.3 m in diameter. Analyses by Browell and Kirkby (1866), Garwood (1891) and Trechmann (1914) showed that the calcite concretions contain up to about 22% of dolomite and that the dolomite between the concretions is slightly calcitic; thin sections (Trechmann, 1914, p. 237 and plate 36, fig. 5) reveal that the dolomite in the concretions forms inclusions in the calcite and that the 13% of calcite in the dolomite of the Great Marl Bed is concentrated in narrow veins. Most of the limestone beds display traces of a complex diagenetic history, and evidence of leaching, stylolitization and partial auto-brecciation is widespread; infiltrated and crystalline cavity-fill commonly have reduced the secondary porosity created by leaching and brecciation.

The site contains several faces of which four are especially noteworthy (1–4 in Figure 3.19); three are in the former Southwick Quarry and one in an unnamed quarry (not Carley Hill Quarry) on the east side of Carley Hill.

Figure 3.20 Mutually interfering subspherical calcite concretions ('cannon-balls') in a matrix of fine-grained dolomite. Note the parallel bedding traces preserved on the surface of some of the concretions. Loose specimen from floor of Southwick Quarry. Reproduced by permission of the Director, British Geological Survey: NERC copyright reserved (NL 130).

Face 1. This upstanding rock-face is about 10 m high and 70 m long; it is overlain by quarry spoil which obscures much of the south-western part of the exposure. Strata (about 18.6 m in total) dip north at up to about 20° and comprise several thick to very thick beds of hard concretionary limestone separated by relatively continuous, but generally thinner, beds of softer cream dolomite; the base of the section is probably slightly above the top of the Flexible Limestone and the sequence exposed comprises most of the 7.6 m unit and the lower beds of the 21 m unit. The limestone beds are mainly finely laminated and feature a wide range of highly complex calcite concretions, including many reticulate types; strong contemporaneous slump contortion is a feature of one of these beds a few metres above ground level on the south-east corner of the face. Most of the dolomite beds contain subspherical non-radial calcite concretions and some contain particularly fine examples of radially-crystalline large calcite lobes.

Face 2. This face surrounds a large excavation in the quarry floor. Strata in it dip generally northwards, in continuation of Face 1, and comprise about 25.5 m of beds of which the uppermost 13 m lie in the 21 m unit and the remainder form the 7.6 m unit and the Flexible Limestone; only the extreme top of the Great Marl Bed is exposed. Interest in this face centres on the Flexible Limestone exposed near the south-west end of the face, and the sparingly fossiliferous Fulwell Fish-Bed near its top.

Face 3. Strata in this face dip southwards at about 5° and comprise about 16 m of much the same sequence as Face 1; finely plane-laminated crystalline limestone predominates, with some calcite crystals exceeding 0.2 m in length. A 5.5 m bed of cream dolomite in the middle of the sequence is atypically thick, however, and is probably the 'Great Marl Bed' of Kirkby (1867); it should not be confused with the bed at the base of the sequence that was accorded the same name by Woolacott (1912) and Trechmann (1914).

Face 4. This comprises two main parts, a small section along the northern side of the old quarry and a long and somewhat overgrown face on the east side. Interest focuses on the small northern face, which furnishes one of the best exposures of complexly reticulate calcite concretions in which the spatial influence of cracks, joints and bedding planes is especially clear. This face is in massive limestone near the top of the 7.6 m unit. The long eastern face is mainly in this unit together with basal beds of the 21 m unit (total about 14 m), and displays both the rolling character of the strata and the great lateral variability of the limestones; the intervening soft cream dolomite beds are more uniform and persistent, as in faces 1-3. In common with many faces in the quarry complex, rockhead in the north of this face features evidence of cryoturbation and large slabs of limestone are embedded in the overlying drift.

Interpretation

The calcite concretions of Fulwell Hills are renowned worldwide for their complexity and variety, and the quarry faces preserved there contain a unique blend of unusual concretionary forms on a wide range of scales; equivalent strata are widely exposed in quarry and coastal cliff sections between South Shields and Whitburn and also at Hendon (a southern district of Sunderland), but none exhibit quite the range exhibited at Fulwell. The Flexible Limestone at Fulwell is generally less flexible than at Marsden, where its flexibility was first noted by Nichol (reported by Winch, 1817), but it is generally richer in fish remains. The cause of the flexibility is not known, but the fineness of the lamination and the preservation of bituminous films and fish remains points to slow accumulation under anoxic conditions well below wave-base.

All the strata exposed in the Fulwell Hills Quarries lie wholly within the Concretionary Limestone Formation as redefined by Smith (1971a), though the 'Great Marl Bed' at the base of the sequence was formerly classified (e.g. Woolacott, 1912; Trechmann, 1914) as Middle Magnesian Limestone. Records of the Carley Hill Well (NZ 3872 5951) in the south of the quarry complex show interbedded marl (i.e. soft dolomite) and laminated limestone to 26 m below the quarry floor, suggesting that at least 22 m of the Concretionary Limestone Formation lies below the Flexible Limestone here. The base of the 0.1 m Marl Slate lies in the well at a depth of 58 m (about -3.4 m O.D.) and the Yellow Sands lie at 95.8 m (about -41.2 m O.D.).

The Concretionary Limestone (up to 120 m) is the thickest carbonate formation of the English Zechstein sequence and its land outcrop spans most of the area between the reef and the present coast (Woolacott, 1912, fig. 5, which mistakenly includes the Seaham Formation); coal exploration boreholes have shown that the formation also crops out on the sea floor (or beneath drift) for several kilometres east of the present coast (Smith, 1994, fig. 34), and extends farther eastwards in the subsurface.

Interpretation of the stratigraphical relationships of the Concretionary Limestone Formation is complicated by its great lateral variability, lack of outcrop continuity and complex foundering, but the writer (Smith, 1970a, 1971a, 1980a, b, 1994) believes that it is co-extensive with the slope facies of the Cycle EZ2 carbonate unit and passes upwards and westwards into a shelf facies represented by the Roker Dolomite; the beds exposed at Fulwell Hills probably were formed on about the middle of the slope, in anoxic water below a basin-wide oscillating pycnocline. The proximity of Fulwell Hills Quarries to the reef at West Boldon shows that the shelf facies may have been less than 2 km wide here. The foundering of the Concretionary Limestone results from the dissolution of the formerly underlying thick (?100 m+?) Hartlepool Anhydrite and is undoubtedly the cause of most of the folds, fractures and brecciation seen in the Fulwell Hills Quarries. Other sedimentological and subsidence features of the formation are further discussed in the account of the coast sections at Marsden Bay, and it will suffice to note here that a slope origin for the Fulwell Hill strata is consistent with the evidence of submarine slumping seen in Face 1 and of the prevalence of thin unlaminated beds (many graded) that are interpreted (Smith, 1970a, 1971a, 1980a, b, 1994) as turbidites.

The origin of the calcite concretions in the Concretionary Limestone has been the subject of endless speculation, but remains uncertain. Sedgwick (1829) recognized that the concretions were secondary and that most of them were formed after much of the rock was partly or fully lithified, and Garwood (1891) showed that the chemical changes could have taken place in a closed system and resulted from a major redistribution ('segregation') of components rather than from the large-scale introduction or removal of matter. Later workers (e.g. Woolacott, 1912, 1919a; Trechmann, 1914; Holmes, 1931; Tarr, 1933) have pondered on the detailed chemistry of the profound mineralogical changes and speculated on the possible involvement of former organic matter, calcium sulphate and other salts; Shearman (1971) and Clark (1980, 1984) considered the possible role of sulphate-reducing bacteria. The petrography, geochemistry and evolution of the Concretionary Limestone in general (but not specifically at Fulwell Hills Quarries) was reviewed in detail by Al-Rekabi (1982) and Braithwaite (1988), who give full references.

Future research

The regional distribution, stratigraphical relationships and sedimentology of the Concretionary Limestone Formation are reasonably well known, although Fulwell Hills Quarries yield only a small part of the evidence on which this understanding is based. The quarries, however, remain a unique repository of almost the full range of concretions in this enigmatic formation, and the exposures here are an essential part of any further studies of the origin of these bewildering structures and of the rocks as a whole. Now that extraction of limestone has ceased, this is no longer a good site for the study of fossil fish, though good specimens are still to be found by the lucky or diligent searcher.

Conclusions

The Fulwell Hill Quarries complex is an internationally important GCR site. The site is justly famous for the enormous range of spectacular calcite concretions which characterize the aptly named Concretionary Limestone, as well as for the Fulwell Fish-Bed for which this is the type locality. It also features a range of structural features caused by the dissolution of the underlying Hartlepool Anhydrite. The origins of the concretions are still poorly understood and therefore this site provides essential exposures for their future study.

HYLTON CASTLE CUTTING (NZ 3594 5862–3611 5888)

Highlights

This small road cutting provides the best and most readily accessible exposure of the lower core of the shelf-edge reef of the Ford Formation. The rock is almost entirely of dolomite and comprises an apparently haphazard crudely-bedded assemblage of masses of hard reef-rock separated and surrounded by shelly rubble; ?algal encrustations are widespread in the hard reef-rock, and steeply dipping ?algal sheets occur in a number of places.

Introduction

The cutting accommodates Rotherfield Road near Hylton Castle, in the north-western outskirts of Sunderland, and was dug in 1955; it exposes about 15 m of exceptionally varied reef dolomite of the Ford Formation, mainly of the lower reef-core facies, and was recorded in detail by the writer (Smith, unpublished Geological Survey fieldnotes, 1955). Preliminary excavations below the floor of the cutting and nearby sewer trenches revealed that the reef-rocks in the cutting were underlain by a typical highly fossiliferous basal coquina.

Large collections from the excavations and cutting led to long fossil lists by Pattison (unpublished Geological Survey report, 1966). More recently, additional collections from the cutting by Hollingworth (1987, table 4) were used (with other data) in his reconstruction of the lower reef-core palaeocommunity (1987, fig. 6.12) which was reproduced by Hollingworth and Pettigrew (1988, fig. 8) in a palaeontological account of the cutting (their locality 1, pp. 40–44). This reconstruction was also reproduced by Hollingworth and Tucker (1987, fig. 5).

The GCR site is at the side of a public highway and great care should be taken both to minimize the risk of personal injury and to avoid causing inconvenience or danger to passing traffic.

Description

The position of the site and of its geological features (including some that are now no longer visible) is shown in Figure 3.21. The rock face in the cutting is about 120 m long and up to 6 m high and its base rises north-eastwards by about 15 m; in

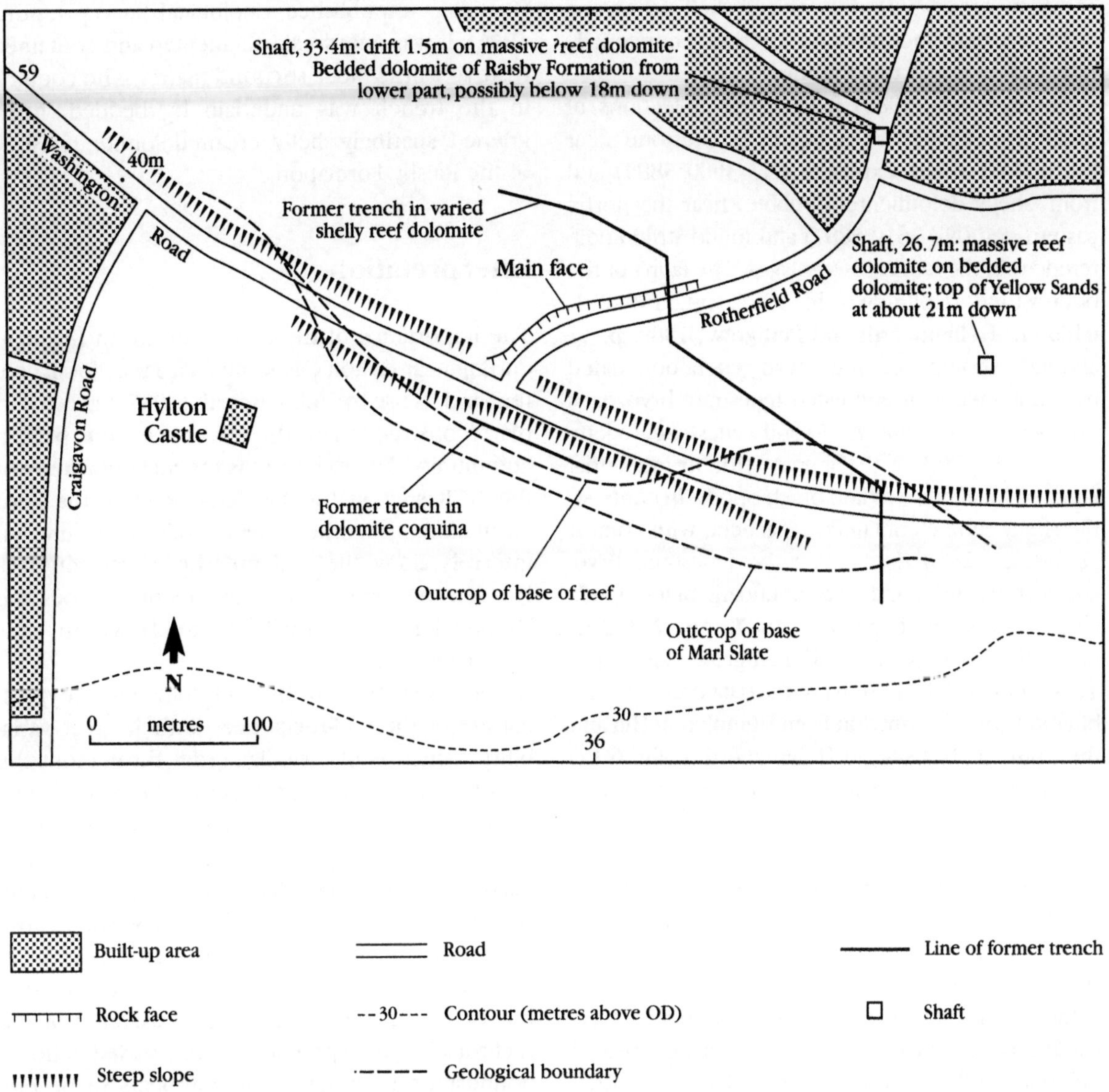

Figure 3.21 Hylton Castle (Rotherfield Road) Cutting and its immediate surroundings, showing the main features of geological interest.

conjunction with former excavations below road level, a section about 20 m thick was exposed, and further information was gained from a number of deep manhole shafts.

The present exposures are mainly in the north wall of the cutting and comprise an extremely varied complex of masses of autochthonous dolomite boundstone, many draped by laminar (?algal) sheets, separated by lenses and pockets of shelly, rubbly reef debris; from a distance the rock has a crude, low-dipping bedded appearance of a type generally found in the reef-flat sub-facies high in the reef and uncommon at this low level in the reef. The boundstone masses are irregular to compressed subspherical in shape (though many are poorly defined) and are commonly 0.5–1 m across (exceptionally 2.5 m); some are concentric; they are composed of dense, hard, cream-buff, finely crystalline dolomite with a sparse bryozoan–brachiopod fauna and widespread (but patchy) concentric ?algal encrustations which in places make up most of the rock. The laminar dolomite sheets are up to 0.3 m thick and some are vertical for 1 m or more; most appear to be contemporaneous coatings of boundstone masses, but some have hints of a bilateral structure and could be fissure-fill. The rubbly pockets between the boundstone masses make up a small to large proportion of the reef according to locality and comprise cream-buff, dolomitized shell debris and fine bryozoan boundstone detritus; concentric ?algal coatings are uncommon.

Hollingworth (1987) made large collections of fossils from a patch of 'laminated boundstone' near the middle of the exposure (NZ 3600 5885) and from 'shelly dolomicrite' (?rubble) near the north-eastern end (NZ 3610 5887) and found striking differences in the faunal assemblages. The fauna of the boundstone (Hollingworth, 1987, pp. 198–201, table 4A; Hollingworth and Pettigrew, 1988, p. 41 and fig. 7) comprised only seven genera dominated by a framework of encrusted fenestrate bryozoans (*Synocladia, Acanthocladia* and *Fenestella*, together comprise 70% of the fauna) with an interstitial fauna of small bivalves and brachiopods. In contrast, the shelly rubble contained 15 genera, with conical *Fenestella* (24%) dominating the fenestrate bryozoans (total 36%) and with abundant brachiopods and bivalves (Hollingworth, 1987, pp. 201–202, table 4B; Hollingworth and Pettigrew, 1988, pp. 41–44 and fig. 7). These were the data used, in combination with information from Humbledon Hill and the Tunstall Hills site, in Hollingworth's (fig. 6.12) reconstruction of the lower reef-core palaeocommunity.

The petrography of the reef-rock in the cutting has been examined by G. Aplin (pers. comm., 1990), who reports evidence of early marine botryoidal cements that have since been dolomitized, and hints of possible primary dolomite cements; Dr Aplin also notes local evidence of brecciation associated with patches of calcite-replaced dolomite, and some uplift-related calcite cements.

The great lateral and vertical variability of the reef dolomite in the cutting was also a feature of reef dolomite temporarily exposed in nearby sewer trenches (see Figure 3.21 for location) which exposed up to 4 m of unpredictably mixed boundstone and shelly rubble (Smith, unpublished Geological Survey fieldnotes, 1955). As in the main cutting, algal encrustations and lamellar drapes were seen to be extremely patchy.

The basal reef coquina is no longer exposed in the Rotherfield Road Cutting, but was seen in 1955 in temporary excavations below road level at the junction of Rotherfield Road and Washington Road and also, at a slightly lower level, in a WNW–ESE sewer trench (NZ 3593 5879–3601 5875) on the opposite (south) side of Washington Road (Figure 3.21). The rock exposed in these excavations was a pale cream accumulation of well-preserved brachiopods and bivalves, with relatively few bryozoans and no recognizable algal encrustations (Pattison, unpublished Geological Survey report, 1966); it was only weakly cemented and contained relatively little bioclastic sand matrix. The coquina in the trench was underlain by bedded, finer-grained, sparingly shelly, cream dolomite, possibly of the Raisby Formation.

Interpretation

The importance of the reef exposure in the road cutting near Hylton Castle stems from its fossil content and its use by Hollingworth (1987, fig. 6.15) in his reconstruction of the lower reef-core palaeocommunity. The exposure is the most northerly of the GCR sites in the shelf-edge reef of the Ford Formation, the others being located at uneven intervals along the reef outcrop as far south as Horden Quarry. See the accounts of the Humbledon Hill, Tunstall Hills and Hawthorn sites for further discussion.

The most noteworthy and somewhat atypical feature of the reef-rocks here is their great variability, which is of a similar order throughout the length of the cutting and also in the nearby temporary trench exposures. The tendency of parts of the reef to be composed of small to medium-sized masses of encrusted algal–bryozoan boundstone separated and surrounded by shelly rubble has been noted at a number of exposures, especially in the road cutting adjoining the Humbledon Hill site, but is unusually clear here; the variability of the reef-rock is an expression of the varied ratio of boundstone to rubble, which appears to be random apart from the strong hint of roughly horizontal bedding. The abundance of ?algal encrustations on bryozoan and other frame elements is also unusual at this relatively low level in the reef, and prompts the question of whether a concentrated reef sequence might be present here; this in turn poses the question of whether the faunal assemblages here might not be fully representative of low-reef faunas as a whole.

Future research

The Rotherfield Road Cutting, having been examined by Hollingworth (1987), appears to offer little immediate scope for further detailed research. It remains an excellent and convenient place where the internal structure of the shelf-edge reef may be examined, and from which the abundant and varied fauna may be collected.

Conclusions

The Hylton Castle Cutting GCR site exhibits the varied lithology of the lower reef-core facies of the Ford Formation, which overlies a highly fossiliferous dolomite (coquina) that is now covered. The site is significant in that it provides an excellent exposure of the internal reef structure, a feature of which is that the frame elements of the reef have ?algal encrustations at an unusually low level in the reef. The site has yielded a large collection of fossils, studies of which have allowed the reconstruction of a lower reef-core palaeocommunity.

CLAXHEUGH ROCK, CLAXHEUGH (FORD) CUTTING AND FORD QUARRY (NZ 3657)

Highlights

This unique complex of large exposures (box 4 in Figure 3.2) provides much of the evidence on which the occurrence of massive, late Permian, submarine slumping may be inferred and is the only readily accessible place where the shelf-edge reef of the Ford Formation is seen to pass into equivalent strata on its western (landward) side. Claxheugh Rock was the site of a well-documented major rock-fall in 1905.

Introduction

Claxheugh Rock (formerly Clack's Heugh = a crag in Mr Clack's property) and the adjoining cutting and quarry lie on the south side of the River Wear in the western outskirts of Sunderland; together the three exposures reveal parts of the Basal Permian (Yellow) Sands, Marl Slate, Raisby Formation and Ford Formation. The sections are the type locality of the Ford Formation, the shelf-edge reef of which was formerly almost completely exposed in cross-section.

The rock faces of the Claxheugh complex of exposures, plus the nearby, but now filled, Claxheugh and Ford (old) quarries, have received more attention in the literature than any other late Permian Marine GCR site. The references range from brief mentions (e.g. Sedgwick, 1829; Howse, 1848, 1858; King, 1850; Howse and Kirkby, 1863; Kirkby, 1866; Lebour, 1884, 1902; Trechmann, 1931) to longer accounts of one or more aspects of the various exposures (e.g. Browell and Kirkby, 1866; Woolacott, 1903, 1905, 1912, 1918; Trechmann, 1925, 1945, 1954; Logan, 1967; Smith, 1969b, 1970a, c, 1981a; Pryor, 1971; Pettigrew, 1980; Hollingworth, 1987; Holling-worth and Pettigrew, 1988). The earlier references were confined to Claxheugh Rock and the railway cutting, but most later workers, were also able to discuss the geology of Ford Quarry which opened in the late 1920s; the south-east and north-east faces of the quarry were specially preserved for geologists by the Sunderland Borough (now City) Council after quarrying ceased in 1971.

Description

The Claxheugh exposures lie entirely within a fault-bounded trough near the core of the NNW–SSE Boldon Syncline. The position of the site is shown in Figure 3.22, together with the locations of the main features of geological interest; the largest rock exposures are Claxheugh Rock and the south-east face of the quarry, but full understanding of the complex facies relationships present stems only from an assessment of all the faces and other available data. The general geological sequence exposed at Claxheugh (including the quarry) is shown below.

	Thickness (m)
Drift deposits, mainly boulder clay (thickest in east)	up to 6
----- unconformity ------	
Ford Formation, reef and backreef facies; at least	55
Raisby Formation, slope facies, with slide-plane and patchy slide-breccia (0–2 m) at top	0–8
Marl Slate	0.77–0.90
Basal Permian (Yellow) Sands (about 18 m seen)	up to ?58
----- unconformity -----	
Upper Coal Measures (Westphalian C)	

The relationships of the several stratigraphical units in the main faces are summarized in Figure 3.23, which was based on drawings made before

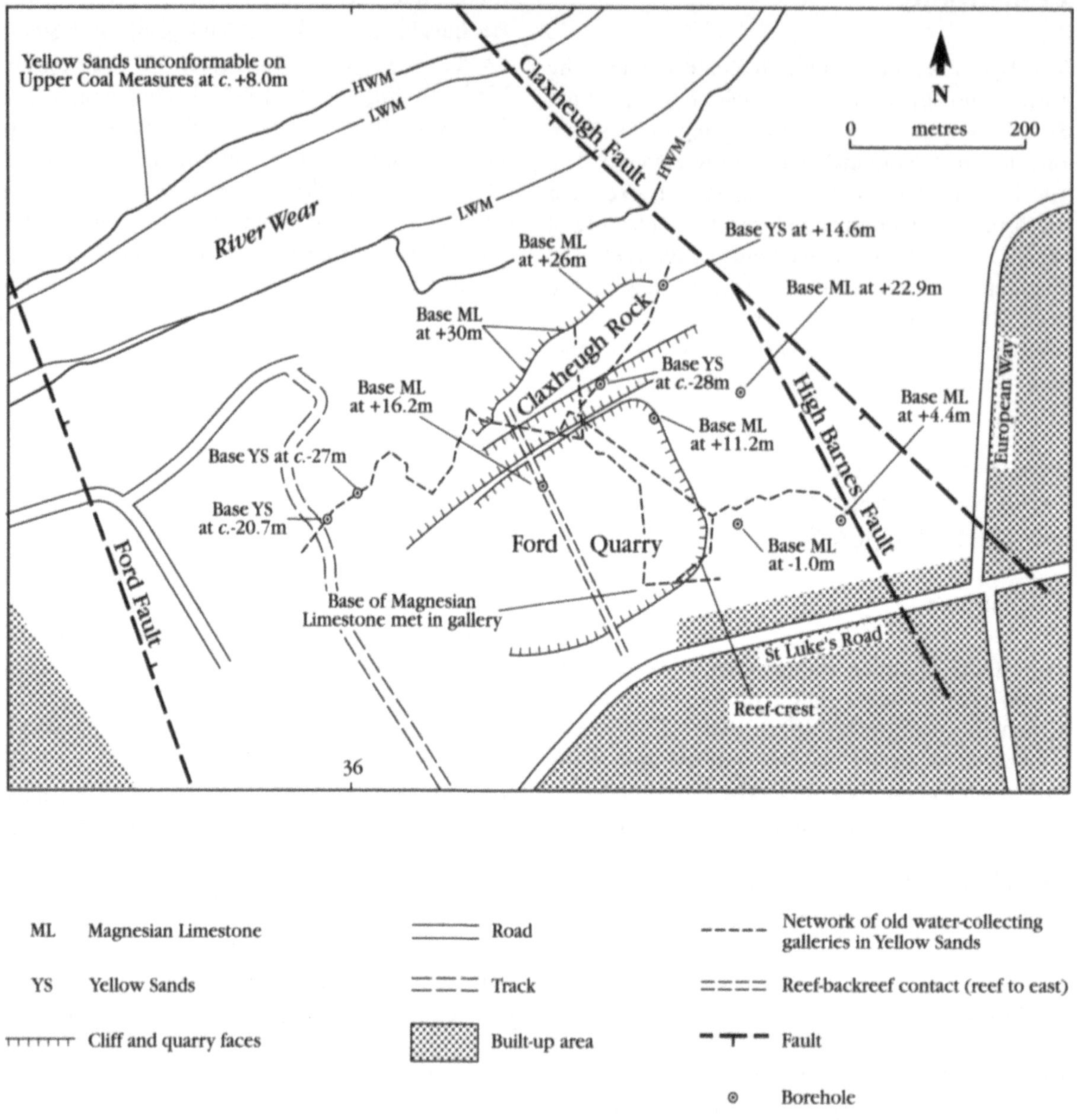

Figure 3.22 Claxheugh Rock, Cutting and Ford Quarry, showing the position of the main features of geological interest.

the partial filling of the quarry. Details of the main Permian units follow.

Basal Permian (Yellow) Sands

The ?early Permian aeolian Yellow Sands are almost at their thickest (about 58 m) beneath Claxheugh Rock; upper parts of the formation here were investigated in detail by Pryor (1971), who classified the weakly-cemented sand as a fine- to medium-grained subarkose with scattered to abundant coarse grains and large-scale, tangential trough cross-stratification. Pryor noted quartz and potassic feldspar contents of 88% and 8%, respectively, and determined a mixed cement of dolomite and calcite with a little illite.

Boreholes have shown that about half of the Yellow Sands at Claxheugh lie below river level. The formation is of special interest here because it was exploited for water by a local factory from which a complex system of galleries was driven south-eastwards deep into the hill along the local water table (Figure 3.22). Tests by Browell and Kirkby (1866) showed that the sand is remarkably porous, with one cubic foot (0.028 m^3) of the deposit able to hold 10 pints (3.41 litres) of water.

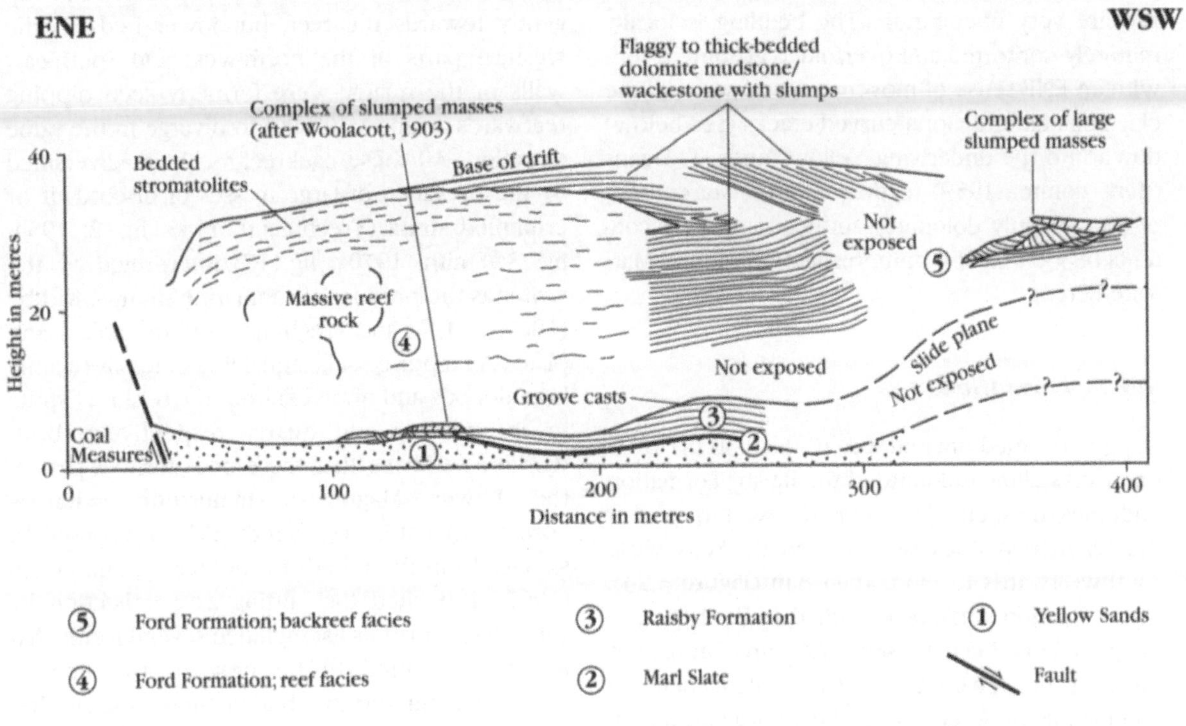

Figure 3.23 Section of strata at Claxheugh Rock, Cutting and Ford Quarry, based on Smith (1970a, fig. 17).

The top of the Yellow Sands at the north-eastern end of the Claxheugh Rock section slopes steeply south-southeastwards (Trechmann, 1954) and boreholes in the quarry floor show that this slope continues for some distance and takes the reef/sand contact from about +26 m O.D. at outcrop at the north-east end of Claxheugh Rock to about +11 m only 150 m farther south, and to about -3 m O.D. in the south of the quarry (Figure 3.24); because of this decline, the base of the limestone was intersected in the most southerly part of the water-gathering gallery.

Marl Slate

This comprises finely laminated, buff and grey, slightly carbonaceous, dolomitic shale with laminae, lenses and thin beds (some partly contorted) of brown, carbonaceous, plastic clay; scales of

Figure 3.24 Steeply south-south-eastwards sloping contact between Basal Permian Sands (pale) and dolomite boundstone of the Ford Formation reef. The slope is interpreted as the northern flank of an east-northeastwards trending submarine slide canyon (Smith, 1971c). The field of view is about 18 m high. (Photo: D.B. Smith.)

palaeoniscid fish are scattered sparingly, but whole fish are very uncommon. The bedding is locally minutely contorted and overfolded at some levels, with an ENE sense of movement, and some of the clay films extend along curved cracks (see below) down into the underlying Yellow Sands. T. Deans (pers. comm., 1959) records a carbonate content of 33% (mainly dolomite) and lead and zinc contents of 330 and 350 ppm, respectively in the Marl Slate here.

Raisby Formation

Flaggy- to medium-bedded (0.05–0.30 m) very finely crystalline dolomite of the Raisby Formation underlies the shelf-edge reef in the western part of the Claxheugh Rock section, and thickens west-southwestwards to more than 8 m (Figure 3.23). The formation was not identified in Ford Quarry, but may have been present and unrecognized in the deep north-west corner (now filled), beneath comparable back-reef strata of the Ford Formation. Some bedding planes bear a thin layer of brown, dolomitic clay and others are slightly stylolitic; a sparse foraminifer–bivalve–brachiopod fauna is present. The truncated top of the formation at the south-western end of the section bears long sub-parallel and divergent WSW/ENE grooves up to several metres long (Woolacott, 1912, fig. 11; Trechmann, 1945), which are interpreted by Smith (1970c) as score marks (groove-casts). Angular fragments and blocks of the formation are complexly intermixed in a coarse breccia lying between the reef and the Yellow Sands in the centre of the section (Woolacott, 1903, section 2, reproduced by Smith, 1970c, fig. 6). Woolacott (1903, section 3) also recorded a system of intersecting curved cracks and minor movement planes that cut the Raisby Formation and underlying beds at the south-western end of the Claxheugh Rock section, but do not extend into the overlying reef.

Ford Formation

This is the main marine stratigraphical unit in the Claxheugh–Ford Quarry complex, and comprises reef and backreef facies; the reef is slightly more than 200 m wide and comprises several sub-facies.

The backreef facies is exposed in the south-western part of the quarry and of the adjacent cutting and comprises at least 55 m of buff, very finely crystalline dolomite in relatively even beds commonly 0.1–0.25 m thick. Only the uppermost 18–20 m of these beds are now visible, and dip gently towards the reef, but lower beds in the western parts of the north-west and south-east walls of the quarry were formerly seen dipping reefwards at up to 30° and to diverge in the same direction. All these backreef-rocks are diversified by the presence of large masses of discordant or crumpled strata (Trechmann, 1945, fig. 2; 1954, fig. 3; Smith, 1970a, fig. 17) interpreted by the writer as the product of submarine slumps. Kirkby (1867, p. 197) and Trechmann (1945, table 1 and plate XV) reported substantial lists of fossils (mainly brachiopods and bivalves) from the backreef strata in the cutting and quarry respectively, both workers believing that these beds formed part of the Lower Magnesian Limestone (=Raisby Formation); Pattison (in Smith, 1970a) recorded 14 species from these beds in the quarry, and Logan (1967) also cited the cutting as a collection locality. Trechmann's list included several forms that he then claimed to be new to the English Zechstein, but the number of 'new' species has declined with more recent studies by Logan (1967), Pattison (unpublished Geological Survey reports, 1969) and Hollingworth (1987). Hollingworth recorded a patch of crinoid debris in the cutting and remarked that the quarry fauna listed by Trechmann indicated an unlithified soft substrate (see also Hollingworth and Pettigrew, 1988, p. 46). Trechmann (1954) noted that the 'Lower Limestone' (= the Ford Formation backreef strata of this account) in Ford Quarry contained nodules and layers of chert that are 'full of Foraminifera and fragments of bryozoa that are not seen in the enclosing dolomite'. Logan (1967), using specimens of shelly 'Lower Magnesian Limestone' (probably the Ford Formation of this account) from the Trechmann collection from Ford Quarry, described and figured hypotypes of *Streblochondria? sericea* (de Verneuil) and *Cleidophorus? hollebeni* (Geinitz).

The reef facies of the Ford Formation forms the bulk of Claxheugh Rock and the easternmost two-thirds of the cutting and quarry; the contact between the reef and backreef beds is only slightly gradational (Smith, 1981a, figs 5, 6) and was formerly seen to be almost vertical for the whole original height (about 35 m) of the south-east face of the quarry (Figure 3.23). Almost all of the reef-rock in the quarry is of reef-core sub-facies, with a passage into reef crest and uppermost reef slope sub-facies along the north-eastern fringe; a few metres of roughly bedded rock at the top of the main faces may be of reef-flat sub-facies.

The reef-core sub-facies comprises a great mass of buff and brown dolomitic bryozoan boundstone (framestone); it is essentially unbedded, but the reef in the 20 m high cliff of Claxheugh Rock features a number of major sub-concordant partings (Woolacott, 1914, fig. 1; Pettigrew, 1980, fig. 7). Most of the rock is finely crystalline, turbid dolomite in which, because of complex diagenesis, fossils are generally poorly preserved (Trechmann, 1945; Aplin, 1985; Hollingworth, 1987). Despite this, Pattison (unpublished Geological Survey report, 1969, and in Harwood *et al.*, 1982, p. 21) recognized some 16 invertebrate species including several species each of bryozoans, brachiopods and bivalves. Trechmann (1931, 1954) commented that the earliest part of the usual reef sequence is missing at Claxheugh Rock, where the reef rests on a deeply scoured surface of Raisby Formation, Marl Slate and Yellow Sands and no coquina is present. Hollingworth (1987) and Hollingworth and Pettigrew (1988) considered that the abundance of epifaunal genera in the reef is consistent with a lithified substrate and reported that bryozoans such as *Acanthocladia* were stiffened and given extra bulk by laminar algal encrustations. A faunal transect of the upper part of the reef-core in the cutting was given by Hollingworth (1987, fig. 6.35) and Hollingworth and Pettigrew (1988, fig. 9).

The reef-core and ?reef-flat dolomite boundstone in the cutting contains a number of vertical tension fissures up to 0.5 m across and 3 m deep (Aplin, 1985, pp. 90–94), that have not been recognized in the nearby quarry. Aplin reports that the fissures are lined with laminated dolomite of possible algal origin, and that some have cores filled with bioclastic debris; he infers from this fill that the fissures were opened whilst the reef was growing.

In the easternmost part of the cutting and quarry, the crudely bedded uppermost part of the reef-core is seen to increase in dip from almost horizontal to up to 50° (Smith, 1981a) as it passes through the reef crest into the uppermost part of the reef slope. The rock here is a crumbly, saccharoidal dolomite boundstone and comprises a highly varied mixture of *in situ* and detached masses of bryozoan boundstone (framestone), complexly anastomosing laminar ?algal sheets and cavity-fill, and pockets of shelly rubble that are locally rich in gastropods, the brachiopod *Dielasma* and the nautiloid *Peripetoceras*; most organisms in the rock are algal-encrusted, and encrustations probably comprise more than 70% of the whole. Many of the laminar sheets bear laterally-linked stromatolite hemispheres, and outward and upward-elongated columnar algal stromatolites up to 0.15 m high and 0.05 m across occur in places at least 5 m downslope from the reef crest (Smith, 1981a, fig. 18).

Interpretation

The complex of faces at Claxheugh Rock and in the adjoining cutting and quarry provide vital links in the chain of evidence favouring massive submarine slumping and sliding during Raisby and Ford formation times and also yield key evidence on the structure, shape and composition of the shelf-edge reef of the Ford Formation and equivalent backreef strata. Except for their involvement in submarine slumping, the rocks of the Raisby Formation, Marl Slate and Yellow Sands here are normal for the district and require no special comment.

Evidence bearing on submarine slumping and sliding

The unusual relationships of strata at the base of the Claxheugh Rock section have been commented on by most of the authors listed in the introduction to this account and were formerly the subject of lively debate. Most of the early authors recognized that the absence of the Raisby Formation and the Marl Slate at the north-east end of the section was a secondary feature caused by their removal after deposition and several inferred an erosional unconformity beneath the 'Shell Limestone' (= Ford Formation reef). A briefly-held alternative explanation by Woolacott (1903) invoked the collapse of a large cave, but was superceded by Woolacott (1912) who envisaged massive destructive east-northeastwards thrusting of the reef over and into the underlying strata. Trechmann (1945) accepted the evidence of thrusting, but clearly had reservations and, in 1954, diffidently suggested that the missing 30 m or so of the lower part of the reef might have been represented by anhydrite, since dissolved. None of these explanations fully accounted for all the facts, however, and this shortcoming led to a new interpretation (Smith, 1970c) in which the Raisby Formation, Marl Slate and the uppermost part of the Yellow Sands here were thought to have slid away downslope (east-northeastwards), leaving a deep slide-canyon which was subsequently filled and buried by the Ford Formation. This interpretation accounts for the scoremarks and matching ridges (= groove casts) between the Raisby and Ford formations, the breccia of Raisby Formation, Marl

Slate and Yellow Sands debris and the overall relationships, but leaves unexplained the enigmatic later growth of a major shelf-edge reef at right angles to the trend of a large, linear sea-floor hollow. Similar features and abnormal stratigraphical relationships were formerly visible in Downhill Quarry (NZ 348601) 3 km NNW of Claxheugh, and large allochthonous masses of Raisby Formation strata, interpreted as slide-blocks (olistoliths), are present downslope at the coastal site at Trow Point and Frenchman's Bay, South Shields. The whole event and its field expression constitute the Downhill Slide.

Woolacott's (1903, Section 2) faithful recording of a particularly complicated part of the Claxheugh Rock section underlines the importance for geologists of making full records of what they see, even if they cannot interpret or understand it; the exposure was covered by debris from the 1905 landslip, but Woolacott's drawing and description were invaluable in the reinterpretation by Smith (1970c) of the history of the area late during Raisby Formation time. Similarly, Trechmann's (1945, fig. 2; 1954, fig. 3) sketches of the south face of the quarry accurately record discordances that he interpreted as evidence of tectonic thrusting, but which are now regarded as submarine slide-planes overlain by allochthonous slide-blocks. These features are still visible, however, and furnish eloquent evidence of the effect of submarine sliding and slumping on partly-consolidated carbonate muds; the exposures of slumped beds in the Ford Formation in the cutting are even more spectacular and convincing (Smith, 1994, plate 10).

Ford Formation

Although advanced diagenesis has obscured many primary details of the shelf-edge reef of the Ford Formation, and made it an indifferent locality for the study of its biota, the south-east face of Ford Quarry is the only place where a fairly complete cross-section of the reef may be seen and is the only place where the reef–backreef transition is readily accessible. The reef is seen to be at least 200 m wide at this point, but the seaward margin is not exposed.

The shelf-edge reef is also a feature of the GCR site at Hylton Castle to the north, and, to the south, of the sites at Humbledon Hill, Tunstall Hills (north and south), Stony Cut (Cold Hesledon), Hawthorn Quarry and Horden Quarry. Aspects of reef distribution, structure, fabric, biotas and diagenesis at these localities are discussed in the relevant accounts. The aspect for which the reef at Ford Quarry is especially noteworthy is that the north-east face is the best exposure of algal-dominated, nautiloid-rich reef crest and high reef slope dolomite and is the only reef exposure containing columnar stromatolites. The presence of large, rolled, detached blocks here is indicative of phases of high energy.

The exposure of the reef–backreef contact is important for a number of reasons. Firstly, it shows that the landward edge of the reef was sharp, with very little relief, proving that the reef surface was barely higher than the backreef sea floor and that backreef carbonates accumulated at the same rate as the reef grew upwards. Secondly, the virtual absence of reef debris in backreef beds implies little erosion and sediment transport landwards across the top of this part of the reef. Thirdly, the general verticality of the contact shows that the landward edge of the reef remained geographically static whilst the seaward margin prograded; the reef therefore became wider with time.

Finally, the exposed backreef strata themselves are unique in being composed of altered carbonate muds and in containing a fairly varied and abundant invertebrate fauna that includes bryozoans and brachiopods. All the other exposures of backreef strata of this age in north-east England are predominantly of altered oolite grainstones with a sparse bivalve–gastropod fauna. The presence of reefwards-displaced allochthonous slide-blocks in these beds is also unique to Ford Quarry and the cutting, and shows that the backreef sea floor sloped towards the reef and was repeatedly unstable. The reasons for these various differences are not known, but Smith (1994) has speculated that they may have arisen because the reef here grew across the floor of a WSW/ENE submarine slide-canyon at least 30 m deep and the sea here may therefore have been deeper than in most other places.

Future research

Most aspects of the several formations exposed at this complex of exposures have been subject to detailed research in recent years and there seems little scope for further detailed studies in the immediate future; an exception is the backreef facies of the Ford Formation, which presents a number of anomalies noted in the text.

Conclusions

This GCR site is the type section of the Ford Formation and is of international importance in that it shows an almost complete section through the late Permian shelf-edge reef and its passage into equivalent backreef strata to the west. It is particularly important in displaying evidence of penecontemporaneous late Permian submarine sliding and slumping on a large scale. This site has been extensively described in the literature and has been the subject of recent research. It is essential therefore that exposures at the site be preserved for further study and for other educational purposes.

DAWSON'S PLANTATION QUARRY, PENSHAW (NZ 3355 5464–3375 5487)

Highlights

This quarry (box 5 in Figure 3.2) contains a superb exposure of a submarine debris flow that lies a few metres above the base of the Raisby Formation. Though generally less than a metre thick, the debris flow displays great lateral variation; it is part of an extensive, but discontinuous, thin sheet of disrupted strata that became unstable and moved east-northeastwards down the marginal slope of the Zechstein Sea. One or more earthquakes may have triggered the movement. The quarry also features many curved joints and minor movement planes, similarly possibly caused by contemporary, but slightly later earth movements.

Introduction

Dawson's Plantation Quarry, Penshaw, exposes about 7 m of limestones and dolomites of the Raisby Formation and contains a thin disturbed sequence interpreted as a submarine proximal turbidite or debris flow. The disturbed bed lies low in the face and perhaps 5–8 m above the (unexposed) base of the formation; it was first reported by Smith (Geological Survey fieldnotes for 1:10,560 Sheet NZ 35 SW, 1953) and later described more fully and interpreted as the product of a complex episode of downslope movement of partly-lithified sediment (Smith, 1970c). The deposit was re-examined in greater detail by Lee (1990), who recognized evidence of three closely-spaced pulses of downslope movement.

Description

The quarry is about 300 m long and lies along the south-east margin of Dawson's Plantation (Figure 3.25); it is cut into the north-west facing escarpment of the Raisby Formation and most beds dip regularly and gently east-northeastwards. The general sequence is given below.

	Thickness (m)
Dolomite, buff, finely crystalline, in irregular beds 0.05–0.30 m thick, gradational base	*c.* 1.0
Calcite mudstone, buff, in slightly irregular beds 0.03–0.1 m thick except in lowest 0.6 m where several more regular beds are 0.1–0.15 m thick; apparently barren	*c.* 2.5
Limestone breccio-conglomerate, grey, and associated grey wackestones, packstones and grainstones; very varied, locally shelly; base sharp and conformable	up to 0.9
Interbedded (thinly in lowest 0.4 m) buff, finely crystalline, dolomite and subordinate grey calcite-mudstone, sparingly shelly	0.9
Very thinly (0.002–0.02 m) unevenly interbedded grey calcite-mudstone and buff finely crystalline dolomite; sparingly shelly	1.6+

The breccio-conglomerate (proximal turbidite or debris flow) may be traced in the quarry face for about 250 m (Figure 3.26); it thins and becomes less pebbly south-westwards. The lithology of the deposit was investigated by Lee (1990) who identified six main rock types that are present in a roughly consistent, but laterally varied sequence:

6 (at top) Calcarenite, upwards-fining, interbedded with host calcite mudstones
5 Fine calcirudites and pebbly calcarenites, grading up into calcite- and then dolomite mudstones
4 Coarse, poorly-sorted, calcirudite, containing subspherical to tabular clasts
3 Calcirudite/calcarenite, upwards-fining
2 Slightly to severely deformed, interbedded calcite- and dolomite mudstones
1 Calcirudite, clast-supported, well-rounded calcite mudstone clasts

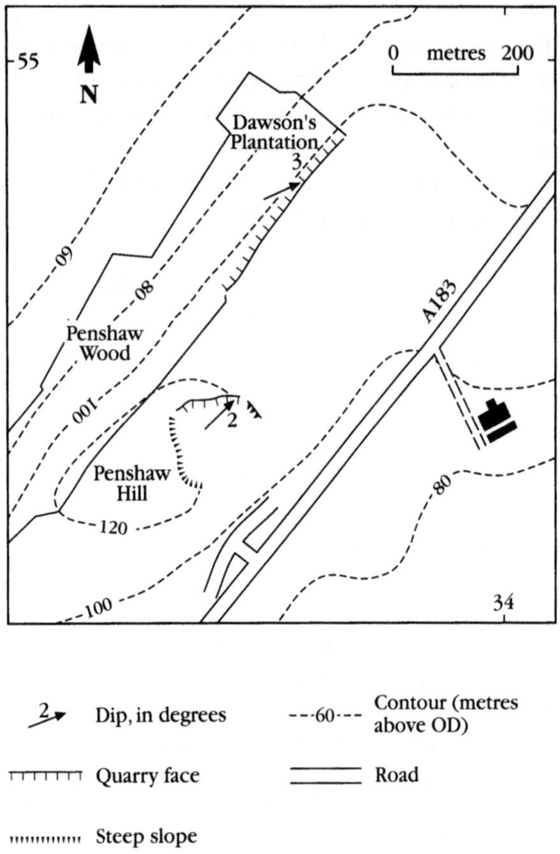

Figure 3.25 Dawson's Plantation Quarry, Penshaw, and its environs.

Lee noted penecontemporaneous erosion surfaces within the deposit, particularly below units 3 and 5, and carefully documented its lateral variability (Figure 3.27). He interpreted this variability as at least partly caused by complex channelling and reworking, the trend of the channels (and therefore the direction of sediment transport) being difficult now to determine; fresher surfaces in 1953 had previously revealed clast imbrication and deformation patterns suggestive of south-west to north-east sediment transport (Smith, 1970c, p. 7). Bioclasts, especially productoids, are much more abundant in the debris flow than in enclosing strata and, though some are deformed, are commonly unusually well preserved. Smith (1970c) has speculated that this good preservation may have resulted from rapid burial of the whole animals rather than slow accumulation of more fragile disarticulated valves and empty shells. Lee (1990) noted a good size correlation between lithoclasts and bioclasts and commented that skeletal remains are most abundant in the finer-grained calcirudites and calcarenites.

In addition, to the debris flow, the Raisby Formation in Dawson's Plantation Quarry features abundant, intersecting, curved low-angle joints and minor rotational movement planes (Smith, 1970c). Most of these are concave-upwards, with a tendency to grade downwards into bedding-plane slips; they cut all strata, including the debris flow.

Interpretation

Dawson's Plantation Quarry contains, without doubt, the best-exposed and most impressive proximal turbidite or debris flow in the Magnesian Limestone. Related disturbed strata are widespread (although not ubiquitous) at about the same stratigraphical level in north-east Durham; they vary greatly in character from place to place, ranging from graded turbidites, as at the former Downhill Quarry (NZ 348601), to coarse breccias composed of slide-blocks up to several metres across as at the High Moorsley Quarry site (Smith, 1970c). Most of the disturbed strata are more fossiliferous than beds above and below, presumably because of rapid deposition, and almost all yield evidence of early sea-floor lithification of the carbonate muds. The rarity of disturbed strata at most other levels in the formation argues against inherent sediment instability through natural oversteepening of the deposition slope and may point to a brief phase of instability caused by contemporary local earth movements.

The low-angle curved joints and minor movement planes in the Raisby Formation at Dawson's Plantation Quarry are similar to others at many northern exposures of these strata, including the Claxheugh Rock site and sea cliffs in Sector 1 of the Trow Point to Whitburn Bay site. At these two localities the joints are truncated upwards at the base of the late Raisby Formation submarine slide sequence, suggesting that they too may have resulted from contemporary earth movements.

Future research

The sedimentology of the Raisby Formation has recently been investigated by Lee (1990, 1993) and there is little immediate scope for further research on this aspect of Dawson's Plantation Quarry. The transported fauna in the disturbed bed and related fall-out deposits, however, are likely more closely to represent the total contemporary benthos than the sparse, selectively-preserved fauna of undisturbed Raisby Formation strata and could repay further study.

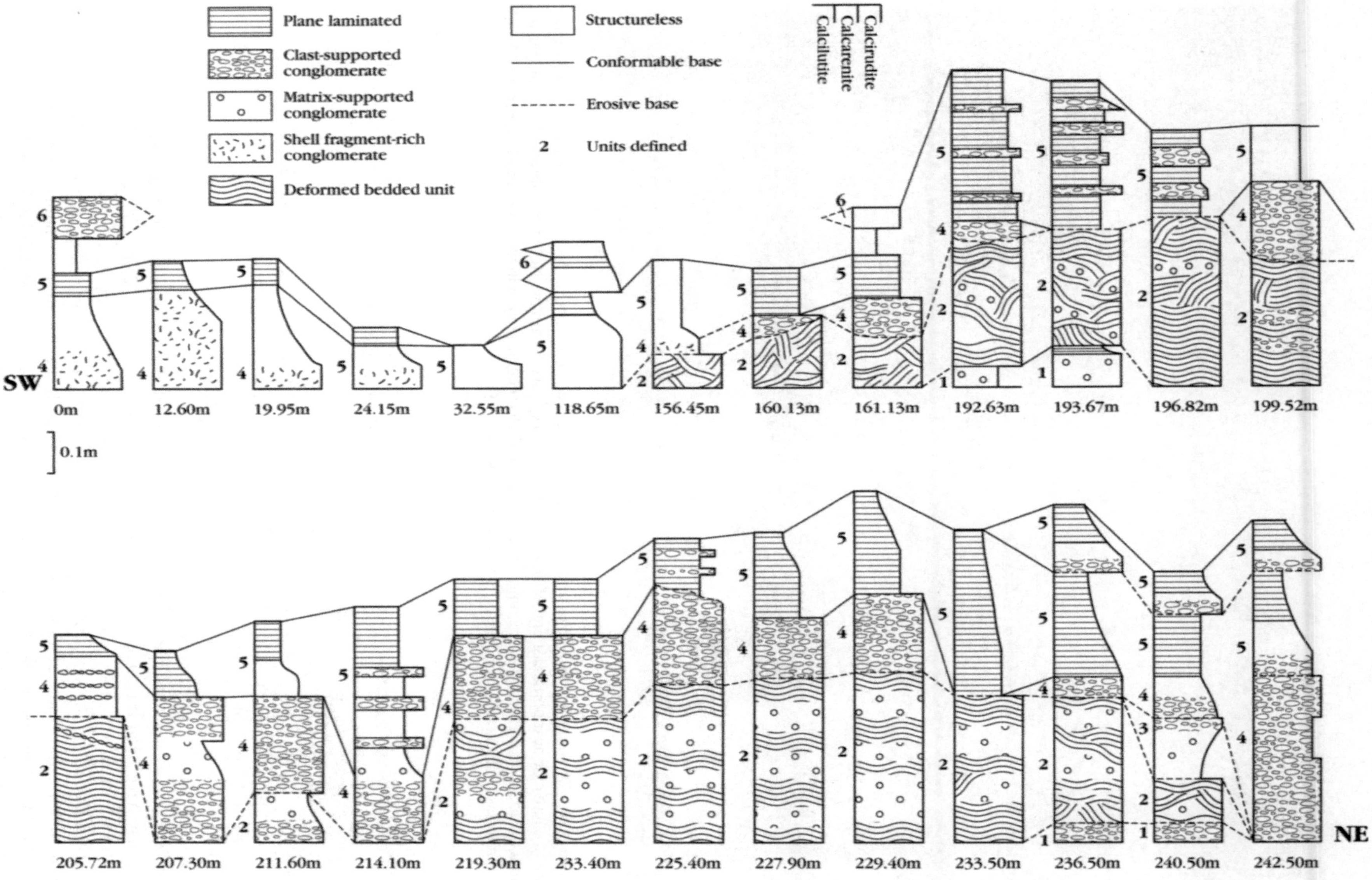

Figure 3.27 Disturbed strata in the Raisby Formation at Dawson's Plantation Quarry, showing their lateral variation. After Lee (1990, fig. 2.21).

Figure 3.26 Debris flow of dolomite- and calcite mudstone a few metres above the base of the Raisby Formation. View to south-east near north-east end of Dawson's Plantation Quarry, Penshaw. Note the imbrication in the partly rounded clasts towards the top of the unit, and the mollusc shells seen in section in the uppermost fine-grained bed, interpreted as a fall-out tail. Hammer: 0.33 m. (Photo: D.B. Smith.)

Conclusions

This site exposes the lower part of the Raisby Formation and is unique in that it is the best exposed example of a debris flow in the marine Permian of the Durham Province. The debris flow is thought to be part of a more extensive sheet of disrupted sediment that moved ENE down the depositional slope near the western margin of the Zechstein Sea. Such downslope movement of sediment may have been triggered by an earthquake. The transported sediment contains a better-preserved fauna than the strata below and above, which may be the result of rapid burial of the shelly organisms on the sea floor. The retention of this site is important for sedimentological study and for future research on the fauna of the disturbed sequence.

HUMBLEDON HILL QUARRY (NZ 381552)

Highlights

Humbledon (formerly Humbleton) Hill Quarry (box 6 in Figure 3.2) is cut into the lower part of the core of the late Permian shelf-edge reef and is world famous as one of the most prolific sources of English Zechstein reef faunas; this locality has yielded more than 40 marine Permian invertebrate type, figured and cited specimens including the brachiopod *Stenoscisma humbletonensis* (Howse) which has been recorded at only one other locality in the region, and the cyclostome bryozoan *Stomatopora voigtiana* (King, 1850) which is thought to be unique world-wide. The quarry contains the best exposure of the contact between the reef and underlying bedded dolomite and, in conjunction with newer roadside sections on the north side of the hill, it provides a transect about 200 m long through the lower part of the reef.

Introduction

Humbledon Hill is a well-known local landmark in the south-western inner suburbs of Sunderland; it is smoothly rounded, roughly circular in plan and 250–300 m across. The hill forms a prominent link in a chain of grassy knolls that mark the position of the shelf-edge reef of the Ford Formation, and the quarry is cut into the steep ENE-facing slope of the hill, not far from the presumed seaward face of the reef; it exposes a thickness of about 15 m of rock in the main excavation and scattered smaller excavations below and above increase the exposed thickness to more than 25 m. Most of the exposed rock is reef dolomite, but the lowest few metres are of sparingly fossiliferous dolomite of uncertain stratigraphical affinity; an apparent erosion surface separates the two main rock units.

Humbledon Hill Quarry has existed for more than 160 years, though the total amount of rock removed is not great and working must have been limited and perhaps intermittent; it was mentioned as a fossiliferous exposure by Sedgwick (1829) - part of his 'Shell-Limestone' - and yielded large numbers of fossils to those great rivals and most dedicated of collectors Richard Howse (1848, 1858) and William King (1848, 1850). Kirkby (1857, 1858) reported on the ostracod fauna from the quarry and Logan (1967) described several of the bivalves.

No additional genera have been found here since the days of Howse, King and Kirkby, but the lithology and fauna were described briefly by Trechmann (1945) and more fully by Hollingworth (1987).

This account is based mainly on the writer's observations of the quarry since 1953 and includes data on parts of the quarry now overgrown, filled or otherwise inaccessible.

Description

The main face of Humbledon Hill Quarry is about 90 m long and more than 15 m high; it, and a little of the adjoining hillslope, comprise the site. The location of the main features of geological interest are shown in Figure 3.28. Parts of the quarry face and much of the hillslope are obscured by vegetation and most of the former quarry floor has been enclosed in private gardens.

The geological sequence in and around the quarry is shown below.

	Thickness (m)
Ford Formation, reef-facies	25+
?Erosion surface	
Ford or Raisby Formation, in south of quarry	2+
Gap	?4–10
Magnesian Limestone (probably mainly Raisby Formation) with Marl Slate at base. Proved beneath drift in nearby well and borehole (two differing records)	82 or 87
Yellow Sands (proved in borehole)	3+

The disposition of the lithological units exposed in the southern part of the main quarry face is shown in Figure 3.29.

Ford or Raisby Formation

The oldest rocks in Humbledon Hill Quarry are exposed in the south-east of the main face and underlie the supposed erosion surface. They comprise about 2 m of evenly level-bedded saccharoidal cream-buff porous dolomite (possibly an altered oolite) with scattered to abundant shell debris; a bed near the middle of the exposed sequence is extremely shelly. Trechmann (1945, p. 341) recorded seven invertebrate genera apparently from this exposure (though his wording is slightly ambiguous), including a bryozoan, brachiopods, a gastropod and two species of bivalves.

The ?erosion surface

This undulate surface has a visible relief of about 0.5 m (Figure 3.29); no erosion products were

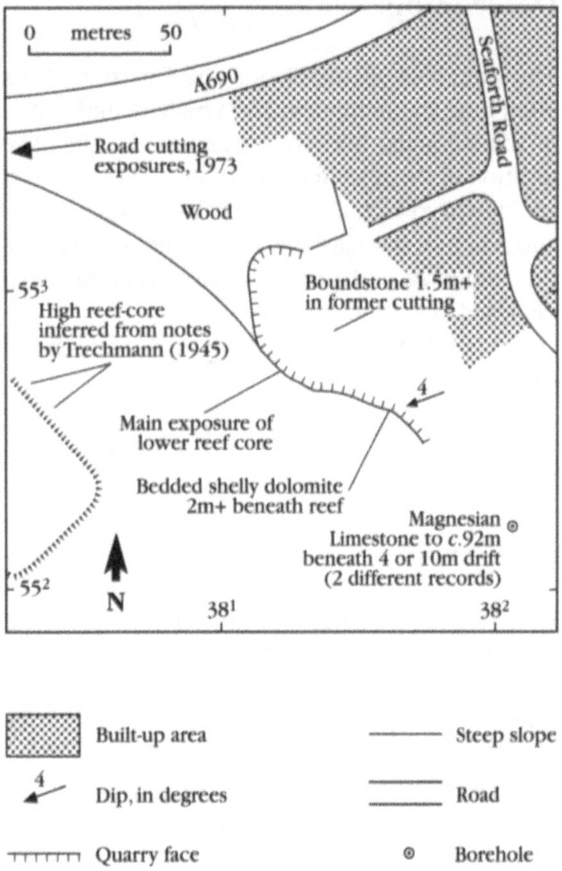

Figure 3.28 Humbledon Hill Quarry and its immediate surroundings, showing the position of the GCR site and the main features of geological interest.

noted on the surface, but overlying dipping rubbly shelly dolomite displays strong onlap. A former 2 m cutting in the quarry floor revealed no bedded dolomite, indicating that the erosion surface in the quarry as a whole has a minimum relief of 2 m.

Ford Formation, reef-facies

Details of the reef-rock in the quarry are somewhat obscured by vegetation and quarry waste, but two main rock types are present: (a) massive hard dolomitized autochthonous bryozoan boundstone (framestone/bafflestone) in ovoid bodies up to several metres across, and (b) tongues, sheets and pockets of crudely bedded dolomitized shelly rubble. The two rock types appear to be randomly distributed relative to each other and both contain scattered cavities after former secondary sulphates.

The biota for which this quarry is renowned is divided unequally between the boundstone and the shelly rubble. Work by Pattison (unpublished British Geological Survey report; Pattison, 1978) on fossils from the adjoining road cutting to the north-west showed that comparable boundstone bodies there contain a relatively low-diversity fauna dominated by *in situ* pinnate bryozoans and small pedunculate brachiopods (Figure 3.30), and Hollingworth (1987, pp. 211–213) broadly confirmed this from smaller collections made in the

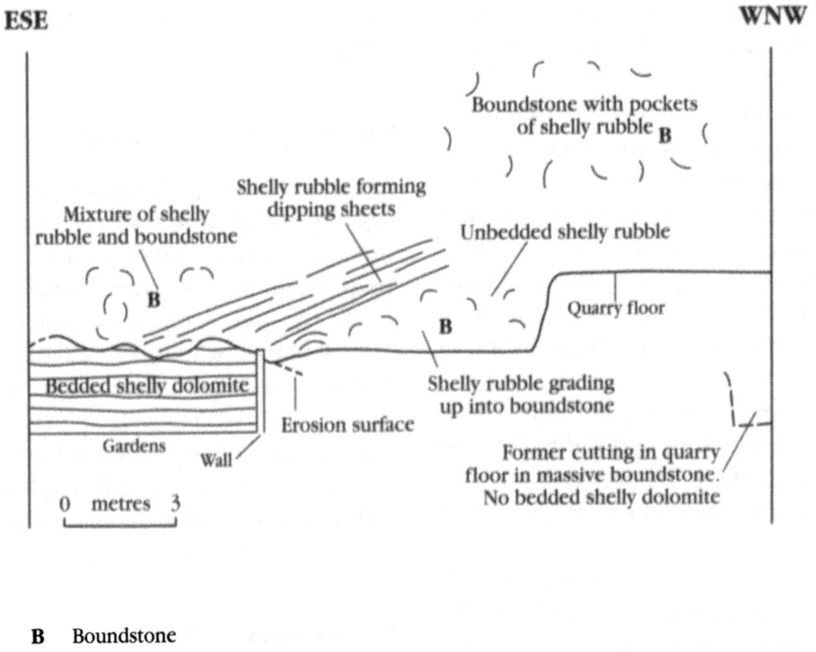

Figure 3.29 Sketch of the stratal relationships in the southern part of the main face of Humbledon Hill Quarry, based on an unpublished drawing made by the writer in 1953.

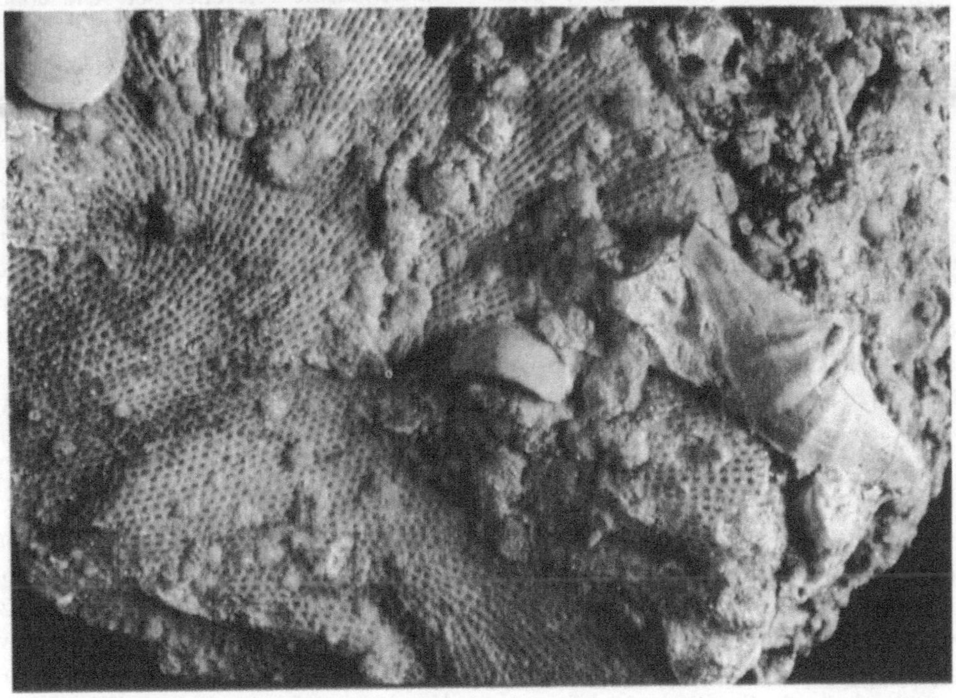

Figure 3.30 Typical elements of the fauna of the boundstone bodies in reef dolomite in Humbledon Hill Quarry, comprising the pinnate bryozoan *Fenestella retiformis* and the pedunculate brachiopods *Dielasma elongatum* and *Pterospirifer alatus*. Field of view about 67 × 100 mm. (Photo: N.T.J. Hollingworth.)

quarry. In contrast, Pattison showed that the enveloping rubble exposed in the cutting contains a highly diverse assemblage of bryozoans, brachiopods, bivalves, crinoids and gastropods. Hollingworth (1987, pp. 212, 216) recorded a similar biota in reef rubble in the quarry, and noted that none of the bryozoan colonies in the rubble were in life position. He inferred that many organisms clearly lived here and were not originally inhabitants of the boundstone bodies. Trechmann (1945) recorded large rolled productids in 'mushy and porous' rock (i.e. rubble) near the base of the reef, and noted that the overall biota remained roughly constant throughout the main face, but became considerably less diverse in small exposures towards the top of the hill behind and above the quarry.

Interpretation

Humbledon Hill Quarry is especially important (a) as the prime source of more than 40 late Permian type, figured and cited marine invertebrate fossils and (b) as one of the best exposures of lower reef-core rocks in the shelf-edge reef of the Ford Formation. Because of the current restrictions on access, the historical aspect is the more important, but the record of the reef-core has assumed additional significance following the unique light thrown on the reef structure by exposures created during the widening of the adjoining road to the north-west in 1973 (see below).

The erosion surface and underlying strata

The significance of the inferred erosion surface beneath the reef at Humbledon Hill Quarry is unknown, and its assessment is complicated by the uncertain stratigraphical age of the underlying bedded dolomite. Similarly uneven inferred erosion surfaces separate reef and underlying beds at an exposure (NZ 3805 5489) some 370 m south of the quarry and also at the Gilleylaw Plantation Quarry site, Silksworth, at both of which the surface has a local relief of about 2 m. Truncation surfaces, indeed, underly reef-rocks at most places where the base of the reef is exposed; in several such exposures the truncation has been ascribed to massive pre-reef submarine slumping (Smith, 1970c), but the thickness of underlying Zechstein strata in these exposures is much less than that at

Humbledon Hill and Silksworth and this explanation may not be appropriate.

The uncertainty regarding the age of the strata immediately beneath the reef in Humbledon Hill Quarry precludes complete understanding of the local stratigraphy and geological history and needs to be resolved; Kirkby (1870) and Woolacott (1912) attributed these beds to the 'Compact Limestone' and 'Lower Limestone' (both now Raisby Formation), respectively, but Trechmann (1945) clearly considered them to be part of the reef; Smith (1969a, 1971b, 1994) tentatively classes them as pre-reef Ford Formation. The argument for a Raisby Formation age presumably was based on their superficial lithological similarity to known Raisby Formation strata beneath the reef at Down Hill, Hylton Castle (not the SSSI) and Claxheugh Rock, but a high content of shell debris is very unusual in upper parts of the Raisby Formation; the faunal list given by Trechmann has little bearing on the problem because of the high species overlap between the Raisby and Ford formations. The argument for a Ford Formation age is based mainly on the height of these strata above the base of the Marl Slate; this is shown by the well and borehole at the adjoining Humbledon Hill Pumping Station to be about 82 or 87 m (two differing records), which greatly exceeds the maximum proved thickness of about 50 m of undoubted Raisby Formation strata in the environs of Sunderland and is comfortably thicker than the maximum exposed thickness of about 45 m of Raisby strata in the general area. Though arguments based on thickness alone are unlikely to be conclusive in a sequence so demonstrably variable as the Magnesian Limestone, the author believes that the thickness and faunal abundance evidence together support a Ford Formation age for the subreef strata at Humbledon Hill Quarry. The thickness of these strata is not known, but they exceed 5 m at their main exposure in the sides of Newport Dene (NZ 385542), 1.2 km SSE of Humbledon Hill; they may have been more than 50 m thick in parts of the Sunderland area (Smith, 1994). Farther south, the reef overlies at least 20 m of lagoon-type dolomite in Castle Eden Dene (NZ 4339).

Ford Formation, reef-facies

The importance of the collections of fossils from Humbledon Hill cannot be over-emphasized; they formed a disproportionately large part of King's (1848, 1850) source material and are now housed at University College, Galway, where they provide an invaluable reference set. The collection was fully curated and catalogued by Pattison (1977). None of the genera named by King was unique to Humbledon Hill, though Howse (1848) erected the species *Terebratula humbletonensis* (later referred to *Camarophoria* by Howse (1858) and now *Stenoscisma humbletonensis* by more recent authors), which at that time had not been recorded elsewhere, but which was later reported from Tynemouth by King (1850). Humbledon Hill (presumably the quarry) was listed by Howse (1848, 1858) as a source (though not the only source) of almost 40 genera of Permian marine invertebrates and Trechmann (1945) listed 46 genera from there. Ostracods from Humbledon Hill were described and figured by Kirkby (1857, 1858) and the quarry supplied material to the Kirkby Collection of other Magnesian Limestone fossils, housed at the Hancock Museum, Newcastle upon Tyne. Many specimens of bivalves from Humbledon Hill, including some from both the King and Kirkby Collection (Hancock Museum), were cited and illustrated by Logan (1967). More recently, the quarry was cited as the only source in the world of the bryozoan *Stomatopora voigtiana* (King, 1850) by Taylor (1980) and as one of the sources of the crinoid *Cyathocrinites ramosus* (Schlotheim) described by Donovan *et al.* (1986). The precise source within the quarry of the type, figured and cited specimens is not clear from the literature, but it is probable that most were collected from the rubbly parts of the main face.

The variability of the reef-rock at Humbledon Hill Quarry was first noted by Howse (1848), who mentioned 'hard somewhat crystalline' and 'earthy and rubbly' varieties; Trechmann (1945) referred to both massive and mushy varieties. Presumably these equate with the autochthonus boundstone and shelly rubble mentioned earlier. The widening of the adjoining road exposed a 125 m transect through the reef a short distance west (i.e. landward in a palaeoenvironmental sense) of the quarry (Smith, 1981a, 1994; Hollingworth, 1987) and revealed that the boundstone there forms discrete to grouped ovoid masses up to several metres across (but generally 1–2 m) that are embedded randomly in the shelly rubble; they increase in proportion towards the ENE (i.e. towards the quarry and reef crest). This key exposure, taken in conjunction with that of the quarry, shows that this part of the reef, when formed, comprised patchily distributed bryozoan thickets or compound colonies, each with a low-diversity specialized associated biota and a relief of a few decimetres, lying on and in (and subsequently covered by) a variable mosaic of bioclastic debris derived from and supporting a highly diverse invertebrate community.

Many of the bryozoans in the boundstone masses are thickly invested with lamellar encrustations, which doubtless contributed bulk and stiffening (and possibly cement), but no evidence of contemporaneous cementation of the shelly rubble has been recorded. Some of the boundstone masses are themselves coarsely concentrically layered, presumably as a result of intermittent growth.

The fauna of the reef-rock at Humbledon Hill Quarry is typical of that of the lower and middle parts of the reef-mass, and was assigned by Hollingworth (1987) to the lower reef-core; the Humbledon Hill exposures, together with that at Hylton Castle site, provide the basis for his portrayal of a typical lower reef-core community (Hollingworth, 1987, fig. 6.12; Hollingworth and Pettigrew, 1988, fig. 8). The reef-base coquina, commonly found elsewhere beneath the reef, is absent at Humbledon Hill, and Trechmann's (1945) record of faunal impoverishment at the top of the hill suggests that high reef-core rock is present and may imply that the reef is unusually thin (and perhaps condensed) here. The lack of evidence of strong contemporaneous erosion in the reef and fossils at Humbledon Hill and similar exposures, and of sedimentary structures in the rubble, points to accumulation in relatively low-energy conditions below wave base (Smith, 1981a; Hollingworth, 1987) and is consistent with Hollingworth's assessment of the living conditions of the faunal community.

Neither the biota nor the structure of the reef dolomite in the quarry provide evidence of proximity to the reef slope, which therefore probably lay at least 30 m east of the main exposure.

The isolation of Humbledon Hill from other known areas of late Permian reef-rock, coupled with its rounded outlines, invited speculation that it might be a link in a chain of reef knolls (Trechmann, 1913, 1925, 1945); the newly-exposed reef-rock along the north side of the hill shows no evidence of lateral passage into bedded rocks or of an approach to a reef margin, however, and it now seems more likely that the knoll-like form of the hill is an erosional feature in an otherwise relatively continuous shelf-edge reef. Similar doubts regarding two rounded hills of reef-rock between West Boldon and Hylton Castle were resolved in 1959 when temporary excavations in the floor of the intervening valley revealed almost continuous reef-rock. The reef is known to form an east-facing NNW/SSE belt extending from West Boldon to Hartlepool, and may once have extended farther; it marks the seaward margin of the carbonate rocks of Sub-cycle 2 of English Zechstein Cycle 1 (Figures 1.4, 3.1 and 3.2). Other GCR sites in rocks of the shelf-edge reef are at Hylton Castle Cutting (Sunderland), Ford Quarry, Cutting and Claxheugh Rock (Sunderland), Tunstall Hills (Sunderland), Stony Cut, Hawthorn Quarry and Horden Quarry; each reveals aspects and parts of the reef that are different from those seen at Humbledon Hill, though the fauna at Humbledon Hill and Tunstall Hills have much in common.

Future research

Restrictions on access now hinder research on all aspects of the reef-rocks in Humbledon Hill Quarry, though many features formerly seen in the quarry may still be seen and investigated in the adjoining road cutting. Opportunities for fossil collecting are now severely limited, but large numbers of fossil specimens from the quarry are available for study at the British Museum of Natural History, the Hancock Museum (Newcastle upon Tyne), Sunderland Museum, University College (Galway) and the British Geological Survey, Keyworth.

Conclusions

The site is internationally famous for its fauna. It has yielded a rich variety of invertebrate fossils characteristic of the English Zechstein reef; in particular, the bryozoan *Stomatopora voigtiana* (King, 1850) is considered to be unique worldwide. The site was formerly one of the best exposures of lower reef-core rocks in the Ford Formation and of the erosion surface immediately beneath the reef. Although access to the site is now restricted, it remains one of major importance for the study of late Permian reef faunas in the Durham Province.

TUNSTALL HILLS, SUNDERLAND; MAIDEN PAPS AND THE TUNSTALL HILLS (ROCK COTTAGE EXPOSURE) (NZ 3954)

Highlights

The twin mounds of Maiden Paps at the north-west end of Tunstall Hills (box 7 in Figure 3.2) are probably the best-known topographical expression of the shelf-edge reef of the Ford Formation and also contain a number of the most important exposures of fossiliferous reef-rock; these include a large exposure of reef-core in the more southerly mound, which also features several steeply-dipping laminar sheets and tension gashes. The nearby 'Rock Cottage' exposure is unique in revealing highly fossiliferous primary limestones at the base of the reef, which bear evidence of contemporary sea-floor cementation. More than 50 type, cited and illustrated fossils are from Tunstall Hills, and at least two species of fossil invertebrates carry the name *tunstallensis*.

Introduction

Maiden Paps are two prominent local landmarks (see Hollingworth, 1987, appendix D, fig. 3) in the southern outer suburbs of Sunderland; they form part of a chain of knolls that mark the position of the shelf-edge reef of the Ford Formation between West Boldon (NZ 3460) and Horden (NZ 4341). The hilltops are excellent vantage points for viewing the close relationship between the local geology and scenery and afford a superb view to the west of the low-lying floor of the former glacial Lake Wear and its associated drainage channels.

Reef-rocks, mainly of limestone, but some of dolomite, make up all the exposures at Maiden Paps and total perhaps 35 m in thickness. The more northerly of the mounds is mainly grass-covered and now bears only a single small quarry exposure of shelly dolomite near its northern extremity; in contrast, the southern mound bears many exposures including a north-facing quarry cliff of reef limestone (8 m+). Slightly farther south is the famous 'Rock Cottage' exposure of reef-base limestone coquina.

The shelly rocks of Tunstall Hills have been known since the writings of Winch (1817) and Sedgwick (1829) and, together with those at Humbledon Hill, were the main collecting ground for Howse (1848, 1858), King (1848, 1850) and Kirkby (1857, 1858, 1859); few additional species have been found here since 1860, though long lists were given by Trechmann (1945), Logan (1967), Pattison (internal Geological Survey report, 1966) and Hollingworth (1987). Hollingworth also gave a faunal list from the 'Rock Cottage' exposure, where some of the gastropods have retained part or all of their original colour (Hollingworth, 1987; Hollingworth and Tucker, 1987; Hollingworth and Pettigrew, 1988).

Analyses of the reef-rocks were given by Aplin (1985), who also illustrated (p. 378) the main face and discussed the complex diagenetic history of the reef. Burton (1911), Woolacott (1912) and Trechmann (1945) noted the presence of coarse, frosted, quartz sand grains in fissures in the main face and the origin and filling of the fissures was discussed by Smith (1981b) and Aplin (1985). The diagenesis of the coquina at the base of the reef in the 'Rock Cottage' section was detailed by Tucker and Hollingworth (1986) and Hollingworth and Tucker (1987), who also drew conclusions on some general aspects of early reef history.

Working of the quarries ceased more than a century ago, but the discovery of the 'Rock Cottage' reef-base limestone by Hollingworth in 1983 was followed by NCC-funded excavation to help in GCR assessment and the exposure has since been enlarged and fenced by the Sunderland Borough (now City) Council.

Description

Rock exposures at and near Maiden Paps include the main face cut into reef-core limestone on the north side of the southern mound, a small quarry (NZ 3910 5472) in reef slope dolomite at the foot of the north slope of the northern mound and the small exposure of basal coquinoid limestone (NZ 3910 5432) high on a wooded slope about 70 m south of 'Rock Cottage' (shown on Ordnance Survey maps as 'Tunstall Hills Cottage'). The position of the above exposures is shown in Figure 3.31 which also shows the positions of several minor exposures.

The north face of the main quarry (a) is about 8 m high and 30 m across, but the exposure also extends a short distance south-eastwards on the north-east side of the hill and rather farther southwards on the west side (Figure 3.31). Although at first sight almost unbedded, close inspection reveals hints that parts of the rock, like that at Humbledon Hill and Hylton Castle Cutting, may comprise rounded 1–3 m masses of bryozoan boundstone that are separated and surrounded by vaguely bedded shelly rubble; in places small *Dielasma* shells in life position form dense swarms attached to boundstone masses that were, by inference, already lithified.

The abundance of fossils for which Tunstall Hills was noted in the past is less noteworthy now,

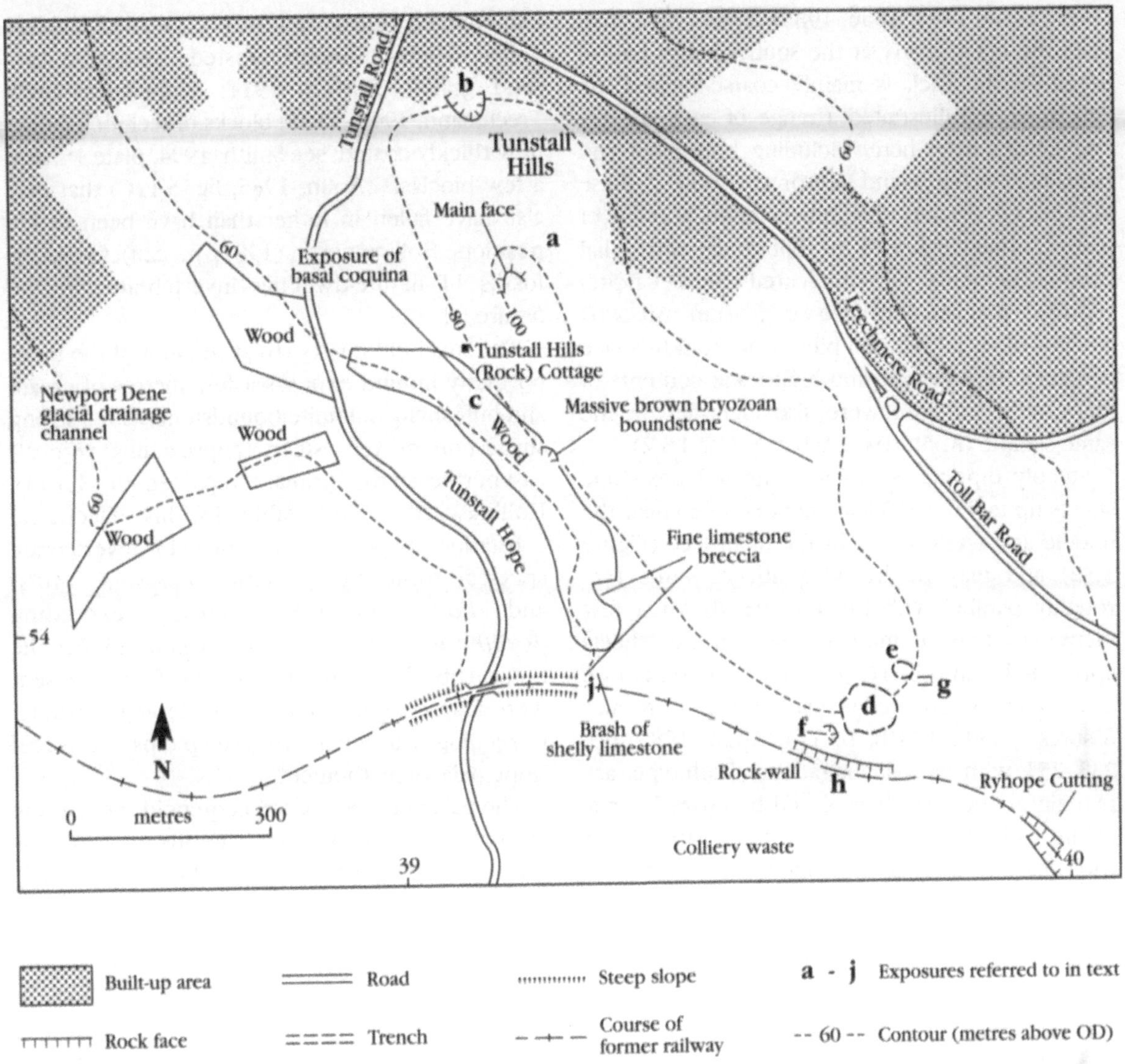

Figure 3.31 Tunstall Hills (north and south), showing the position of the main exposures and features of geological interest.

though locality citations in the early (and some of the later) works were generally limited to 'Tunstall Hills' or 'Tunstall' and the precise point of origin cannot now be identified. These localities, presumably, though not undoubtedly, referring to Maiden Paps, were a source of more than 50 type, cited and/or illustrated individuals for King (1848, 1850), almost 40 for Howse (1848, 1858) and more than 40 for Trechmann (1945) who gave some precise locality details; Logan cited the hills as a source of almost 30 species of bivalve and illustrated several from here. The main face (a) yielded 16 species to Hollingworth (1987, table 7), with *Dielasma* forming 38% and *Bakevellia (Bakevellia)* 21% of the fauna; he described the assemblage as of low diversity, with less than 10% of bryozoans, and combined the data from here with that from the main quarry at the south-eastern end of the hills and from Humbledon Hill, to reconstruct a model of a reef-core palaeocommunity (Hollingworth, 1987, fig. 6.15, reproduced as fig. 6 in Hollingworth and Tucker, 1987 and as fig. 16 in Hollingworth and Pettigrew, 1988). Fossil preservation ranges from poor to extremely good, even in single specimens. In several substantial parts of the face, particularly near the centre-top, bryozoan and other frame-building elements are thickly coated with concentric ?algal encrustations that locally form almost all of the rock.

Most of the rock in the main face and adjoining exposures is of hard, crystalline, brown, ferruginous limestone with abundant coarse radial calcite

(Trechmann, 1945; Aplin, 1985). Aplin noted that, as in the main quarry at the south-eastern end of the hills, the rock is mainly coarse-grained; he described and illustrated a range of early diagenetic fabrics from here, including botryoidal and radial-fibrous types, and interpreted some of these as replaced, primary aragonite cements possibly of marine origin; many of the botryoidal and radial fibrous cements are nucleated onto calcite-replaced nodular anhydrite or gypsum. An XRD analysis of algal bryozoan bindstone from this face revealed 100% calcite, and high trace contents of manganese and iron were also identified in the same sample (Aplin, 1985, tables 5.1 and 5.2).

Steeply dipping, sinuous, laminated limestone sheets up to 0.6 m thick are concentrated near the middle and western end of the main face (Figure 3.32) and may be traced south-eastwards (i.e. roughly parallel with the reef trend) for a few metres to tens of metres. Some of the sheets appear to be unilateral (Figure 3.33), but others are bilateral and undoubtedly coat the walls of former fissures (Smith, 1981b, p. 174; Aplin, 1985, pp. 233–253, with several illustrations); both types are patchily to extensively replaced by coarsely crystalline radial calcite (see Aplin, 1985, table 5, for XRD analyses from here). Some of the fissures are incompletely filled and retain irregular median voids; others contain frosted, coarse quartz sand grains (Burton, 1911; Woolacott, 1912; Trechmann, 1945), fallen blocks of reef-rock (some now thickly coated; see Smith, 1994, plate 15) and a few bioclasts (Aplin, 1985, fig. 5.12C) that may also have fallen in rather than have been in life position. Hollingworth (1987, p. 220) found no fossils of fissure-dwelling invertebrates in the fissure fill.

The small old quarry (b) at the foot of the more northerly mound exposes a few metres of cream and buff shelly dolomite boundstone with a strong suggestion of very steep (approaching vertical) east-northeasterly ?primary dips (Smith, 1981a). Hollingworth (1987, table 13, his 'Electricity Substation' exposure) recorded 14 invertebrate species from here, with *Dielasma* (40%) and *Bakevellia* (18%) greatly exceeding *Acanthocladia* (7%) and *Synocladia* (5.5%); he noted (pp. 279–280) that all the forms present were adapted to life on a steeply-sloping substrate, in keeping with the inferred steep dips of this reef slope palaeoenvironment.

The palaeontology of the coquinoid rocks at the important 'Rock Cottage' exposure (c) was investigated in great detail by Hollingworth (1987)

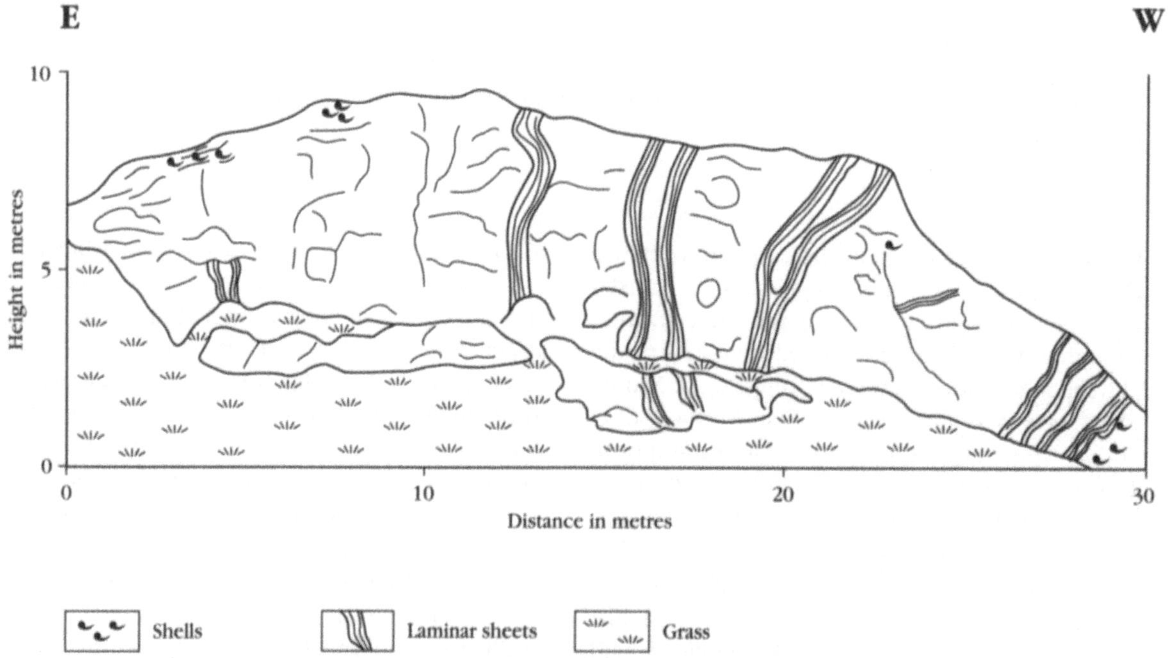

Figure 3.32 The main quarry face on the north side of the more southerly of Maiden Paps, Tunstall Hills. Most of the rock is brown bryozoan boundstone, locally with profuse encrustations. At least two of the tension gashes have partial central voids. Simplified from a sketch by Aplin (1985, p. 378).

Figure 3.33 Laminar to botryoidal (?algal) limestone (possibly dedolomite) lining the footwall of a steeply inclined fissure in boundstone of the shelf-edge reef of the Ford Formation on the north side of the main exposure at Tunstall Hills (north). Hammer: 0.33 m. (Photo: D.B. Smith.)

and their petrology was described by Tucker and Hollingworth (1986) and Hollingworth and Tucker (1987). The base of the coquina is not exposed, but the lowest part of the new excavation lies only a few metres up the hill from temporarily exposed, sparsely fossiliferous, well-bedded 'dolomicrite'; the outcrop of the coquina may be traced south-eastwards for about 100 m, but then ends abruptly at the High Barnes fault which has an estimated downthrow south of 25 m. According to Hollingworth (1987, pp. 170–173), the coquinoid limestone comprises a crumbly basal unit (0.6 m+) of 'pale cream to buff, well-bedded, calcified coquina overlain by dark brown, iron-rich, partially decalcified coquina' (about 2 m) which, in turn, is succeeded by about 3 m of 'slightly more massive buff-brown, crystalline coquina'. The basal unit has a diverse biota of brachiopods, bivalves, gastropods and bryozoans, but the dark brown limestone is composed almost entirely of the articulated valves of *Dielasma*. The uppermost unit contains an exceptionally varied and abundant fauna dominated by *Dielasma*, and also contains teeth of two genera of fish (*Janessa* and *Wodnika*); some of the gastropods retain their original colours. The coquina passes up into 0.4 m+ of bryozoan boundstone (framestone) in which fenestrate bryozoans (mainly *Fenestella*) are dominant, with *Dielasma* persisting in abundance. In total the coquina yielded the remains of 23 invertebrate species, enabling Hollingworth (1987, fig. 6.3) to reconstruct a base-of-reef palaeocommunity; this diagram was reproduced by Hollingworth and Tucker (1987, fig. 4) and Hollingworth and Pettigrew (1988, fig. 15). The coquina/reef-core transitional strata yielded 26 species (Hollingworth, 1987, table 2).

Petrographic examination of reef-base rocks from the Rock Cottage site (Tucker and Hollingworth, 1986; Hollingworth, 1987; Hollingworth and Tucker, 1987) revealed that they had escaped dolomitization and, partly in consequence, retain traces of primary marine cements, now calcite; these include aragonite crusts and botryoids, isopachous layers and fans of acicular calcite, and calcite fans. The cements are interpreted to indicate widespread episodic early cementation of the coquina as it lay on the shallow sea floor, with occasional brief episodes of exposure and cement dissolution.

Interpretation

The shelf-edge reef of the Ford Formation features in, or is the main constituent of, seven GCR sites in north-east England, each exposing different subfacies; taken together they reveal much of the structure, character and history of the reef. The designated sites are scattered unevenly along the outcrop, and comprise (from the north) Hylton Castle road cutting, Claxheugh Rock and adjoining exposures, Humbledon Hill, Tunstall Hills (both ends), Stony Cut, Hawthorn Quarry and Horden Quarry (Table 3.1); in general the more northerly sites expose the low and middle parts of the reef and the southern sites expose the higher parts.

Some information on the distribution of the reef is given in the accounts of the other reef sites herein, and was summarized by Trechmann (1925, plate 15) and Smith (1981a, fig. 9). There are, however, still many places where the information is too poor to allow precise delineation and the maps, accordingly, are locally little more than speculative; this is particularly so in low-lying areas, where the ridge or topographic step that normally indicates the position of the reef is missing and thick drift deposits cover the rock. The correspondence between topography and reef is generally good at Tunstall Hills, however, with the north-west to south-east ridge probably closely following the reef slope on its eastern side; some erosion of the western side has undoubtedly reduced the width of the reef there, but the hills as a whole are only a little narrower than the usual 300-600 m reef width. The inferred correspondence between the eastern side of the ridge and the reef foreslope is strongly supported by the presence of Cycle EZ2 collapse-breccias in the Ryhope Cutting at the south-eastern end of the hills and by the observed juxtaposition of the reef slope with post-reef collapse-breccia in hill-slope exposures at Easington Colliery (NZ 436437) and Horden (NZ 435417).

In greater detail, Tunstall Hills are important as one of the two main source localities of fossils reported on by King, Howse, Kirkby, Trechmann, Logan and Hollingworth, though locality information given by the early authors was generally vague. The exceptionally large collections of well-preserved fossils made by Hollingworth from the precisely located 'Rock Cottage' exposure is especially significant for the light it throws on the palaeocommunity of the reef-base coquina; this exposure is also noteworthy for the evidence of early cementation and the inference of shallow-water deposition of the coquina uniquely preserved here (Tucker and Hollingworth, 1986; Hollingworth and Tucker, 1987). Though lacking a framework and therefore not being a true reef, the cemented coquina nevertheless provided a firm substrate upon which the succeeding reef could be constructed.

The main face (a) of reef-core limestone affords by far the best exposure of steeply-dipping laminar sheets in the Cycle EZ1 reef and is one of the two main exposures investigated by Aplin (1985) in his study of coarsely crystalline partly secondary reef limestones. The origin of the laminar sheets remains somewhat uncertain because, although some are clearly bilateral fissure-fill, others that are lithologically similar lack deposits on a hanging wall and may be primary reef-surface encrustations (Smith, 1981a; Aplin, 1985). The fissures presumably resulted from tension caused by unequal support around the reef crest and the steep (50–90°) upper reef slope, and their presence is an additional indicator of early cementation of the reef; similar fissures occur in comparable parts of many major shelf-edge and barrier reefs, including the famous Capitan Reef of New Mexico and West Texas. Aplin's exhaustive study of the laminar limestone sheets suggested that they were mainly early and of marine origin, but that some could be inorganic subaerial flowstones (speleothems); this latter interpretation, coupled with the presence of ?wind-blown quartz sand grains in at least one fissure, suggests a phase or phases of subaerial exposure of the reef (Trechmann, 1945; Aplin, 1985).

The host limestones at Maiden Paps, like those in the main quarry at the south-east end of Tunstall Hills, were thought by Aplin to have resulted from the calcitization of partly dolomitized limestones and are therefore, in part, dedolomites; many of the fabrics, he commented, are probably neomorphic replacements of primary aragonite and high magnesium-calcite.

The topographically high level of the 'Rock Cottage' exposure (about +78 m O.D.) and the presence of sparingly fossiliferous dolomite a few metres lower provides further evidence of the marked primary relief of the base of the reef (see interpretation of the Humbledon Hill and Gilleylaw Plantation Quarry sites). The base of the 'Rock Cottage' exposure lies an estimated 103 m above the Marl Slate, in an area where the intervening Raisby Formation is unlikely to be more than 50 m thick and the Yellow Sands to be no more than a few metres thick; it follows that at least 50 m of bedded dolomite of the Ford Formation probably underlies the reef-base at 'Rock Cottage'.

Future research

The overall structure and stratigraphical position of the reef-rocks here are still only poorly understood and require further investigation (though this would probably necessitate drilling or the creation of additional surface exposures), and the age and origin of the fissure-fill and other laminar sheets remains uncertain.

Conclusions

The site is one of a series of GCR sites which highlight the shelf-edge reef of the Ford Formation in north-east England. This series includes a number of important exposures of fossiliferous reef-rock, and one exposure of the underlying reef-base coquina. Both reef-core and reef slope rocks are exposed, the former comprising masses of bryozoan-rich rocks surrounded by shelly rubble, and the latter characterized by very steeply-dipping, sparingly shelly dolomites and limestones. The coquina contains a varied and abundant fauna, which has allowed workers to reconstruct a base-reef community. The former shape of this part of the reef and its relationship to the surrounding rocks is still not well understood, and further study is required. Although exposures are now limited, the preservation of this site is important for the overall understanding of late Permian reef development.

TUNSTALL HILLS (SOUTH-EAST END) AND RYHOPE CUTTING (NZ 395538–399537)

Highlights

The group of exposures (boxes 8a and 8b in Figure 3.2) at the south-eastern end of Tunstall Hills is unique in including a readily accessible large exposure of debris-rich foreslope beds of the shelf-edge reef of the Ford Formation, here overlain by massive *in situ* reef. Slightly east of the reef is an excellent exposure of the younger residue of the Hartlepool Anhydrite Formation, which is in turn overlain by rocks of the Roker Dolomite Formation; the latter were almost totally brecciated (broken up) by foundering for at least 50 m when the anhydrite was dissolved by percolating groundwater.

Introduction

The exposures of reef-rock here form an integral part of the ridge that extends south-eastwards for almost a kilometre from Maiden Paps; they comprise one large and several small quarries cut into the hill itself and a rock-wall cut into the base of the southern extremity of the hill to accommodate a former railway. The quarries are all in massive reef dolomite and limestone with a relatively varied fauna dominated by bryozoans, brachiopods, gastropods and bivalves, and the rock-wall is in eastward-dipping reef debris and interbedded low reef slope deposits, the latter with their own distinctive assemblage of fossils.

The off-reef exposure is in a former railway cutting some 220–270 m ESE of the easternmost exposures of reef-rock; it comprises 2 m+ of bedded reef-equivalent dolomite overlain by 1–6 m of powdery and brecciated dolomite which, in turn, is succeeded by foundered collapse-brecciated Roker Dolomite. The north-western part of the cutting was preserved and re-excavated by the local council for research and teaching purposes when the remainder of the cutting was filled in 1981.

Early references to the reef-rocks at Tunstall Hills are too imprecise for locations mentioned to be identified now, though it is possible that all the various exposures were grouped together for reporting purposes; if this were so, fossils from the south-eastern end of the hills could have been included in the fossil lists published by, for example, King (1848, 1850), Howse (1850, 1858) and Kirkby (1857, 1858, 1859). Analyses of reef-rock by Trechmann (1914, p. 241) seem likely to be from the large quarry at this end of the hills, however, and brief comments on the reef-rock in this same quarry were made by Trechmann (1945, p. 344). Logan (1967, plate 4) later cited the quarry as the source of a hypotype of ***Bakevellia* (*Bakevellia*) *ceratophaga*** (Schlotheim) and Smith (1971b) summarized the main features of the geology. Further details were given in a series of excursion guides and reports (e.g. Smith, 1973a, 1981d) and by Smith (1981a, 1994). Finally, the petrology and diagenesis of the reef-rocks here were investigated in depth by Aplin (1985) and the faunal communities in the reef were analysed and discussed by Hollingworth (1987) and Hollingworth and Pettigrew (1988).

The earliest detailed reference to rocks in the railway cutting was by Trechmann (1954, p. 198), who listed the fauna in the lowest beds exposed and interpreted the overlying powdery dolomite breccia as 'a sort of mylonite' associated with a thrust plane; this dolomite was later reinterpreted as the dissolution residue of the Hartlepool Anhydrite by Smith (1971a, 1972) and the overlying breccia was interpreted as the foundered remains of the Cycle EZ2 carbonate unit.

The nomenclature used in this account follows traditional practice except that, to avoid confusion, the term 'Ryhope cutting' is restricted to the former railway cutting at Ryhope and the former half-cutting starting some 220 m farther WNW is referred to as the 'rock-wall'.

Description

The several parts of the site at the south-east end of Tunstall Hills and the Ryhope Cutting are shown in Figure 3.31, which also shows the main features of geological interest.

Exposures of reef-rocks of the Ford Formation

These comprise the main quarry ('d' in Figure 3.31), two smaller quarries (e, f), a trench (g) and a large artificially steepened rock wall (h); the position of a number of minor additional exposures (j) is also shown in Figure 3.31. The mutual relationship of exposures (d to h) is shown in Figure 3.34.

The main quarry (d) is about 80 m across and exposes up to 15 m of massive buff to brown algal-bryozoan boundstone (framestone/bafflestone) with several sinuous anastomosing thin sheets of complexly laminar bindstone that dip eastwards at 15-70°; western parts of the quarry are interpreted as lying in the reef-core with eastern parts approaching the mid reef slope (Smith, 1981a, fig. 3; Hollingworth, 1987, fig. 1.6; Hollingworth and Pettigrew, 1988, fig. 4). The fauna of the main quarry was listed by Pattison (Geological Survey internal report, 1966) and Hollingworth (1987, table 8), and was discussed by Hollingworth (1987) and Hollingworth and Pettigrew (1988, their locality 7, 'Tunstall Hills East'); Hollingworth showed that the fossil assemblage is broadly typical of the reef-core, with *Dielasma* being numerically dominant (34%) and fenestrate bryozoans totalling 16%. Many of the fossils are unusually well preserved in the limestones (Smith, 1981a; Aplin, 1985; Southwood, 1985; Hollingworth, 1987; Hollingworth and Pettigrew, 1988) and Smith (1981a) and later authors have speculated that the structures in these partly secondary rocks might have been inherited from primary fabrics that are less clear in the intervening dolomite stage.

Most of the rock in the main quarry is of coarsely crystalline, iron-stained limestone with very coarse replacive calcite spherulites concentrated along the western side (Trechmann, 1914, 1945; Aplin, 1985). Trechmann (1914, p. 241) quoted three analyses of limestone from this quarry and one photomicrograph (plate 36.4) and Aplin (1985, table 5.1) presented two analyses and several photomicrographs (figs 5.4, 5.4, 5.6, 5.8, 5.9, 5.17C and 5.18B); calcite was shown to be predominant in all five rocks analysed, with less than 4% of dolomite in each, but extensive petrographic examination by Aplin of these and other rocks

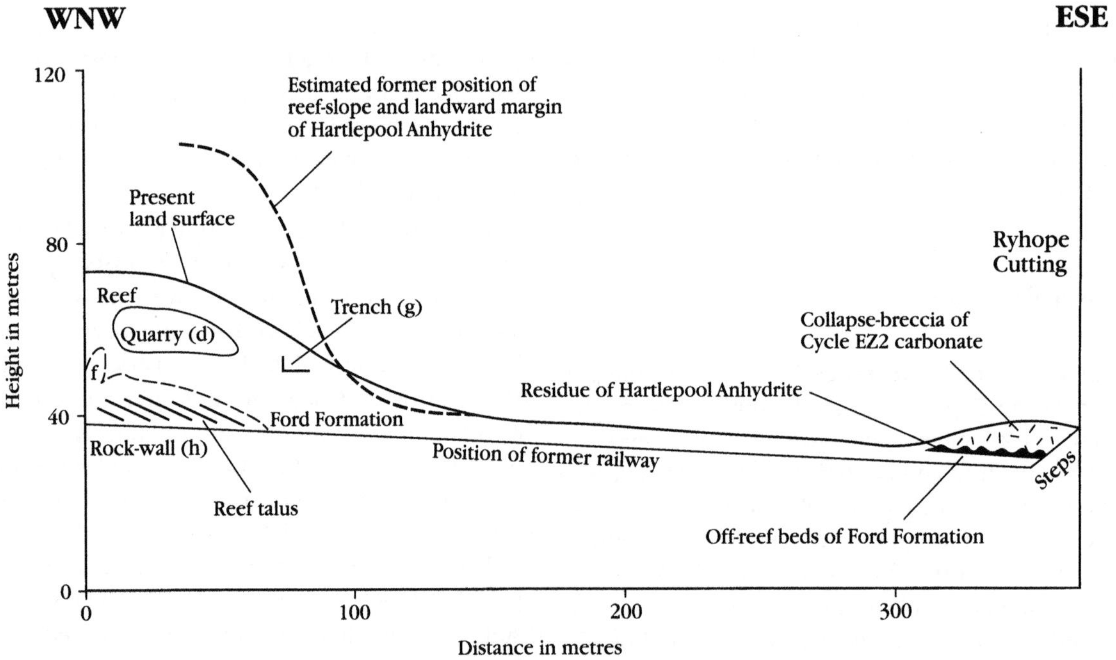

Figure 3.34 Relationships of reef and off-reef (including post-reef) strata at the south-east end of Tunstall Hills and in the Ryhope Cutting (diagrammatic). See Figure 3.31 for the location of the various exposures.

from the quarry revealed a complex diagenetic history and at least some calcite replacement of dolomite and gypsum/anhydrite. Preservation of primary fabrics is generally good in much of the limestone and reveals evidence of former marine botryoidal and fibrous ?aragonite cements (Aplin, 1985); complex ?algal encrustations coat many of the bryozoan frame elements (Smith, 1981a, fig. 26; Aplin, 1985, fig. 5.6) and in places form most of the rock (Figure 3.35). Boundstone masses such as characterize the reef-core at Humbledon Hill are fewer here, though an excellent example of a 1.8 m concentric algal-bryozoan subspherical mass lies in the extreme south-east corner of the quarry. The sinuous laminar bindstone sheets are mainly less than 0.2 m thick and some extend for the full height of the quarry face; some appear to be unilateral, having accreted only basinwards, but others appear to be bilateral and presumably coat the walls of fissures in a fully lithified host rock.

The small old quarry ('e' in Figure 3.31; NZ 3975 5394) situated a short distance north-east of the main quarry, is cut mainly into hard, massive, brown, crystalline, slightly dolomitic limestone similar to that in the main quarry and is interpreted as reef-core facies approaching the middle of the reef slope. The rock appears to be formed mainly of hard, crystalline bryozoan boundstone (framestone/bafflestone) with pockets and irregular sheets of shelly rubble; the rubble dips predominantly south-eastwards at 10–25°, but local westward dips also occur. The fauna of the quarry has not been specially studied, but crinoid debris is abundant locally. Very coarse calcite spherulites form much of the rock near the south-east corner of the quarry.

The small old quarry ('f' in Figure 3.31; NZ 3964 5386) immediately south-west of the main quarry lies at a lower topographical and stratigraphical level and exposes massively rubbly high-foreslope reef talus. The major clasts in the talus are of algal-bryozoan boundstone superficially similar to that in the main quarry, but work by Aplin (1985) shows that the limestone is generally more finely

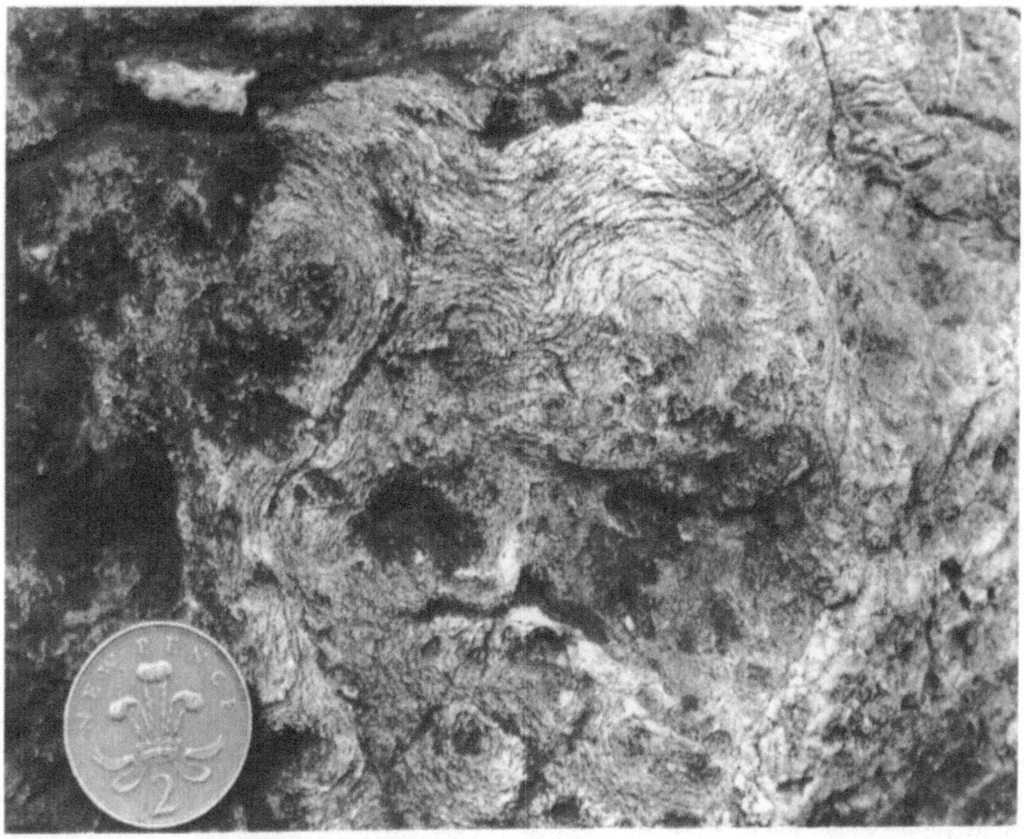

Figure 3.35 Dense ?algal encrustations in reef boundstone in the main old quarry (d) at the south-east end of Tunstall Hills. Delicate bryozoan frame elements form less than 5% of the rock. Coin: 26 mm across. (Photo: D.B. Smith.)

crystalline and that, paradoxically, primary fabrics are less well preserved; the limestone contains abundant dolomite relics. Aplin states (p. 260) that the talus interdigitates westwards into *in situ* reef-rock, here represented by unbedded dolomite algal–bryozoan boundstone in a large face that extends westwards from the small quarry.

The trench ('g' in Figure 3.31; NZ 3978 5392) lies about 40 m north-east of the main quarry and topographically slightly lower. It is about 1.5 m deep and is cut into brown (iron-stained) crystalline bryozoan boundstone (framestone/bafflestone) with hints of a gentle north-easterly dip and much coarsely spherulitic calcite. Hollingworth (1987) and Hollingworth and Pettigrew (1988, their locality 8) fully describe and discuss the brachiopod-dominated fauna and doubtfully assign the trench rocks to the reef crest sub-facies (see Interpretation); some of the bryozoans are especially well preserved and were thought by Hollingworth (1987) to be in life position. A full faunal list was given by Hollingworth (1987, table 9).

The rock wall ('h' in Figure 3.31; NZ 3964 5382–3971 5380) is about 70 m long and up to 8 m high and is cut in dolomitized reef talus deposited on the lower part of the reef-foreslope; the talus comprises a crudely eastward-dipping (20–35°) accumulation of debris derived from higher parts of the reef slope (including the crest), together with the remains of an indigenous fauna (Smith, 1981a). Tumbled blocks of bryozoan and bryozoan–algal boundstone form more than half the rock (Smith, 1981a, figs 8, 23) and lie in a matrix of finer rubble. The blocks, some more than 3 m across, contain a mid- to high-slope fauna dominated by brachiopods and bryozoans, but the finer rubble contains a much more varied fossil suite; careful collecting by Hollingworth (1987) from a representative pocket of shelly rubble here enabled him to distinguish the allochthonous from the indigenous brachiopod and bivalve fauna, and allowed reconstruction of a reef talus palaeocommunity (Hollingworth, 1987, fig. 6.34, reproduced as Hollingworth and Tucker, 1987, fig. 8 and Hollingworth and Pettigrew, 1988, fig. 17). Faunal lists were given by Pattison (Geological Survey internal report, 1966) and Hollingworth (1987, table 19).

The remaining substantial reef-related exposures ('j' in Figure 3.31, NZ 3926 5395–3929 5394) lie in a separate group some 400 m WNW of the rock wall and are of several different rock types; diagenetic changes and extensive brecciation have obscured much of the primary character of the rocks, but they include both dolomite and limestone. Fragments in the brecciated rocks are mainly less than 0.1 m across, though some exceed 0.3 m, and algal-type lamination is present in some. Extremely fossiliferous calcitic dolomite (possibly basal coquina) is abundant in hillside brash (NZ 3929 5395) at the eastern end of this group of exposures.

Exposures of rocks of the Ford, Hartlepool and Roker formations in Ryhope Cutting

This preserved remnant (NZ 399537) of a former 300 m-long railway cutting reveals the following sequence.

	Thickness (m)
Roker Dolomite Formation (foundered):	
Breccia of subrounded to angular fragments up to 0.30 m across (but mainly less than 0.05 m) of grey and buff calcite- and dolomite-wackestone/?grainstone in a matrix of fine-grained carbonate; much laminar cavity-fill, especially in lower part; base markedly uneven, relief 2–4 m	7+
Inferred residue of Hartlepool Anhydrite Formation:	
Breccia of small, soft, angular fragments of cream finely saccharoidal dolomite in a cream powdery matrix; strong flow-lines in places and several laccolith-type intrusions into overlying breccia (most now covered); sharp rolling base (relief 3 m), partly truncating underlying strata	1–6
Ford Formation, off-reef beds:	
Dolomite brachiopod-bryozoan wackestone, cream-buff, in gently folded thick beds; partly dedolomitized at top at north-west end of cutting (T.H. Pettigrew, pers. comm., 1981)	2+

The bedded dolomite at the base of the section is sparingly to moderately fossiliferous and is notable for containing *Neochonetes davidsoni* which is typical of off-reef Ford Formation strata; faunal lists were given by Trechmann (1954, p. 198) and Hollingworth (1987, table 20), and T.H. Pettigrew (pers. comm., 1981) also records ?*Astartella* and *Permophorus costatus*. No reef-derived debris other than bioclasts has been noted in this dolomite. Judging from the record of strata encountered in Ryhope Colliery shaft nearby (NZ 3989 5353), Cycle EZ1 strata are here about 85 m thick, of which the Raisby Formation is unlikely to exceed 50 m.

The top of the bedded dolomite in the preserved faces has some of the characteristics of an erosion surface, with slopes reaching 70° (T.H. Pettigrew pers. comm. 1981); in the cutting as a whole, before filling, the surface was generally gently and unevenly rolling and had a total relief of at least 4 m in all directions.

The overlying soft breccia, mainly 1-2 m thick in the preserved exposures, but thickening to 6 m where intruded into the overlying hard breccia (see Smith, 1994, fig. 44), was seen in the former cutting to thin south-eastwards to 0.1-0.3 m; it was probably from here that Trechmann (1954) recorded stellate gypsum clusters. This was the bed regarded by Trechmann as marking a thrust plane, and later reinterpreted (Smith, 1972) as the residue of the Hartlepool Anhydrite. The harder calcitic breccia is interpreted as the foundered remains of the Cycle EZ2 carbonate formation, here probably of Roker Dolomite facies.

Interpretation

The advantage of the exposures at the south-east end of Tunstall Hills and in Ryhope Cutting lies in their close grouping, which provides not only a convenient view of the rocks themselves, but also of their mutual relationships; the reef-rocks furnish unambiguous and unique evidence that the reef-core prograded basinwards (i.e. eastwards) over earlier reef talus and the rocks east of the reef show that the succeeding 50 m+ Hartlepool Anhydrite formerly approached to within 200 m of the reef foreslope.

Reef-rocks

The relationship between the reef and the local topography has been discussed in the account of the exposures at the north-western end of Tunstall Hills and the correspondence is particularly close at the south-eastern end; here, notwithstanding the probable erosion of some of the landward margin of the reef, it is clear that the group of quarries lies in the core of the reef, towards its seaward margin; equally clearly, the exposures of bedded reef talus show that bed by rough bed, the front of the reef migrated basinward as it built upwards, so that the reef-core extended out over the former reef foreslope.

Relationships between reef, topography and stratigraphy are less clear on the south-west side of the valley of Tunstall Hope, where the main shelf-edge reef undoubtedly forms much of the high ground at Tunstall village and to the south, but is difficult to delineate. Similarly, the presence of Tunstall Hope makes it impossible to relate the shelf-edge reef directly to the reef-rocks exposed in the nearby old railway cutting (NZ 388538) at High Newport, though these were probably formed as a patch-reef in the Ford Formation lagoon (Smith, 1981a; Hollingworth, 1987).

In greater detail, rocks in the main quarry (d), the small quarry (e) and trench (g), together comprise a partial transect from mid reef-core to near the middle of the reef foreslope, though Hollingworth's tentative identification of reef crest lithofacies in the trench is difficult to reconcile either with some aspects of the biota or with the lithology. The main quarry is doubly important in that, together with the exposures of reef-core at the northern end of Tunstall Hills and at Humbledon Hill, it yielded the fossils upon which Hollingworth (1987, fig. 6.15) was able to reconstruct the reef-core palaeocommunity.

The main quarry is also important in being one of the two main exposures (the other being at the northern end of the hill) where the reef is of coarse-grained limestone, rather than the more usual dolomite, and in which primary fabrics are well preserved; from his detailed studies Aplin (1985, p. 305) concluded that the coarse-grained limestone was produced by the calcitization of partially dolomitized limestone and is thus partly dedolomite. 'Some fabrics' he wrote, 'were regenerated during dedolomitization, but many of the fabrics observed are thought to be secondary replacements of primary aragonite and a high-Mg calcite'.

The small quarry (f) is also important mainly because of Aplin's work; he identified it as one of the main places in the shelf-edge reef where the rock is composed of finely crystalline limestone in which primary fabrics are poorly preserved. Aplin (p. 305) concluded that these limestones are dedolomites and resulted from the reaction of the former dolomite with meteoric fluids during or after Mesozoic/Tertiary uplift.

The rock-wall (h) is important in being the only large surface exposure of the talus aprons of the shelf-edge reef in north-east England. An underground exposure of the reef talus has been recorded in the walls of tunnels through the reef at Easington Colliery 10 km to the SSE (Smith and Francis, 1967, pp. 136-137 and 169). The lithology and biota of the rock at the two exposures is generally similar.

The presence of rocks resembling collapse-breccias at exposure (j) is enigmatic, and the brecciation could be entirely diagenetic in origin; if they are collapse-breccias it implies the former presence here of abundant soluble rocks, presumably secondary anhydrite, which would be difficult to account for by present views of the local palaeogeography.

Off-reef Ford Formation and later rocks

Ryhope Cutting, in addition to providing evidence of the former proximity to the reef of Hartlepool Anhydrite, is important in being one of only four places where the basin-floor bedded equivalent of the reef is known to be exposed. The apparent absence of reef-derived detritus other than bioclasts is particularly noteworthy here in view of the short distance (less than 200 m) between the exposure and the toe of the reef slope. The base of the reef-equivalent strata is not exposed, so that their thickness is not known, but comparable strata proved in the Easington Colliery tunnels are 9-15 m thick (Smith and Francis, 1967, p. 137), indicating a very sharp eastward thinning of the off-reef strata. Farther east, at Frenchman's Bay and Trow Point (South Shields), reef-equivalent strata appear to be absent.

The other exposures of basin-floor rocks of probable reef age are in the floor of Ryhope Dene (NZ 4131 5170) 2.5 km SSE of the cutting and at Dene Holme (NZ 454404), Horden; all are different from each other, both lithologically and faunally, though a link is the common presence of chonetoid brachiopods and nodosariid foraminifera. A number of exposures of sparingly shelly, bedded dolomite in the East Boldon area, 8 km NNW of the cutting, may also be in reef-equivalent strata basinward of the reef, but stratigraphical relationships cannot be proved; if these rocks are of the same age as the reef, the reef foreslope may not be as high in the northern part of its course as the 100 m or so inferred from the spatial relationships of reef and succeeding strata in the Hawthorn-Easington area farther south (Smith, 1981a).

Future research

The structure, palaeontology, ecology and petrology of the reef have recently been investigated in considerable detail, so that there is presently little scope for further broadly based research into these aspects. There remains, however, considerable doubt regarding the relationship of the reef to enclosing and underlying strata, especially in the High Newport and Tunstall areas, and these aspects are worthy of further investigation; delineation of the reef-backreef boundary, for example, and identification of the age of subreef strata (Raisby Formation or Ford Formation?) would help to reduce the present uncertainties, and the possibility that collapse-breccias may be present at exposure (f) deserves investigation in view of its considerable palaeogeographical implications.

East of the reef, scope for further research undoubtedly exists into the sedimentology, distribution and biota of basinal strata of reef age, and determination of the age of bedded, shelly dolomite east of the reef in the East Boldon area would be extremely helpful in reconstructing the contemporary palaeogeography. Further research is also required on the petrography and origin of the supposed residue of the Hartlepool Anhydrite in Ryhope Cutting, and on the overlying inferred collapse-breccias.

Conclusions

The varied rocks in these two readily accessible groups of excavations at the south end of Tunstall Hills include *in situ* reef limestone and dolomite that can be seen to have prograded over its own earlier detritus and, in the cutting, one of the few exposures of fossiliferous basinal rocks equivalent in age to the shelf-edge reef. Detailed study of the reef-rock in the western group of excavations has revealed a complex history of mineralogical changes, and study of the abundant fauna in the reef detritus (talus) has yielded vital clues to the assemblage and lifestyle of the many invertebrate organisms that lived there.

The presence of an inferred residue of Hartlepool Anhydrite directly overlying the reef-equivalent beds in the cutting, is important in showing how close the anhydrite must once have approached to the steep seaward slope of the reef, although the anhydrite was almost certainly wholly younger.

The reef-rocks have been researched in detail but there remains the question of the relationship of the reef to the surrounding strata, the petrography of the residue of Hartlepool Anhydrite, and the origin of the collapse-breccias, presently considered as part of the Roker Dolomite Formation.

For these reasons, the site is an important link in the understanding of reef growth and progradation in the late Permian, and the style of sedimentation along the western margin of the Zechstein Sea.

GILLEYLAW PLANTATION QUARRY (NZ 375537)

Highlights

Though small, the old quarry (box 9 in Figure 3.2) in Gilleylaw Plantation, Silksworth, is the best remaining exposure of a late Permian marine patch-reef in the Durham Province of north-east England. The reef forms part of the Ford Formation and rests discordantly on bedded dolomite also probably of the Ford Formation; the quarry is cut into the northern end of the patch-reef and is the source (or one of the sources) of more than 20 type, figured or cited genera of marine invertebrates and of several growth-forms of supposed marine algae.

Introduction

Gilleylaw Plantation Quarry lies amongst trees at the northern end of a long low north–south hill at Silksworth in the south-western outer suburbs of Sunderland. The hill is probably roughly co-extensive with the patch-reef exposed in the quarry and has also been quarried along much of the northern part of its western side. Quarrying ceased long ago.

A quarry at Silksworth was mentioned by both Howse (1848, 1858) and King (1848, 1850), but no details were given and its exact site is unknown; it seems likely, however, that this quarry is that in Gilleylaw Plantation because both authors quote substantial fossil lists and Gilleylaw Plantation Quarry is much more fossiliferous than the others. Silksworth was also the source of many fossil specimens in the Kirkby collection, housed at the Hancock Museum, Newcastle upon Tyne. The quarry was not mentioned again in the literature until Smith (1958, 1981a) briefly described the section and illustrated oncoids and several other algal growth-forms from there, and Logan (1967) illustrated lectotypes of several species of late Permian bivalves from Gilleylaw. Most recently, a detailed faunal and ecological analysis was given by Hollingworth (1987).

Strata exposed in Gilleylaw Plantation Quarry comprise a lower unit of unevenly-bedded soft saccharoidal dolomite, a median unit of varied reef dolomite (a mixture of algal-bryozoan boundstone and shelly rubble) and a thin upper unit of thin-bedded pisoidal (oncoidal) dolomite. Early authors (e.g. Howse, 1848; King, 1850) apparently believed the reef-rock to be an outlying erosional relic of the main 'Shell-Limestone' reef, then thought to be 1.3-5 km wide; the present view that it is more likely to be a patch-reef in the lagoon landward of a much narrower main reef was proposed by Smith (1981a) and supported on palaeontological grounds by Hollingworth (1987).

After lying unused for over a century, the quarry was filled with builders' rubble during the early 1980s; it was subsequently re-excavated following representations by staff of the Planning Department of the Sunderland Borough (now City) Council acting in consultation with staff of the Nature Conservancy Council and Sunderland Museum and Art Gallery. The floor of the quarry is now occupied by a house and garden, but the face remains available for study by prior permission of the occupants.

Description

The position and shape of Gilleylaw Plantation Quarry are shown in Figure 3.36, which also shows the location of the main features of geological interest. The faces of the quarry total about 60 m in length and the main face is up to 11 m high; parts of the face are obscured by vegetation and soil.

The general geological sequence in the quarry is shown below.

	Thickness (m)
Ford Formation, oncoid facies	up to 1.3
Ford Formation, probable patch-reef	up to 5.5
?Erosion surface	
?Ford Formation, backreef (lagoonal) facies	up to 5.2

The disposition of the lithological units exposed in the south and east faces of the quarry is shown in Figure 3.37.

?Ford Formation, backreef (lagoonal) facies

Strata beneath the ?erosion surface at Gilleylaw Plantation Quarry comprise unevenly thick-bedded cream and buff porous saccharoidal dolomite that

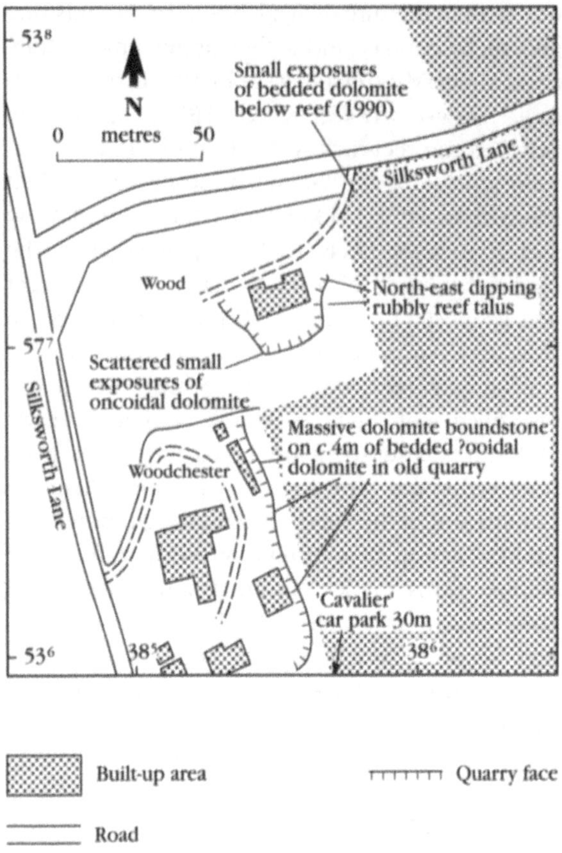

Figure 3.36 Gilleylaw Plantation Quarry and its immediate surroundings, showing the position of the main features of geological interest.

contains sparingly scattered molluscan bioclasts and many empty or thinly calcite-lined cavities up to 10 cm across after secondary anhydrite; the rock is probably an altered ooid grainstone. These strata may be divided into two main units, the lower (*c.* 2.8 m+) of which comprises rock that is variably bedded and rather vuggy, and the upper (3-3.5 m) comprises clearly bedded rock with continuous beds 0.1-0.4 m thick and only local wedging-out. Beds just above the base of the upper unit feature two mound-like structures (Figure 3.37); the larger of these is about 3 m across and 0.7 m high and has a confused (possibly brecciated) mainly dolomite core, and the other (Figure 3.38) is about 1 m across and 0.4 m high and has a core of dense grey limestone (?dedolomite). The mounds contain no obvious organic framework, but are clearly contemporaneous and they may be small algal patch-reefs.

The ?erosion surface

The inferred patch-reef is separated from the underlying bedded dolomite by a sharp break interpreted as an erosion surface (Figure 3.37). This has a relief in the quarry of more than 2 m, and sharply truncates beds beneath it on the east face where it dips north-eastwards at about 20°; its precise position and relief on the remaining (mainly inaccessible) faces is less clear. Hollingworth (1987) noted no evidence of contemporaneous cementation beneath the ?erosion surface.

Ford Formation, probable patch-reef

The inferred patch-reef at Gilleylaw Plantation Quarry occupies most of the upper part of the main face and comprises a varied mixture of massive dolomitized algal-bryozoan boundstone and subordinate dolomitized shelly rubble (Figure 3.37). The boundstone is a dense hard rock (bafflestone/framestone) in which scattered pinnate and straggling bryozoans are partly to thickly encrusted by fine concentric ?algal coating; in a massive 1.5 m bed in the south face such encrustations exceed 90% of the bulk of the rock. Only a restricted range of invertebrates is found in the boundstone masses but the shelly talus, exposed mainly in the east face of the quarry, contains a varied and abundant invertebrate fauna. The biota includes some genera that are absent or very uncommon in the main shelf-edge reef and lacks some common reef forms such as *Horridonia* and *Cyathocrinites* (Hollingworth, 1987); a full list by Hollingworth shows the striking difference between the faunal assemblages of the two rock types in the Gilleylaw reef and also between the overall biota of the inferred patch-reef and that of the main shelf-edge reef.

The association of massive bryozoan-algal boundstone and shelly rubble in the Gilleylaw reef is not unlike that in parts of the main shelf-edge reef, at the Humbledon Hill site for example, and it must be assumed that most of the type, figured and cited genera listed in the early works were specimens collected from the shelly rubble.

Ford Formation, oncoid facies

Up to 1.3 m of irregularly thin-bedded grey and buff oncoidal dolomite was formerly exposed at the top of the slope above the south face of the quarry where it rested with pronounced onlap on the uneven surface of the underlying patch-reef (Figure 3.37). These beds are now mainly covered, but many small exposures and loose blocks may still be found; they appear to contain no invertebrate remains.

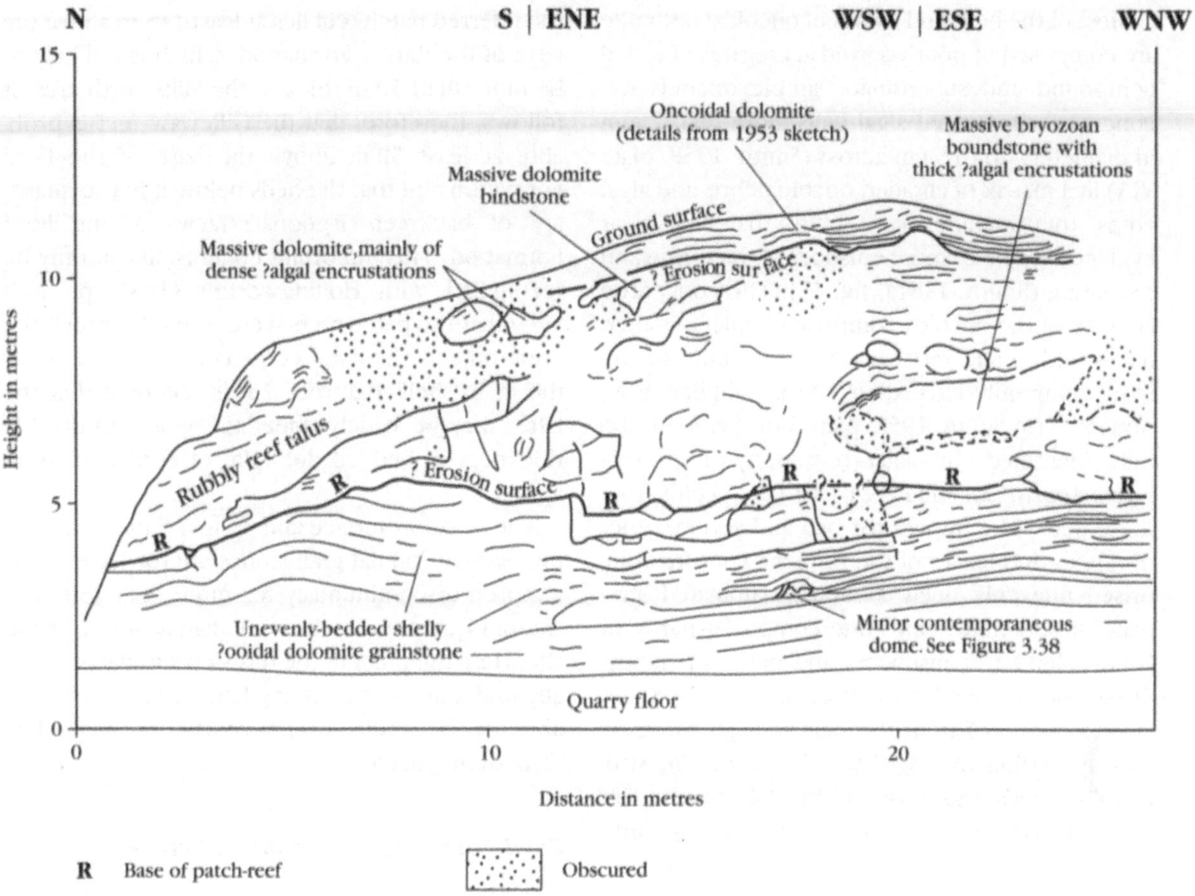

Figure 3.37 Sketch of patch-reef in the east and south faces of Gilleylaw Plantation Quarry, incorporating some details of strata formerly exposed, but not now visible.

Figure 3.38 Contemporaneous minor mound-like structure in backreef dolomite of the Ford Formation in the south face of Gilleylaw Plantation Quarry. Bar: 0.32 m. (Photo: D.B. Smith.)

Most of the beds and lenses of oncoidal dolomite are composed of poorly-sorted aggregates of rolled compound and subordinate simple oncoids (ie. concentrically-layered algal balls more than 2 mm in diameter) up to 5 cm across (Smith, 1958, plate VIA) in a matrix of abraded oncoid debris and algal chips; they include many grains that bear clear evidence of one or more episodes of fracturing and re-coating (Smith, 1981a, fig. 17). The concentric laminae of the oncoids comprise couplets of alternately turbid and relatively clear dolomite microspar, commonly exceeding 100 in number. F.W. Anderson (in Smith, 1958, plate VIII, fig. 3) doubtfully identified the algal (cyanophyte) growth-forms *Aphralysia* and *Bevocastria* in oncoids from these beds. The uppermost bed and several other thin beds and lenses of the former exposure comprised unevenly finely sinuously laminated stromatolitic dolomite bindstone composed partly of laterally-linked hemispheres and partly of tightly-fitted oncoids similar to those figured by Smith (1981a, fig. 12c) from the nearby High Newport railway cutting (NZ 388538). These *in situ* stromatolitic beds and lenses yielded the mammillar algal growth-form, cf. *Bevocastria conglobata* Garwood (F.W. Anderson in Smith, 1958, plate VII, figs. 1, 2). The visible relief of the base of the oncoidal dolomite was about 0.6 m.

Interpretation

Gilleylaw Plantation Quarry is of major importance as (a) the most accessible and complete exposure of a late Permian inferred marine patch-reef in the Durham Province and (b) the prime source of more than 20 type, figured and cited specimens of late Permian marine invertebrate specimens and algal growth-forms. It is also the first locality in Britain from which late Permian marine oncoids were illustrated, though some of the beds from which these were obtained are no longer fully exposed.

The ?erosion surface and underlying strata

These are closely comparable with the erosion surface and underlying strata exposed in Humbledon Hill Quarry in the main reef 1.7 km NNE of Gilleylaw Plantation Quarry and comments made in the Humbledon Hill account apply equally here. Judging from the record of strata proved in Silksworth Colliery South Shaft (NZ 3766 5404) located some 350 m NNE of the quarry, the base of the inferred patch-reef lies at least 115 m above the base of the Raisby Formation, which is unlikely to be more than 50 m thick in the Silksworth area; it follows, therefore, that the Gilleylaw reef is probably at least 50 m above the base of the Ford Formation and that the beds below it in the quarry are of backreef (lagoonal) facies of the Ford Formation. This important conclusion can only be reconciled with Hollingworth's (1987, p. 367) view that the patch-reefs were formed at much the same time as the basal coquina and lower core of the main shelf-edge reef if it is accepted that the latter may be widely underlain by a considerable thickness of bedded dolomite of Ford Formation age.

The ?erosion surface and up to 4.5 m of underlying bedded ?ooidal grainstones are (or have been) exposed discontinuously for more than 150 m in the old quarries almost immediately south of the site. Here the relief of the ?erosion surface is generally low, but the underlying beds are indistinguishable from their counterparts in Gilleylaw Plantation Quarry.

Ford Formation, probable patch-reef

Gilleylaw Plantation Quarry and adjacent sections provide the most readily accessible exposures of a late Permian inferred patch-reef in the Durham Province of north-east England; such bodies are concentrated in the Silksworth area and are not known farther south. The only other permanently exposed large inferred patch-reef is in an abandoned railway cutting (NZ 387538) at High Newport, about 1.2 km farther east; it differs from the Gilleylaw reef in a number of respects and is, on balance, less varied. Both bodies are only a few metres thick. Other, smaller, reef bodies have been noted in temporary excavations around Silksworth by Smith (1971, 1981a, 1994) and Hollingworth (1987) and it is probable that they number some scores or perhaps hundreds in total; some are wholly embedded in shelly lagoonal ooid grainstones.

The doubts about whether the reef-rocks around Silksworth are truly patch-reefs stem from the generally poor quality of most of the exposures and from the lack of exposure between Silksworth and the main shelf-edge reef complex. The early view that the reef-rock at Silksworth is part of the main reef was presumably based on lithological and faunal similarities, and in the absence of firm evidence to the contrary, cannot yet wholly be refuted; however, the discovery by Smith (1981a) that the main reef is generally much narrower than previously thought made this view difficult to sustain and the visible bilateral symmetry and fringing talus of some of the temporarily exposed, small reef bodies made interpretation as patch-reefs seem almost unavoidable.

This interpretation is strongly supported by the faunal evidence advanced by Hollingworth (1987), especially his discovery that infaunal bivalves and the bryozoan *Kingopora* are relatively much more abundant in the rubble of the inferred patch-reefs than in the main reef, and that crinoids are absent.

The Gilleylaw patch-reef is also exposed for more than 150 m in a series of old quarry faces stretching southwards from the private grounds of Woodchester (Figure 3.36) into the upper car park of 'The Cavalier' public house where the reef-rock is readily accessible. Here the reef is a coarsely and very unevenly bedded body dominated by hard buff dolomite boundstone composed of complex algal laminites, encrustations and ovoid masses and, in places, including tilted blocks up to 0.5 m across of algal laminite and bryozoan-algal boundstone; contemporaneous lithification and considerable energy levels are indicated. Most of the bryozoans and shelly fossils are tightly cemented into the rock, which lacks talus sheets, but scattered pockets and lenses contain many small gastropods and bivalves. The overall appearance and composition of the reef-rock in the 'Cavalier' car park is much like that of the reef-flat sub-facies of the main shelf-edge reef at Townfield Quarry (NZ 434438, Easington Colliery) and at the Hawthorn Quarry site, and similar shallow-water deposition seems likely.

In summary, the evidence suggests that whilst construction of the main shelf-edge reef was actively proceeding a short distance to the east, the shallow ooid-dominated floor of the backreef lagoon was, in the Silksworth area, dotted with bun-shaped patch-reefs ranging from less than 1 m to (exceptionally) several scores or hundreds of metres across. In time, perhaps through building up to sea level, slight sea-level fall and/or a salinity increase, the tops of the larger patch-reefs evolved to become inhospitable (?hypersaline) algal flats.

Although patch-reefs in the Durham Province are restricted to the Ford Formation in and around Silksworth, hundreds of patch-reefs occur in dolomitized open shelf ooid grainstones of the Wetherby Member of the Cadeby Formation in the Yorkshire Province (Smith, 1974a, b, 1981b, 1989); striking examples of the Yorkshire patch-reefs are exposed in the GCR sites at Cadeby Quarry (SE 5200), Newsome Bridge Quarry (SE 379514), South Elmsall Quarry (SE 483116), Ashfield Brick-clay Pit (SK 515981) and Wood Lee Common (SK 5391) and the mutual relationships of reefs and enclosing grainstones are particularly well seen in the picturesque lanes of Hooton Pagnell (SE 4808). The patch-reefs in Yorkshire differ from those in the Durham Province in having cores composed of sack-like masses ('saccoliths') of straggling bryozoans almost without laminar encrustations, and, in the larger examples, having an upper unit of coarsely domed algal stromatolites.

Ford Formation, oncoid facies

The importance of this thin unit rests partly on its role as the prime source of most of the small number of figured late Permian marine oncoids and partly on the light it throws on contemporary depositional conditions. The oncoidal bed is also poorly exposed at the top of a disused quarry (NZ 3759 5354) some 160 m farther south, where it is closely comparable with that in the GCR site. Elsewhere in the Permian marine sequence of north-east England, lithologically similar rocks have been recorded only in the reef-flat sub-facies of the main shelf-edge reef in Stony Cut (Cold Hesledon) Cutting (NZ 418473) (Smith and Francis, 1967, p. 133) and in exposures of reef-flat rocks at Yoden (NZ 4315 4176) (Smith and Francis, 1967, p. 139 and plate XB); superficially similar pisoids at the top of the Boulder Conglomerate of the Hesleden Dene Stromatolite Biostrome at the Hawthorn Quarry and Blackhalls Rocks sites may have a different origin from those at Gilleylaw, Cold Hesledon and Yoden.

Modern oncoids comparable with those at Gilleylaw are formed in a range of peritidal and shallow-water environments near the hypersaline margins of tropical seas such as the Persian Gulf and the Red Sea. The reworked Gilleylaw oncoids presumably accumulated as lenses and sheets of fine gravel on a shallow or peritidal reef-flat in comparable latitudes; the abundant evidence of fracturing, abrasion and re-cementation points to some contemporaneous lithification and to at least moderate energy levels at times, and the apparent absence of a shelly fauna is consistent with an atypical (either high or low) salinity. These inferences on the palaeoenvironments of the oncoids suggest a sharp and considerable change of conditions from those under which the underlying patch-reef was formed, although a shallowing water level and partial exposure may have sufficed.

Despite the presence of algal influences on the formation of the oncoids, the lamination of many of them lacks undoubted organic growth-forms and at least partial inorganic precipitation (as in some modern pisoids formed in unusual environments, such as splash-cups and surge pools) cannot be wholly excluded.

Future research

There are uncertainties and substantial gaps in our knowledge and understanding of most aspects of the Gilleylaw patch-reef and the light it throws on the late Permian sedimentary and stratigraphical evolution of the area. Aspects in particular need of detailed research are listed below.

1 The lithology and diagenetic history of the three main rock units exposed in Gilleylaw Plantation Quarry and also in the car park of 'The Cavalier' public house.
2 The stratigraphical position of the Gilleylaw reef and its relationship (if any) to the main shelf-edge reef and the significance of the presumed erosion surface. If it can be proved that the reef is both well above the base of the Ford Formation and roughly synchronous with the basal coquina and lower core of the main shelf-edge reef, present interpretations of the local late Permian stratigraphy will need to be reconsidered.
3 The nature and origin of the apparently contemporaneous minor mounds in beds underlying the reef.

Conclusions

The massive and rubbly dolomite of this small historic quarry contains an abundant and varied marine fauna, and much evidence of the former presence of marine algae. The massive rock is interpreted as a patch-reef and the rubbly dolomite is thought to be talus at the margin of the reef. The reef overlies a minor erosion surface of unknown significance and is surrounded by shallow-water lagoonal oolites which contain many other small patch-reefs in the Silksworth area. The upper surface of the reef is also an erosion surface, and is overlain by a thin deposit of marine pisoliths.

The detailed sedimentation and stratigraphical position of the patch-reefs are still relatively unknown, and further research is needed, as outlined above. The preservation of Gilleylaw Plantation Quarry is essential both to achieve this aim and to safeguard an example of a patch-reef formed in the late Permian backreef lagoon.

SEAHAM (NZ 4349)

Highlights

This coastal site (box 10 in Figure 3.2) is the type locality and by far the best surface exposure of both the Seaham Formation and the Seaham Residue, and is one of the best places in Britain for observing the effects of evaporite dissolution; it is also the best surface exposure of the highest beds of the Roker Dolomite Formation. The Seaham Formation here is unusual in its content of several thick units rich in calcite spherulites, some exceptionally large, and the Seaham Residue, the dissolution residue of the Cycle EZ2 (Fordon) evaporites, is at its thickest and most spectacular.

Seaham harbour is of interest in being an artificial anchorage created between 1828 and 1831 by the hollowing-out of a former limestone headland. An original Stephenson locomotive was used until the late 1950s on harbour maintenance work and a paddle tug, *The Eppleton Hall*, for many years plied the South Dock of the harbour before becoming an exhibit at the maritime museum in San Francisco.

Introduction

The south-eastward dipping rocks at Seaham are the highest Permian strata exposed on the Durham coast. They occupy a gentle asymmetric downfold on the north side of the Seaham Fault, a major west to east dislocation that has a northerly downthrow of about 200 m in Coal Measures rocks beneath the harbour, but perhaps only a small fraction of that in the overlying Permian strata. The sequence exposed comprises the uppermost part of the Roker Dolomite Formation (8 m+), the Seaham Residue (up to 9 m) and, at the top, the Seaham Formation (about 31 m). Breccia-gashes in the eastern part of the north wall of the North Dock contain foundered debris of higher strata including red 'marl' (mudstone/siltstone) and blocks of cellular limestone that may be the remains of calcitized evaporites. Fossils abound in the Seaham Formation (locally in rock-forming proportions) and comprise two species of bivalve and the supposed alga *Calcinema* (formerly *Filograna*) *permiana* (King); the same two species of bivalve have been found in the local Roker Dolomite beds.

The complex of exposures at Seaham was virtually ignored in the literature until the formal definition of the Seaham Beds and Seaham Residue by Smith (1971a); the name Seaham Beds was subsequently changed to Seaham Formation. The rocks of the Seaham Formation (Smith *et al.*, 1974) were informally known by Trechmann (1954) as the *Filograna* beds and were generally regarded as part of the Concretionary Limestone Formation (e.g. Woolacott, 1912; Trechmann, 1925). This attribution was accepted by Smith (in Smith and Francis, 1967), but was shown to be incorrect by Taylor and Fong (1969); these authors discovered calcite concretions in cores of late Permian marine limestones at two different levels separated by evaporites in boreholes in North Yorkshire and disclosed that those at the higher level were associated with the diagnostic biota of the Seaham Formation. Magraw (1975) proposed the term 'Upper Nodular Beds' for concretion-bearing limestones at about the level of the Seaham Formation in a number of partly cored, offshore coal exploration boreholes, but this name has found little favour.

The exposures of Roker Dolomite at Seaham lie towards the northern end of the designated area (Figure 3.39), and the top of the formation rises gradually northwards. These rocks were classified by Trechmann (1931) as Post-reef Middle Magnesian Limestone (=Ford Formation of modern usage), but their position immediately below the Cycle EZ2 Seaham Residue and above the inferred residue of the Cycle EZ1 Hartlepool Anhydrite in the nearby Seaham Borehole (Smith, 1971a) shows that this view cannot be correct.

The inferred presence of the residue of the Hartlepool Anhydrite in the Seaham Borehole, if correct, implies that all the Permian strata exposed at Seaham have foundered as a result of the dissolution of perhaps 120 m of underlying evaporites.

Description

Seaham (NZ 4349) lies on the east coast of County Durham some 8 km south of Sunderland; the position of the GCR site there is shown in Figure 3.39, together with the geological boundaries and the position of the main features of geological interest.

The general geological sequence in the designated area at Seaham is:

	Thickness (m)
Drift deposits, mainly boulder clay and beach deposits	up to 8
------unconformity------	
Red mudstone/siltstone (Rotten Marl) occupying breccia-gashes	up to ?4
Seaham Formation (EZ3Ca), seen mainly in the harbour walls and at Red Acre Point	30-31.5
Seaham Residue (of the Fordon Evaporites, EZ2E), seen mainly in cliffs from the war memorial northwards	up to 9
Roker Dolomite (EZ2Ca), seen mainly from Red Acre northwards	8+

Roker Dolomite Formation

This formation lies at the base of the sequence exposed at Seaham, its top first appearing at beach level about 300 m SSE of Featherbed Rocks and rising cumulatively until it crops beneath drift about 350 m north of the designated area; beyond this the cliffs for several hundred metres are composed solely of drift on Roker Dolomite.

The Roker Dolomite Formation in the northern part of the site and in the cliffs to the north is composed mainly of fine-sand grade, hollow ooids (Trechmann, 1914, plate 37(5)), and generally is a cream to buff porous rock with less than 2% of calcite. The uppermost 1-5 m of the formation, however, is a hard, finely crystalline, grey limestone (dedolomite) with a network of fractures that in places is so dense that the rock has become a breccia; internal sediment is locally abundant in this breccia. The fractures, which doubtless once were filled with anhydrite, gypsum or salt, diminish in number downwards, and the proportion of dedolomite similarly diminishes until limestone forms only narrow selvedges alongside the downwards-tapering cracks. There is also clear evidence, in the form of irregular to stellate calcite-lined cavities up to 0.1 m across, of the former presence of abundant replacive and intergranular anhydrite.

Where least altered, the Roker Dolomite at Seaham exhibits a range of shallow-water sedimentary structures and has yielded two species of bivalves (*Liebea* and *Schizodus*) (Trechmann,

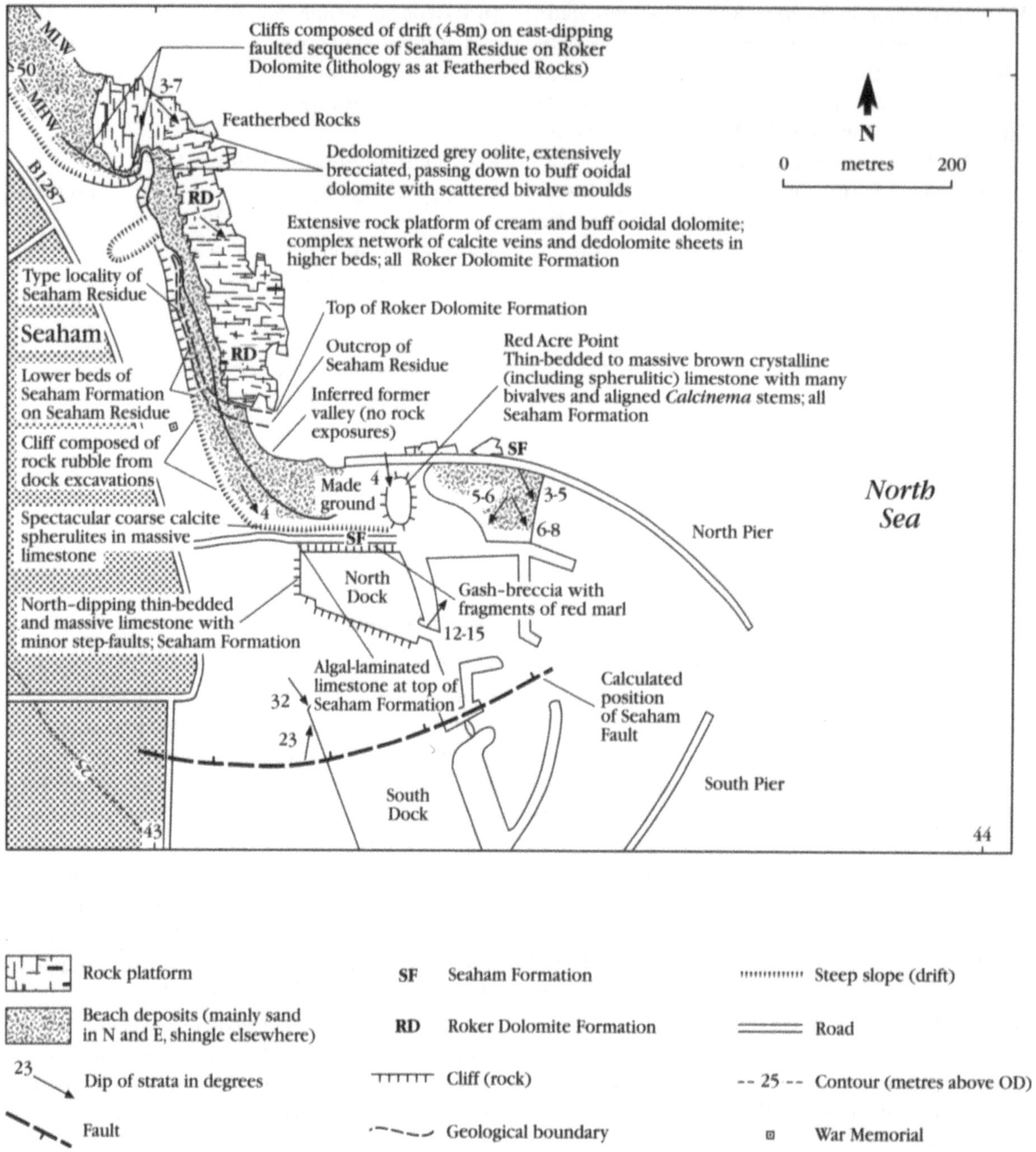

Figure 3.39 The Seaham GCR site, showing the position of the main features of geological interest.

1914, p. 235) from 'a mass of rock isolated from the cliff-section' (presumably Featherbed Rocks).

The contact between the Residue and the Roker Dolomite is generally sharp and smoothly rounded at the southern end of its outcrop and has a local relief of up to 2 m; this relief diminishes towards the northern end of the site, where the contact is generally less sharp and locally interdigitates.

Seaham Residue

The Seaham Residue is not seen in the harbour walls, but its top is exposed at the foot of the cliffs about 200 m NNW of North Dock (dependent on the height of beach gravel) and rises gradually northwards for more than 200 m before cropping out beneath drift at the top of the cliff near

Featherbed Rocks; much of the bed, however, is visible almost as far north as the former Vane Tempest Colliery (NZ 425502), well north of the GCR site, as a result of several minor downfolds and small step faults.

The Seaham Residue was identified and defined by Smith (1970a, 1971a) and further described and illustrated by Smith (1972). It is thickest in the cliffs at Red Acre where it is also strongly contorted (Figure 3.40); despite great lateral variation, the residue has a roughly uniform general sequence exemplified in the cliff face (NZ 4303 4973) some 210 m south of Featherbed Rocks.

	Thickness (m)
Residue, mainly yellow-buff, comprising a weakly layered and partly contorted heterogeneous clayey dolomite or dolomitic clay with scattered small angular fragments of limestone. Top generally sharp, but uneven, relief up to 1 m	1-2.5
Limestone, white, grey and buff, thin-bedded and flaggy, partly contorted, an altered ooid grainstone, with thin beds of yellow-buff clayey ?residue	0.8-1.1
Limestone, off-white, ooidal, finely cross-laminated, with possible bivalve moulds	1.0-1.2
Residue, buff in uppermost part grading down to buff-grey, grey-buff and brown, comprising an upper unit (up to 1.2 m thick) of strongly contorted flaggy and thin-bedded limestone (partly ooidal) passing down to contorted calcareous clay and clayey dolomite with scattered to abundant angular blocks of altered ooid grainstone; some of the latter are cross-laminated and contain flat-pebble conglomerates	1.5-5.0

Traced northwards, the Seaham Residue thins gradually to 2-3 m and is less contorted and varied.

Seaham Formation

The several rock walls of Seaham harbour, Red Acre Point and the cliff section from 155 m to the north, together constitute the type locality of the Seaham Formation (Smith, 1971a); in general, the lower parts of the formation are exposed north of the harbour and the higher parts in the harbour walls. No single continuous section exposes the whole formation and the overall sequence (based on Smith, 1994) has been pieced together from several sections separated by minor faults and stretches of unexposed ground.

	Thickness (m)
Limestone, grey and brown, finely crystalline, hard, thin-bedded, unevenly algal-laminated, with abundant stellate and rectilinear cavities after former sulphate	1.2-1.5+
Limestone, buff, grey and brown, finely crystalline to finely saccharoidal, hard, flaggy to thick-bedded, but with discontinuous beds of calcite concretions and of limestone. Mainly thin-bedded, with symmetrical low amplitude massive coarsely-crystalline ripples, shallow cut-and-fill structures, low-angle planar and tabular cross-lamination and abundant graded bedding. *Calcinema*, *Liebea* and *Schizodus* at most levels, partly in rock-forming proportions	*c.* 27-28.5
Dolomite (exposed in cliffs from about 180 to 300 m NNW of the north-west corner of North Dock), cream and buff, soft, finely saccharoidal, mainly thin-bedded, with *Calcinema*, *Liebea* and *Schizodus*	*c.* 1.8+

Calcite concretions occur almost throughout the Seaham Formation, but are most abundant slightly above the middle where they merge patchily to form massive spherulitic beds individually up to 3 m thick (Figure 3.41); most of the spherulites are only a few millimetres to centimetres across, but in places they exceed 0.2 m in diameter and completely obscure primary sedimentary features. Seaham is the best and most readily-accessible place to study these concretions.

The tiny stick-like tubular remains of *Calcinema* are present in enormous numbers in much of the rock, and form dense swarms aligned roughly WSW/ENE to west to east at some levels; at other levels however, they appear to be disposed randomly or to form complex swirls.

The algal-laminated (stromatolitic) limestone at the top of the Seaham Formation is exposed for only a few metres on the north side of an artificial

Figure 3.40 Strong contortions in the lower part of the Seaham Residue at the type locality, showing detached blocks of cross-laminated ooid grainstone (middle) and the lowest part of the bedded ooid grainstones that here form a median unit in the Residue. Hammer: 0.33 m. (Photo: D.B. Smith.)

terrace, high in the north-west corner of the North Dock; although such algal-laminites are widespread in subsurface provings at this horizon, this small surface exposure is unique in north-east England.

The minor faults seen cutting the Seaham Formation in the dock walls may be partly tectonic in origin, and related to the proximity of the Seaham Fault, but most of them probably resulted from fracturing of the brittle rocks when the underlying Fordon Evaporites were dissolved. Total brecciation such as is seen in the Cycle EZ2 collapse-breccias is uncommon in the Seaham Formation, but step-faults, partial brecciation and contortion are widespread; an excellent section displaying some of these features is seen in the west wall of the North Dock. It is important to bear in mind that all the strata exposed at Seaham have also foundered by at least 100 m because of the dissolution of the Cycle EZ1 Hartlepool Anhydrite and that some dislocation in the Seaham Formation may have resulted from this cause.

Red mudstones/siltstones (Rotten Marl)

The Rotten Marl is well known in boreholes in County Cleveland (e.g. Marley, 1892), and is extensive on the sea bed off the Durham coast (Smith, 1994). Surface exposures of this formation are known, however, only as fragments in a few breccia-gashes in coastal cliffs between Ryhope (NZ 4152) and Crimdon (NZ 4837), and one of these was excellently exposed (NZ 4327 4950) near the eastern end of the north wall of the North Dock. About 4 m of fragmented, soft red-brown, argillaceous siltstone and silty mudstone were formerly seen here; it contained a number of large angular blocks of pale grey cellular limestones that may be the carbonate framework of the Billingham Anhydrite (EZ3A) or (less likely) cellular dolomite from the Sherburn Anhydrite (EZ4A); the former separates the Rotten Marl from the underlying Seaham Formation, but has been dissolved from all surface outcrops, and the latter overlies the Rotten Marl and has also been dissolved.

The uppermost 2–3 m of the breccia-gash were removed when a new road was constructed in 1980, and much of the remaining Rotten Marl has since been obscured by landscaping; the remainder of the breccia-gash, however, may be seen from the footpath along the northern lip of the North Dock, where the gash is about 7 m wide.

Interpretation

The site at Seaham is of national importance both as a reference section and for teaching purposes; several features exposed there are unique either in north-east England or in Britain as a whole, and the exposures together provide an unrivalled expression of the effects on carbonate rocks of the dissolution of interbedded and pervasive evaporites. The main features of interest and importance are discussed on pages 93–5.

Figure 3.41 Typical limestones of the Seaham Formation immediately north of the harbour at the type locality, showing massive secondary spherulitic limestone overlying unevenly, mainly thin-bedded, *Calcinema* bivalve calcite mudstones and wackestones with shallow mega-ripples and cut-and-fill structures. Note the minor step-fault at top right. Bar: 0.32 m. (Photo: D.B. Smith.)

Roker Dolomite Formation

The clear exposures of the uppermost beds of the Roker Dolomite Formation at the northern end of the Seaham site are unique in north-east England and their protection is important on this account alone; there are, moreover, no readily available records of cored boreholes through this part of the sequence, although it has been cored in confidential commercial boreholes at several places in Cleveland. The discovery by Trechmann (1914) of *Liebea* and *Schizodus* in these rocks at Seaham is also important in providing corroborative evidence that they are not part of the unfossiliferous post-reef Middle Magnesian Limestone as had been claimed previously.

Seaham Residue

The Seaham Residue at Seaham is a superb example of an evaporite dissolution residue, certainly one of the best in the British Isles and ranking highly amongst such features in western Europe; poorer exposures occur beneath the Seaham Formation in coastal cliffs north of Crimdon Park (NZ 4837), Blackhalls Rocks (NZ 4638) and Easington Colliery (NZ 4343). Its great lateral variability in thickness and composition, its commonly gradational contacts with the underlying and overlying strata and its local strong contortion are all typical of evaporite dissolution residues, and combine to make the main exposure a most valuable teaching section.

The stratigraphical affinities of the Seaham Residue leave no doubt that it is equivalent to the Fordon Evaporite Formation (Cycle EZ2), which, approximately along strike in North Yorkshire, is about 15–30 m thick and comprises interbedded anhydrite and halite with subordinate carbonate and siliciclastic rocks. The absence of red and grey siliciclastic rocks in the Seaham Residue, plus the presence of underlying Roker Dolomite, shows that it is not the residue of the Edlington Formation into which the Fordon Evaporites pass landwards (i.e. westwards). The unusually great thickness of the residue, although enhanced at Seaham by lateral flow into a minor anticline, presumably indicates that these evaporites had a high content of insoluble material.

Seaham Formation

The Seaham exposures were chosen as the type locality of the Seaham Formation because of the apparent permanence and ease of access of the various faces and the nearly complete coverage of the whole thickness of the formation; no other exposure offers these advantages. The sedimentary features and biota at Seaham harbour are generally typical of the formation in north-east coastal districts (so far as may be judged from limited alternative exposures) and, except for the graded bedding, are comparable with those in the equivalent Brotherton Formation in Yorkshire and the Plattendolomit in Germany. The removal of some important faces and the covering-up of others during the subsequent construction of a road along the north wall of North Dock was particularly unfortunate, but still left the best overall exposure of these rocks. Other exposures of the formation lie in the coastal cliffs north of Crimdon Park (NZ 4837) (the only other large section), near Blackhall Colliery (NZ 4639), Easington Colliery (NZ 4443) and Dawdon Colliery (NZ 4347), and in the valley of Seaton Burn (NZ 413503), Seaham.

The calcitic concretions and massive beds of spherulitic limestone at Seaham have much in common with those of the Cycle EZ2 Concretionary Limestone of the Fulwell and Carley Hill Quarries at Sunderland and also of several other localities in the Cleadon-Marsden-Sunderland area; the main difference is that those at Seaham generally do not feature the fine parallel primary lamination of their Concretionary Limestone counterparts and white crystalline calcite is much less common. Striking exposures of coarse calcite spherulites lie low in the cliffs (NZ 4315 4951) immediately north of the harbour, where much of the rock is a massive coarsely crystalline limestone. Calcite spherulites are also abundant in parts of the Seaham Formation at its other main exposure in coastal cliffs north of Crimdon Park, but are smaller and less striking.

The composition and petrography of the Seaham Formation were investigated as part of wider studies by Al-Rekabi (1982) and Braithwaite (1988), who present illustrations (including photomicrographs) of several of the rock types present. Both authors inferred a late phase of leaching and partial internal collapse. Al-Rekabi concluded (albeit diffidently) that at least some of the radial calcite concretions may have had a sulphate precursor, but this view was disputed by Braithwaite who concluded that most of the late (i.e. post-burial) calcite replaced a range of earlier carbonate rock types without a sulphate intermediary or precursor.

Structural and petrographical effects of evaporite dissolution

The main structural effects of evaporite dissolution at Seaham cannot readily be seen, but their overall effect may be inferred from evidence visible at other exposures along the north-east coast and from reconstructions of regional stratigraphic relationships from boreholes in Yorkshire (Smith, 1974b; Taylor and Colter, 1975); these data show that all the strata exposed on the coast at Seaham must have foundered by more than 100 m as a result of the dissolution of the Cycle EZ1 Hartlepool Anhydrite. Farther north, study of cliff sections around Sunderland, Whitburn and Marsden shows that the fracturing and brecciation caused by such foundering die out upwards in the Cycle EZ2 Concretionary Limestone and Roker Dolomite formations, but that the less obvious effects persist upwards in the form of broad open folds and scattered faults. Presumably some such folds and faults have affected the sequence exposed at Seaham, but cannot now be distinguished from possible tectonic dislocation associated with the formation of the Seaham Fault and from the effects of the removal of the Fordon Evaporites.

The less severe effects of the dissolution of the Fordon Evaporites take the form of numerous minor faults and folds that break up much of the Seaham Formation into blocks commonly only a few metres across; most of the faults are normal and stepped, but a few reverse faults are present and also some minor troughs. Complete brecciation of parts of the Seaham Formation occurs locally, but is not as widespread as that noted in foundered Cycle 2 strata (Smith, 1972). Late stage breccia-gashes are similarly less common than in Cycle EZ2 rocks.

The relative importance of the varied processes leading to brecciation in rocks associated with evaporites varies greatly from place to place and embraces fracture by and the injection of pressurized formation fluids, including those liberated by the dehydration of gypsum, into rocks both above and below the evaporite beds; it is probably this process that accounts for the network of evaporite-filled veins found at depth in many Zechstein carbonate rocks in north-east England and which ultimately accounts for the brecciation of the rocks when the evaporite veins dissolve and leave the fragments unsupported. The partial brecciation of the uppermost part of the Roker Dolomite Formation in the northern part of the site probably results from this process.

The petrographical effects of evaporite dissolution on the carbonate rocks at Seaham have not been studied in detail, but appear mainly to have resulted in dedolomitization. This process here has widely converted the uppermost part of the Roker Dolomite from soft, porous, ooidal dolomite into hard, dense limestone; the effect is general in the uppermost 1-3 m, but dies out downwards. The basal dolomitic parts of the overlying Seaham Formation have also locally been dedolomitized.

The dedolomitization of the uppermost part of the Roker Dolomite Formation presumably was caused by the reaction of the dolomite with pervasive calcium sulphate-rich solutions. These may have been released during burial by the dehydration of primary gypsum, but more probably originated during uplift when the anhydrite was hydrated and the resulting gypsum was dissolved by phreatic groundwaters.

Breccia-gashes

Most of the known breccia-gashes (under a variety of names such as gash-breccia and breccia-pipe) are in the Concretionary Limestone, but a few at Seaham and in cliffs to the north of Crimdon Park are in the Seaham Formation. Work on equivalent Cycle EZ3 strata in Yorkshire (Smith, 1972; Cooper, 1986) has shown that such gashes probably form through the collapse of caves located at the intersection of joints or faults, and propagate upwards until they choke with debris. According to Cooper, this choking occurs when the pipe is 5–10 times the original height of the cave, through the stoping of angular fragments of roof rock. The end results of this process are near-vertical bodies of breccia, some with slight fault displacement, composed of fragments of wall rock, but also commonly including fragments of rocks from strata since removed by erosion; such breccia-gashes are analogous to small faulted-in outliers, and afford valuable information on the former local stratigraphy. The breccia-gash near the east end of the north wall of the North Dock at Seaham, though now degraded, is an excellent example of a structure associated with a minor normal fault and in which are preserved fragments of younger strata, the Rotten Marl, now otherwise eroded away.

Future research

Although the broad outlines of the geology of the natural and man-made cliffs at Seaham are reasonably well known and understood, many details remain uncertain and would amply repay further research. Amongst the more fundamental aspects in need of investigation are (a) the geochemistry and petrology of the Seaham Residue, with the aim of determining the character, thickness and depositional environment of the Fordon Evaporites of which it is the insoluble remains, and (b) the precise thickness and sedimentology of the Seaham Formation, with the aim of accurately documenting this formation at its type locality and of attempting to deduce its depositional environment.

Conclusions

The coastal site at Seaham contains the highest Permian strata exposed on the Durham coast, and is the type locality of both the Seaham Residue and the Seaham Formation. This is a classic site for the study of post-depositional changes in sedimentary sequences. The site exhibits a range of features including unusually large calcite concretions in the Seaham Formation, together with the only known clear exposure of the top of the Roker Dolomite Formation. Of particular significance are the well-exposed remains of former evaporite beds since removed by dissolution. The evaporites resulted from the evaporation of the Zechstein Sea, producing salt and anhydrite concentrates. These were later taken back into solution by invading groundwater, leaving an insoluble residue and causing collapse of the overlying strata. In particular, breccia-gashes, which contain fragments of rocks which have elsewhere been eroded from the area, were created. The site is therefore of major importance in observing post-depositional changes in the Permian rocks of Durham.

STONY CUT, COLD HESLEDON (NZ 4171 4724–4186 4744)

Highlights

This shallow cutting (box 11 in Figure 3.2) uniquely exposes a transect from the reef-flat to the crest of the shelf-edge reef of the Ford Formation. The reef-flat rocks are exposed in the south-west and central parts of the cutting and comprise a partly crudely-bedded mixture of *in situ* and reworked shallow-water reef dolomite; this passes north-eastwards into massive reef dolomite in which successive positions of the reef crest appear to be marked by sharply steepening thin sheets of laminar (?algal) dolomite.

Introduction

Stony Cut is a disused cutting on a former colliery wagonway and exposes up to 3 m of varied reef dolomite of the Ford Formation beneath a thin cover of Late Devensian boulder clay. The reef-rock is exposed for about 260 m (Smith, 1962) and is divisible into a reef-flat sub-facies (about 190 m seen) in the south-west and a reef crest sub-facies in the north-east; the latter is important as one of only four places where the crest of the Ford Formation reef is now exposed.

The cutting gave valuable insight into the disposition of reef sub-facies as originally identified by Smith (1958) and was later described in more detail by Smith and Francis (1967) and Smith (1981a). The palaeontology of the reef here was investigated by Hollingworth (1987), who reported marked differences between the fauna of the reef-flat and reef crest sub-facies and a striking north-eastwards increase in faunal abundance and diversity. Aplin (1985) reviewed the petrology and diagenesis of the reef-rock and discussed the origin of laminar ?algal encrustations and laminar fissure-fill in north-eastern parts of the cutting.

Description

The position of Stony Cut is shown in Figure 3.42. The rock faces are generally only 1–2 m high and are commonly overgrown and obscured in high summer; the floor of the cutting falls gently north-eastwards at about the regional dip in the Magnesian Limestone, and the north-eastern end of the cutting coincides with the edge of the strong topographic bench that marks the basinward margin of the Ford Formation reef.

Dolomitized reef-flat rock in the south-western part of the exposure is buff algal–bryozoan boundstone with lenses and pockets of oncoids (coated reworked algal chips) and skeletal debris. The rock

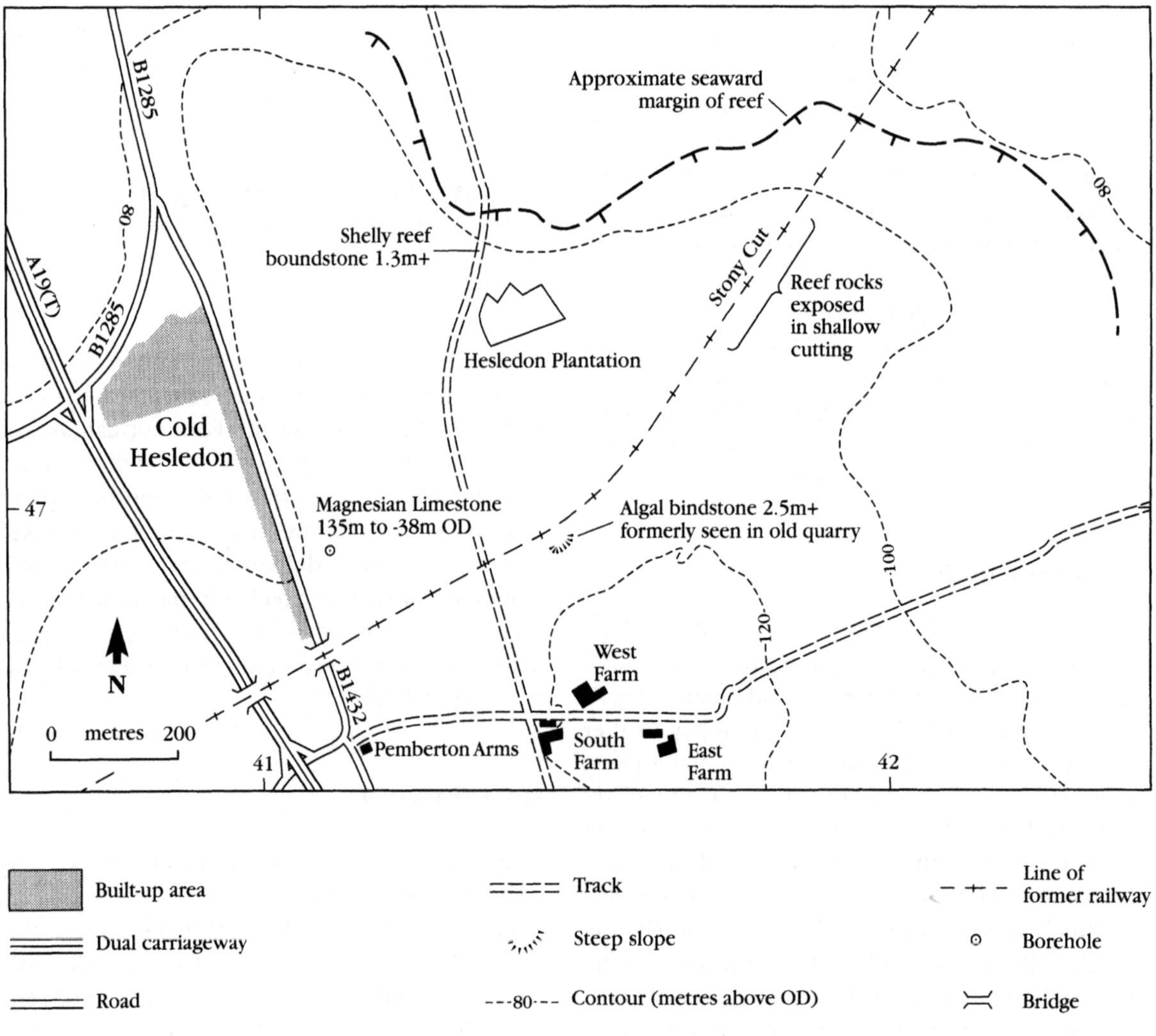

Figure 3.42 Location of Stony Cut, Cold Hesledon.

is generally unbedded and is a heterogeneous assemblage of mutually interfering masses up to 0.5 m across of boundstone (some rolled) and abundant draped sheets of laminar ?algal bindstone; the boundstone masses have a sparse framework of ramose bryozoans (almost exclusively *Acanthocladia* according to Hollingworth, 1987) which are thickly to very thickly coated with concentric ?algal encrustations (Smith, 1958, plate VIB, and 1981b, fig. 12 A,B).

Central parts of the cutting, extending for more than 100 m, are in crude and very uneven thick-bedded dolomitized algal-bryozoan boundstone; this includes many minor primary boundstone domes (some rolled), a wide variety of complex laminar (?algal) sheets and encrustations, and scattered to abundant lenses and pockets of fine boundstone debris, skeletal remains and oncoid rudstone. The overall dip is roughly parallel with the floor of the cutting, but is widely varied and local primary dips of up to 30° in all directions testify to contemporary reef-top relief of up to 1 m.

The dolomitized reef-rock in the north-easternmost 70 m of the cutting is, by contrast, relatively uniform. It contains less skeletal and boundstone debris and mainly comprises massive *in situ* brown-buff algal-bryozoan boundstone divided into steeply dipping panels a metre or more thick by thin finely laminar sheets. The boundstone has a sparse framework of ramose and fenestrate bryozoans, most of which are thickly coated with fine concentric ?algal encrustations that locally form up to half of the rock (Aplin, 1985, p. 385). Some of the thin laminar sheets are subvertical to vertical and were interpreted by Aplin as fissure-fill, but many are gently north-east dipping at the top of the section and steepen sharply to up to 85° below (Smith, 1981a; Aplin, 1985) (Figure 3.43); these appear to be algal coatings of reef masses or successive positions of the reef crest and remains of algal filaments were identified in such laminite by Aplin (1985, fig. 2.16C).

Early selective fossil collections from Stony Cut by Pattison (in Smith and Francis, 1967, p. 133) were augmented by more detailed sampling by Hollingworth (1987); both authors noted a sharp north-eastwards increase in faunal abundance and diversity and Hollingworth (1987, fig. 6.38) convincingly illustrated this trend and showed that the increase is not uniform, but is interrupted by a low-diversity belt in central parts of the cutting where bryozoans are less common. Hollingworth noted that *Acanthocladia* persists across the full width of the reef tract, but is accompanied by

Figure 3.43 Laminar ?algal bindstone sheets with high primary east-northeastwards dip, in reef boundstone of the Ford Formation near the north-east end of Stony Cut, Cold Hesledon. The sheets are thought to mark the temporary position of the upper part of the reef foreslope and grade upwards into the reef-crest and reef-flat dolomite. Hammer: 0.33 m. (Photo: D.B. Smith.)

Dyscritella, Synocladia and *Fenestella* in the north-eastern sector; a number of species of gastropods, bivalves and brachiopods are also confined to this sector. He commented that the absence or rarity of infaunal and quasi-infaunal forms suggested a measure of contemporaneous lithification of the substrate, supporting the evidence of such lithification by the rolled boundstone blocks and reef gravels (Smith, 1981a).

Interpretation

The complex and varied rocks of Stony Cut provide a unique transect across much of the shelf-edge reef of the Ford Formation; this reef extends sinuously from near Sunderland to West Hartlepool and also features in the GCR sites at Hylton Castle, Claxheugh Rock, Humbledon Hill, Tunstall Hills, Ryhope, Hawthorn Quarry and Horden Quarry. Indications from topography, the exposures in the cutting and from other excavations nearby suggest that the reef at Cold Hesledon may be about 400 m wide, compared with estimates of 250–400 m in the Sunderland area and at least 300 m at Hawthorn Quarry. The only other transects are at the Claxheugh Rock site, where the rocks are probably somewhat older, at the Hawthorn Quarry site where the reef-rock is probably of about the same age, but is now partly covered, and in Castle Eden Dene where much of the section is almost inaccessible.

The lithology of the heterogeneous roughly-bedded carbonate rocks in most of the cutting, together with their indigenous and derived fauna, strongly suggests that they were formed on a sub-horizontal, but somewhat rugged reef-flat under water no more than a few metres deep and perhaps at times intertidal (Smith, 1981a; Aplin, 1985; Hollingworth, 1987). Salinity was probably normal to slightly above normal, and energy slight to high according to location and weather conditions. Other, larger, sections in reef-flat dolomite are in the Hawthorn Quarry site and in Townfield Quarry (NZ 4343 4380), Easington Colliery.

The sections at Hawthorn Quarry (now covered) and Horden Quarry, provided the key to understanding Stony Cut, for they showed (a) that massive boundstone at the self-evident reef crest at the two quarries was lithologically and faunally indistinguishable from that in the north-eastern part of the Cold Hesledon cutting and (b) that boundstone at the progradational reef crest, as in the cutting, is divided into steeply-dipping panels by thinner upwards convex steeply-dipping laminar sheets (Smith and Francis, 1967, especially plate IX; Figure 3.47) that strike parallel with the reef foreslope. The origin of these laminar sheets is uncertain and some could fill former tension gashes like those in reef-rock at the Maiden Paps site, Tunstall Hills, Sunderland. The author believes, however, that their upwards-and-outwards convexity is more in keeping with a succession of reef crest ?algal coatings and if this is so, they indicate reef foreslopes approaching vertical and an extremely sharp reef crest. From his analysis of the biota at and near the supposed reef crest here, Hollingworth (1987) inferred that these rocks were formed subaqueously in turbulent water slightly deeper than that covering the reef-flat, and that the vicissitudes of reef growth provided a wide range of ecological niches that were exploited by the abundant and varied invertebrates. The origin of the thick concentric encrustations that freely coat the skeletal framework here and in many other parts of the Ford Formation reef were discussed by Smith (1981a), who concluded that they were probably formed by blue-green algae.

Future research

The palaeontology, ecology and petrology of the rocks in Stony Cut have all quite recently been investigated in detail (Aplin, 1985; Hollingworth, 1987) and there is little immediate scope for further work on these aspects. The curved laminar sheets near the inferred reef crest are worthy of further research, however, because of their importance and probable significance in the interpretation of reef crest and high reef slope morphology and evolution.

Conclusions

This site comprises a unique cross-section from the reef-flat to the crest of the shelf-edge reef of the Ford Formation. It is additionally important in that it is now one of only four places where the crest of the reef is exposed. Reef-flat carbonate rocks in the form of sub-horizontally bedded dolomite, sharply pass into steeply-dipping laminar sheets at this crest. Indications that the reef-flat was subject to very shallow water, possibly intertidal conditions, are the presence of oncoids, of algal encrustations on the reef-building framework and of large rolled boulders of reef-rock that had already become hard.

HIGH MOORSLEY QUARRY (NZ 334455)

Highlights

High Moorsley Quarry (box 12 in Figure 3.2) is typical of many exposures in the much-quarried Magnesian Limestone (Permian) escarpment and exposes a representative section of the lower part of the Raisby Formation (formerly the Lower Magnesian Limestone). In addition, it contains a spectacular submarine debris flow, other evidence of contemporaneous, mass downslope sediment movement and a coarse, mineralized breccia.

Secondary features in the quarry include evidence of markedly widened major joints, the opening of which probably resulted from cambering and/or mining subsidence.

Introduction

High Moorsley Quarry is cut into the west-facing Permian escarpment a short distance south-west of High Moorsley village and exposes about 17 m of the lower part of the Raisby Formation. The rocks exposed are typical of this formation in north-east England, and include an inferred slide-breccia and a debris flow; secondary calcite-marcasite mineralization is a feature of the northern part of the quarry. The general section and the mineralization at High Moorsley Quarry were noted by Francis (1964) and in Smith and Francis (1967, p. 109), and details of the debris flow and slide-breccia were given by Smith (1970c). Lee (1990) presented isotopic analyses of several rock types from the quarry and discussed their diagenetic history.

Description

The position of High Moorsley Quarry is shown in Figure 3.44; the main exposures of the Raisby Formation are in the high, east face of the quarry, but the breccias and debris flow are best exposed in the north of the quarry.

The generalized section in the quarry (from the top) is given below.

Bed		Thickness (m)
7	Dolomite mudstone, coarsely mottled buff and grey-buff, mainly in uneven to lenticular nodular wavy beds 0.15–0.30 m thick in lowest 1.6 m where patchily bioturbated, 0.15–0.20 m beds above; some auto-brecciation	*c.* 5.1
6	Dolomite mudstone, buff with grey-buff patches, in even to wavy beds mainly 0.05–0.15 m thick but becoming thick-bedded in places, partly finely nodular; some beds bioturbated; sharp planar base	*c.* 1.4
5	Dolomite mudstone, finely mottled buff and grey-buff, partly unevenly laminated, strongly bioturbated, one bed, planar base	0.1
4	Dolomite mudstone/wackestone, buff and grey-buff, thick-bedded in the north but mainly in varied uneven to lenticular beds 0.10–0.20 m thick with some boudin-like structures; some beds dip at up to 7° in large cross-sets; ripple-like linen-fold fluting (WSW–ENE, relief 0.10 m, wavelength 0.50–0.80 m) 0.60–0.80 m above base	*c.* 2.3
3	Breccio-conglomerate, buff and grey-buff, comprising ill-sorted sub-rounded to tabular dolomite mudstone clasts up to 0.15 m across (some WSW–ENE imbrication) in a varied matrix of skeletal dolomite mudstone, wackestone and packstone; locally clast-free in uppermost 0.05–0.25 m. Discontinuous, forming at least three discrete 5–10 m-wide lenses (?lobes) in about 60 m along the north-east face of the quarry	0–0.6
2	Dolomite mudstone, buff and grey-buff, mainly in very uneven to lenticular beds 0.05–0.15 m thick, with linen-fold fluting (?nearly symmetrical ripples, relief 0.05–0.10 m) *c.* 0.30 and 0.80 m beneath the scoured top, several to many penecontemporaneous sub-concordant glide-planes, and with scattered large penecontemporaneously folded and brecciated patches (more in north than south).	*c.* 4.0
1	Calcite mudstone, mottled in shades of grey, mainly in finely augen-nodular beds 0.01–0.06 m thick with sub-stylolitic contacts; scattered poorly-preserved molluscan debris; partly broken-up by penecontemporaneous brecciation. Normal top not exposed	1.5+

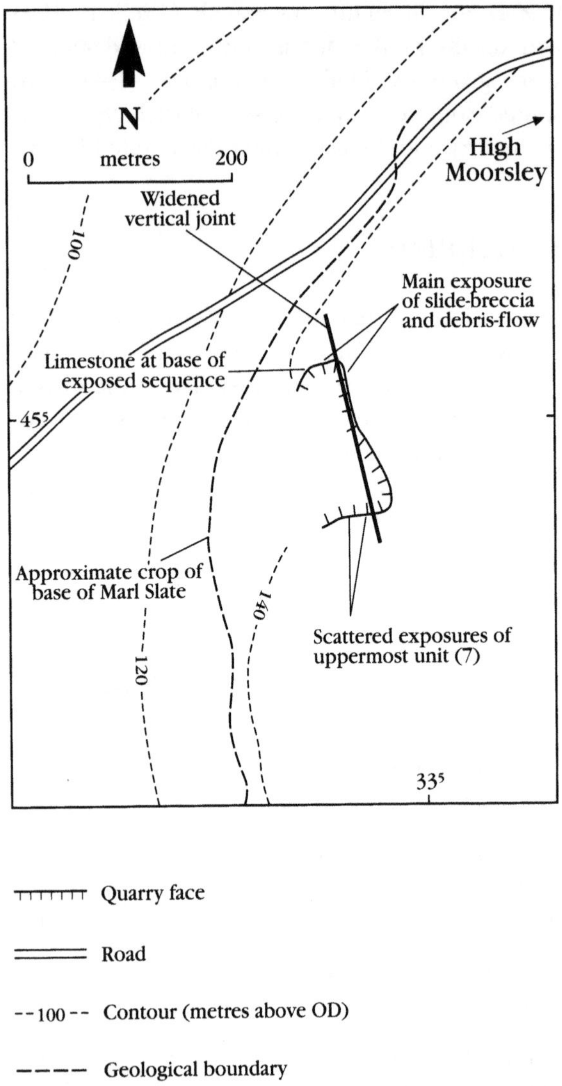

Figure 3.44 High Moorsley Quarry and its immediate surroundings, showing the main features of geological interest.

Judging from small exposures at the north and south ends of the designated area, and from topographical features, the floor of the quarry lies less than 5 m above the base of the Raisby Formation.

All the dolomite rocks in the quarry contain scattered to abundant, irregular to ovoid, calcite-lined cavities after replacive anhydrite, and small calcite laths after anhydrite are also present; manganese dioxide speckles and dendrites are widespread. Most beds are lithologically varied when traced laterally. Beds 1 and 2 are slightly to severely broken-up into irregular blocks, slices and lenses up to 10 m across and 3 m thick, and are interpreted (Smith, 1970c) as comprising a slump- or slide-breccia caused by downslope movement of a mass of fairly-well lithified strata; crumpling and slight truncation of bedding in otherwise apparently undisturbed parts of this disrupted sequence (Figure 3.45) show that some bedding-plane sliding was on an unusually large scale. Heavy calcite-marcasite mineralization of parts of the breccia implies that many inter-block cavities remained at least partly empty of sediment after movement ceased, but M.R. Lee (in letter 1990) believes that the association is coincidental and that the mineralization may be related to later tectonic brecciation. Limestone from bed 1 was analysed for its isotopic composition and strontium content by Lee (1990), who reported lower carbon and oxygen values than is normal for dolomite rocks in this formation, but a higher strontium content. Bed 3 was interpreted as a proximal turbidite (= debris flow) (Smith, 1970c), overlain by a fall-out tail from a suspension cloud. The breccia ranges from clast-supported to matrix-supported and is very variable in thickness; it appears to thin out southwards in the east face. Two clasts from this bed were interpreted by Lee (1990) to have been at one time composed of replacive anhydrite, but their isotopic composition was found to be almost the same as that of normal dolomite rocks in the quarry.

Conspicuously wide, sub-vertical joints in the north and south faces of the quarry are probably a response to massive cambering along the escarpment, but may have been further widened by differential mining subsidence; they cause inherent instability in parts of the east face. Other widened joints trend WSW–ENE in the east face.

Interpretation

High Moorsley Quarry is important because it provides a representative section of the Raisby Formation in this part of County Durham; both the lithology and scanty biota are typical of those in many abandoned quarries in the escarpment, and the debris flow and slide-breccia are characteristic of a disturbed sequence commonly found 3–7 m above the base of the formation. Together they throw much light on the depositional environment of the Raisby Formation.

The Raisby Formation is the first major carbonate unit of the English Zechstein sequence in the Durham Province, and is up to about 73 m thick in some eastern parts of County Durham. Generally, however, it is considerably thinner, and is unlikely to have been more than 50 m thick at High Moorsley. At outcrop in eastern Durham and

Figure 3.45 Crumpled bedding in the lower beds of the Raisby Formation near the north end of the east face of High Moorsley Quarry, with evidence of contemporaneous truncation at the top of the disrupted beds. Hammer (middle top): 0.33 m. (Photo: D.B. Smith.)

adjoining areas it is almost everywhere a carbonate mud rock, and is mainly dolomitized. Judging from the distribution of similar rocks in Yorkshire (Smith, 1974b, 1989, fig. 6), it accumulated mainly on the gentle marginal slopes of the Zechstein Sea and passed westwards into a belt of shallow water dolomitized packstones/grainstones formed on a progradational carbonate shelf. Such shelf rocks have since been eroded from the Durham Province, but may, by selective storm winnowing, have been the source of some or most of the hemipelagic carbonate muds deposited on the slopes (i.e. now the Raisby Formation).

The evidence of sediment instability forms another part of the argument favouring a slope location for the deposition of the Raisby Formation at outcrop in northern Durham and Tyne and Wear, and was summarized by Smith (1970c, 1985); both the debris flow and the slide-breccia are textbook examples of their kind, though the exposure of the latter could be improved by the clearance of debris.

The exposures at High Moorsley Quarry are representative of disturbed strata at about this stratigraphic level for more than 40 km between offshore boreholes east of Blyth (NZ 8231) and the village of Ludworth (NZ 358413) and it seems likely that most of the disturbance was caused by an external stimulus such as an earthquake shock or a closely spaced group of shocks; instability through natural over-steepening seems to be excluded by the limited stratigraphic range and absence of widespread turbidites such as characterize the Concretionary Limestone Formation in, for example, Marsden Bay (Trow Point to Whitburn Bay GCR site). The abundance of bioclasts in the matrix of the debris flow has been attributed to rapid burial and consequent escape from predation.

Future research

The sedimentology and diagenesis of the complex rocks of the Raisby Formation at High Moorsley Quarry have recently been investigated by Lee (1990) and there is little immediate scope for further research on these aspects.

Conclusions

The site is notable for the exposure of the lower part of the Raisby Formation, comprising dolomites and limestone typical of the sequence found in Durham, together with well-exposed interbedded slide-breccia and debris-flow units, a characteristic feature of strata close to the base of the formation. These are indicative of the movement of sediment down an inclined depositional surface as a result of instability, perhaps triggered by earthquake shocks. This site, together with Dawson's Plantation Quarry, is one of the best exposures of evidence of downslope sediment slumping and sliding low in the Raisby Formation, and needs to be preserved for this reason.

HAWTHORN QUARRY (NZ 4346)

Highlights

Hawthorn Quarry (box 13 in Figure 3.2) is one of the largest exposures of late Permian reef-rocks in north-east England and the only exposure in which their contact with overlying strata is seen; unique exposures formerly recorded, but no longer available, revealed the overall profile of the basinward crest of the reef and its juxtaposition with down-faulted or foundered younger strata to the east.

Introduction

Hawthorn Quarry, which ceased working in 1985, exposes the basal beds of the ?Roker Dolomite Formation (15 m+), the whole of the Hesleden Dene Stromatolite Biostrome (?22-26 m, including a basal 0-4 m boulder conglomerate) and the uppermost 15 m of the shelf-edge reef of the Ford Formation. Exposures available up to mid-1958, but now quarried away or covered, showed that the eastern (basinward) crest of the reef crossed the eastern end of the quarry (Smith, 1962), and boreholes drilled in about 1974 proved reef-rock to a depth of at least 44 m below the then quarry floor at about +49 m O.D.

The strata exposed in Hawthorn Quarry dip generally southwards at less than 5° and there are no major folds or faults in any of the strata now visible. The pre-1958 exposures, however, revealed a complex 80° reverse shatter belt abutting and trending parallel with the reef crest, which brought down brecciated, ooidal dolomite on the basinward side.

The rocks at Hawthorn Quarry have been discussed and illustrated by Smith and Francis (1967), Smith (1973b, 1981a), Kitson (1982, petrography), Aplin (1985, petrography of the reef) and Hollingworth (1987, palaeontology of the reef); their interpretation has changed little during the period covered, the main development being the revealing of the boulder conglomerate between the reef and biostromal laminites in about 1980 as the quarry extended gradually westwards. A provisional faunal list for the reef-rock was given by Pattison (in Smith and Francis, 1967, p. 134) and the fauna was analysed in detail by Hollingworth (1987, pp. 258-266).

Description

Hawthorn Quarry (NZ 4346) is cut into an east-facing slope near the Durham coast, about 3 km south of Seaham; the boundaries of the quarry are shown in Figure 3.46, together with the geological boundaries and the position of the main features of geological interest.

The general geological sequence in and around the quarry is given below.

	Thickness (m)
Soil on Durham Lower Boulder Clay	0-5
Gravel, partly calcreted, present only near entrance in east of quarry and in minor rockhead depressions	0-4
------unconformity------	
?Roker Dolomite Formation	up to 15
Hesleden Dene Stromatolite Biostrome, with boulder conglomerate at base in west of quarry	*c.* 22-26
------erosion surface------	
Ford Formation, reef-facies, in floor of quarry	*c.* 15+

The disposition of the various lithological units within the quarry site is shown in Figure 3.47.

Ford Formation, reef-facies

Dolomite rocks of this unit are exposed in the lowest levels of the quarry, where they are up to 15 m thick; a borehole in the quarry floor proved an additional 29 m of reef dolomite (Figure 3.47), but its total thickness here may exceed 100 m. The reef comprises a complex assemblage of autochthonous masses of bryozoan and algal boundstone

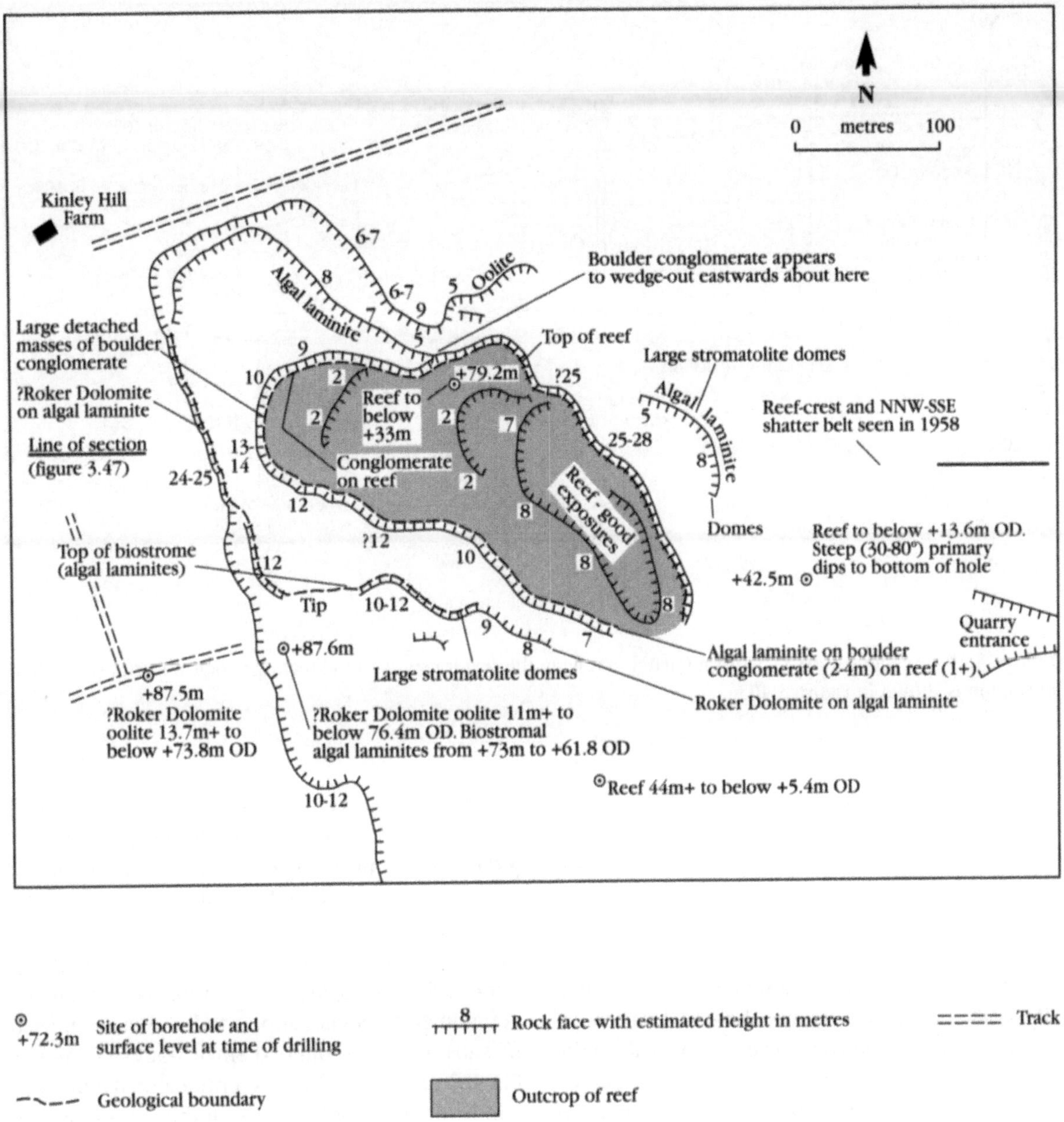

Figure 3.46 Hawthorn Quarry, showing the location of the main features of geological interest.

separated and surrounded by sheets and pockets of shelly detritus (Smith, 1981a, pp. 169–174); laminar (at least partly algal) encrustations and ramifying laminar sheets abound and locally form most of the rock, and a few rolled blocks (some coated) also occur. The reef-rock has a general roughly horizontal thick bedding, with primary dips of up to 30° traceable for a few metres. Pattison (in Smith and Francis, 1967, p. 134) gave a provisional faunal list for reef-rock from Hawthorn Quarry and this has been supplemented by a full faunal analysis by Hollingworth (1987), who distinguished between assemblages in the boundstone masses and surrounding ?algal laminites. Aplin (1985) gave details of the petrography and diagenesis of the rock.

Unique exposures formerly visible (Figures 3.46 and 3.47) near the quarry entrance revealed the reef crest, where successive reef-flat beds bent sharply over to dip east-northeastwards at up to 90° down the reef front (Smith and Francis, 1967, plate 9; Smith, 1973b, 1981a); similarly steep dips were encountered in a nearby borehole, proving that the reef front maintained a high angle to a depth of at least 35 m below the crest. In inaccessible parts of the former exposures, the reef crest

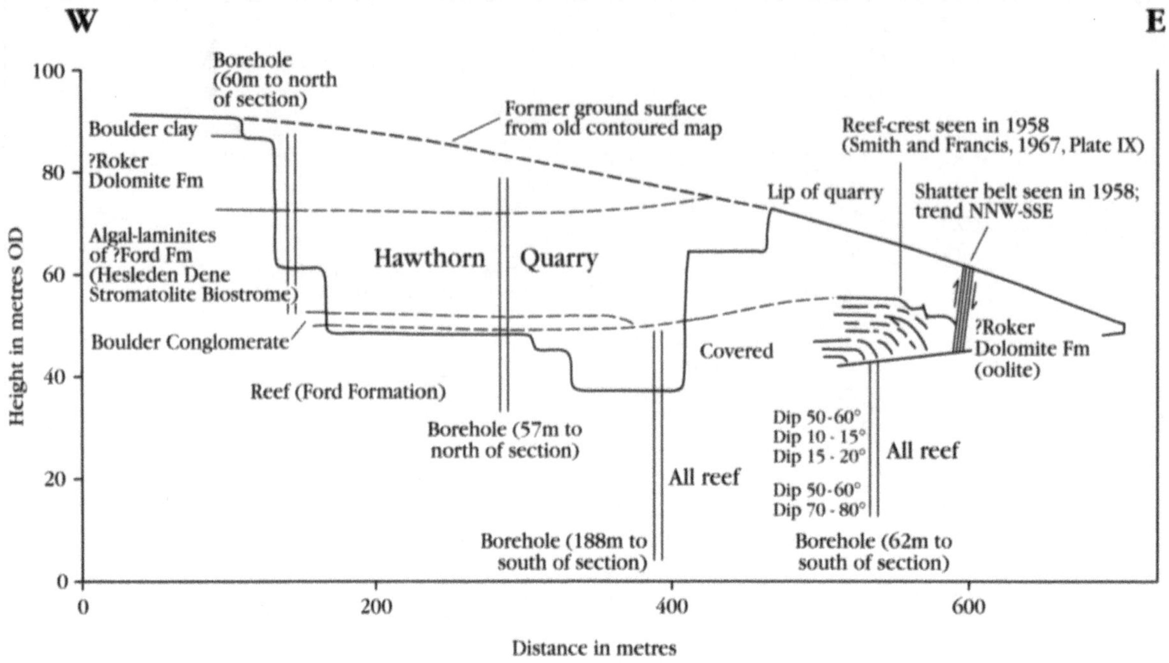

Figure 3.47 Section across Hawthorn Quarry, showing the relationships of the main geological features. The line of section is shown in Figure 3.46.

appeared to decline eastwards in a flight of 1-2 m steps (Figure 3.47).

The erosion surface

This is exposed mainly in the south-west and west of the quarry, but is not well seen because it generally coincides with a quarry bench. In the few places where it is exposed clearly, especially in the middle of the south and west faces of the quarry, the surface is sub-horizontal with an average relief of 0.05-0.10 m. In detail, the surface is diversified by scattered slopes of up to 45° and minor depressions up to 0.15 m deep. Even where well exposed, however, the contact has commonly been blurred by diagenetic changes.

Hesleden Dene Stromatolite Biostrome

This member at Hawthorn Quarry comprises a basal boulder conglomerate up to 4 m thick that is present only in the west and south of the quarry, and a thick (*c.* 22 m) unit of algal-laminated dolomite bindstone which is (or was) recognizable all round the quarry except near the entrance.

The boulder conglomerate at the base of the biostrome is 2-4 m thick in the west and south faces of the quarry, but appears to die out eastwards (or to pass into non-conglomeratic bedded rock) in the inaccessible north face. Generally the conglomerate forms a single bed, but in a number of places boulders are only thinly scattered in the uppermost 1-2 m of the unit which is composed mainly of chaotic, finely laminated dolomite with a wide variety of ?algal growth forms. Below this the conglomerate is clast-supported and is composed of subangular to rounded small boulders (mostly 0.1-0.3 m across, but exceptionally 0.6 m) and subordinate cobbles of dolomite boundstone derived from the underlying reef. Most of the clasts are uncoated and a few bear signs of contemporaneous fracturing and re-cementation. The rock is generally poorly-graded and lacks clear imbrication or cross-bedding; possible crude east-dipping foresets, however, are present where the bed thins out in the north face. Interstices between the cobbles and boulders are filled or partly filled with white to pale cream, unfossiliferous, saccharoidal dolomite or calcite (according to location), and calcite-lined irregular voids are common. The filling comprises an estimated 20-30% of the conglomerate and is mainly faintly finely laminar, the laminae being parallel with the cavity walls; similar material also lines the walls of scattered steeply-inclined cracks, some of which are now more than 1 m deep and may have been contemporaneous.

The algal-laminated dolomite is of cream, pale buff and pale grey, silt- to sand-grade saccharoidal dolomite and calcitic dolomite with abundant calcite-lined, irregular cavities after secondary anhydrite; scattered patches of grey and brown saccharoidal dedolomite occur. Bedding is thin to thick, partly according to the state of diagenesis, but is mainly thin where the rock is least altered. At several levels the beds are disposed in rounded to flat-topped domes up to 20 m across and 3 m high, though most are 3-10 m across and 1-2 m high, and some are quite small (Figure 3.48); dips on the flanks of domes range to almost vertical, and extend down to the basal surface on which the domes lie. In detail the dolomite of the domes, where least diagenetically altered, is finely and slightly unevenly laminated, the laminae featuring widespread delicate crenulation and commonly being domed on a millimetre-centimetre scale. The basal 1.5 m of the algal laminates is generally an almost pure dolomite rock and is composed of distinctively crenulated dolomite (Smith, 1981a, fig. 27) that readily distinguishes it from the remainder of the biostrome; the lamination in this, the informally designated 'Crinkly Bed', is accentuated by slight concentrations of manganese dioxide (Figure 3.49).

Loose fragments of pisoidal dolomite are mixed on the quarry floor with debris from the 'Crinkly Bed' and from the boulder conglomerate, but the pisoidal rock has not been found *in situ*; similar pisoidal rock at Blackhalls Rocks appears to be an uncommon local variant of the 'Crinkly Bed', filling deep pockets between atypically tall algal domes. The pisoids are up to 18 mm across, flattened, simple or compound, with smooth fine concentric coatings; Kitson (1982, fig. 62) illustrated partly silicified pisoids from this bed at Hawthorn Quarry.

?Roker Dolomite Formation

Rocks doubtfully referred to this formation form the upper part of the faces of most of the quarry, including those of the narrow entrance where they have been lowered to present ground level by faulting or foundering (or both); they also underlie most of the surrounding fields, where they are known from small quarries and soil brash.

The base of the formation is taken at a thin, but varied bed of brown clayey dolomite or dolomitic clay which is generally inaccessible, although it is well exposed near the south-west corner of the quarry; here it is 0.15-0.25 m thick and has been partly contorted by plastic flow. The basal

Figure 3.48 Small columnar stromatolites just above the boulder conglomerate of the Hesleden Dene Stromatolite Biostrome near the middle of the south face of Hawthorn Quarry. Bar: 0.32 m. (Photo: D.B. Smith.)

Figure 3.49 Slight concentrations of manganese dioxide coating a bedding plane in the 'Crinkly Bed' near the base of the Hesleden Dene Stromatolite Biostrome near the middle of the south face of Hawthorn Quarry. Note the asymmetry of the ?algal growth-forms, indicating water flow from the right. Coin: 20 mm across. (Photo: D.B. Smith.)

grainstone at this exposure displays slight onlap, perhaps indicative of a depositional hiatus, but there is no unequivocal evidence of truncation or of a hiatus at the top of the underlying biostrome, and no evidence of the former presence of evaporites.

The ?Roker Dolomite at Hawthorn Quarry comprises pale buff and cream, mainly ooidal dolomite grainstone in which abundant irregular calcite-lined cavities up to 0.10 cm across mark the site of former secondary (replacive) anhydrite. According to Kitson (1982), rock in the basal 2 m of the formation is of relatively pure dolomite, but higher beds have a sparry calcite cement; local dedolomitization (?by surface water) has occurred near the base of the drift, where travertine and calcite veins are abundant and the rock has been brecciated in places. Component ooids are of coarse to very coarse sand-grade and have leached centres (Kitson, 1982, fig. 44); most are simple and subspherical, but a few irregular compound grains are present in most hand specimens and compound pisoids up to 5 mm across are locally common. Most of the rock is thin- to medium-bedded, with much of the bedding poorly defined. Preservation of sedimentary structures is similarly generally poor, although traces of ripple lamination, cut-and-fill structures and small-scale planar cross-lamination occur locally, especially in ooidal grainstones exposed near the quarry entrance (NZ 4383 4631).

Interpretation

The exposures at Hawthorn Quarry are of special importance because they furnish a complete sequence from the reef of the Ford Formation well up into the inferred Roker Dolomite Formation. They are unique in being the only place where the contact of the reef with overlying strata may be studied; the record of the former position of the reef crest, now covered, enables the sequence to be precisely located relative to the main late Permian facies belts in the area.

Ford Formation, reef-facies

Hawthorn Quarry contains one of only three remaining substantial exposures of the reef-flat sub-facies of the reef of the Ford Formation, the other being the much smaller Townfield Quarry (NZ 4343 4380) at Easington Colliery and Stony

Cut GCR site. Several small quarries in this facies in the Easington–Hawthorn area have been filled in recent years, but were described by Smith and Francis (1967). The reef-rocks at Hawthorn Quarry (then much smaller than now) and Townfield Quarry were discussed briefly by Smith (in Smith and Francis, 1967) and later by Kitson (1982), Aplin (1985) and Hollingworth (1987); Smith (1981a) gave a general review of the characteristics of rocks of the reef-flat based mainly on the two quarries, and Hollingworth (1987) gave detailed analyses of the fauna from them.

Much of the importance of the exposures of reef-rocks at Hawthorn Quarry stems from their large size, which alone allows a comprehensive overview of the general lithology, faunal distribution and great lateral variability of the reef-flat sub-facies. This sub-facies is shown by a combination of former and present exposures to have been at least 300 m wide, and boreholes in the quarry floor showed that it overlies reef-core and reef slope rocks; for comparison, the reef transect at Ford Quarry, Sunderland, revealed a total reef width there of at least 200 m.

Exposures (NZ 4415 4666) at the foot of present coastal cliffs, about 450 m north-east of the reef crest in Hawthorn Quarry, throw some light on contemporary reef-front relief here; allowing for a dip not exceeding 2°, a contemporary reef-front relief of 40–50 m is suggested. This is comparable with that inferred at the southern end of Tunstall Hill GCR site, but is appreciably less than the inferred relief of at least 80 m at Beacon Hill, about 1 km south of Hawthorn Quarry. The exposures feature a coarse breccia (?talus) of reef-derived boundstone, overlain by up to 1.4 m of bedded cream ?ooidal dolomite that fills hollows in the surface of the breccia and is, in turn, succeeded by the dissolution residue of the Hartlepool Anhydrite. The breccia has yielded shelly reef fossils (Trechmann, 1954) and appears to include rounded masses up to 4 m across of intensely encrusted dolomite similar to that at the summit of Maiden Paps, Tunstall Hills.

The reef as a whole is known to stretch in a somewhat tortuous belt from West Boldon (NZ 3464) southwards to Hartlepool (e.g. Trechmann, 1925; Smith, 1981a), and forms the edge of the Cycle EZ1 carbonate shelf-wedge; it plunges gently southwards so that, in general, lower parts are exposed in the north and higher parts in the south. Smith (1980c, 1981a) recognized and defined several main sub-facies of the reef, and Hollingworth (1987) investigated the faunal distribution and ecology of these. Each of the main sub-facies is exposed in one or other of the several GCR sites in the reef, which include Hylton Castle road cutting, Claxheugh Rock and Ford Cutting and Quarry, Humbledon Hill and Tunstall Hills.

The erosion surface

This is readily accessible only at Hawthorn Quarry, its other known surface exposure being in a gorge (NZ 471370) in Crimdon Dene where the reef is in a different facies; a possible additional exposure at the northern end of Blackhalls Rocks awaits the removal of about 1.5 m of recent colliery waste from the beach there so as to reveal the reef top (suspected from a borehole (NZ 4716 3991)) drilled in 1984 by the University of Durham.

The importance of the erosion surface lies in its bearing on interpretation of the local and regional sedimentary history and on some aspects of the stratigraphy. In particular, it must record an episode of erosion and redeposition of the rocks of the underlying reef-flat, and, by inference, a sea-level fall of at least a few metres. The extent and duration of the sea level bears on the problematical age of the biostrome (Cycle EZ1 or Cycle EZ2?) and the choice of the EZ1/EZ2 boundary.

Hesleden Dene Stromatolite Biostrome

Hawthorn Quarry is one of the three main exposures of this unit, the others being the eponymous type locality (not a GCR site) and Blackhalls Rocks. Most of the features of the main part of the biostrome are common to the three main exposures, except that broad algal domes such as occur here and at Blackhalls Rocks have not been recorded at the type locality. Algal domes of this exceptionally large size (up to 20 m) were first reported from Hawthorn Quarry and Blackhalls Rocks (Smith and Francis, 1967), but have since been recorded by Eriksson (1977) from Precambrian dolomites in South Africa and from Precambrian limestones in north Africa. Nothing comparable has been recorded in rocks of any age in the British Isles or elsewhere in rock of Zechstein age. The striking 'crinkly' algal laminite at the base of the main part of the biostrome is similar in both thickness and lithology at each of the three main exposures and was illustrated by Smith (1981a, fig. 27).

The environmental interpretation of the algal laminites of the biostrome was considered by Smith (1981a, p. 15), who concluded from modern partial analogues that the rocks were formed on the broad reef-flat of the Ford Formation under a few metres of hypersaline water; this view was accepted by Kitson (1982).

The areal extent of the biostrome is poorly documented, with only a few exposures and borehole provings other than those cited. It has not been proved north of Hawthorn Quarry and known surface exposures farther south are restricted to a working quarry (NZ 475345) near Hart and to two small old quarries (NZ 448340) near Whelly Hill Farm (Smith and Francis, 1967, p. 144); in the subsurface, the biostrome was proved above reef dolomite in two boreholes (NZ 465337) at Naisberry Waterworks and, judging from the brief records by Trechmann (1932, p. 170 and 1942, pp. 321-322), possibly also above reef-rocks in boreholes (NZ 507333) at Hartlepools Water Works. It may also have been present in the Mill Hill Borehole (NZ 4122 4248), Easington. With only one exception the biostrome has not been proved east of the main shelf-edge reef and its apparent absence from areas north of Hawthorn Quarry may result from poor exposure and erosion.

The boulder conglomerate at Hawthorn Quarry, like the overlying laminites, is exposed also in Crimdon Dene (NZ 4715 3705) (a downstream continuation of Hesleden Dene) and at Blackhalls Rocks, but is thinner and less diverse at Hawthorn. Here, too, it differs uniquely in having only a partial matrix, the laminar fill in many of the larger interstices having a central void up to several centimetres across. Other, poorer, exposures of the conglomerate are at the base of coastal cliffs between Hive Point (NZ 443458) and Beacon Point (NZ 444454), near Hawthorn, where they appear to form part of the collapse-breccia.

The conglomerate at Hawthorn was first described by Kitson (1982). It is an accumulation of clasts of boundstone derived from the underlying Ford Formation reef; the angularity of many of the clasts shows clearly that they were eroded and transported from an already lithified reef surface, indicating a high-energy environment similar to that of a modern boulder storm beach. The origin of the laminar matrix is problematical, but it much resembles travertine and deposition from marine or partly vadose waters passing through the interstices seems likely; the incompleteness of the filling may indicate early constriction of the 'throats' by fine detritus and contemporaneous cements, but could also have resulted from inadequate time before burial; close proximity to sea level is probably indicated. The period of conglomerate formation was completed by a phase of apparently chaotic ?algal-stromatolite growth before the more uniform subaqueous regime of the succeeding crenulated algal laminites became established.

?Roker Dolomite Formation

?The Roker Dolomite Formation exposed in the quarry is normal for the region and requires no special comment except on the uncertainty of its attribution; this doubt results from its apparent lack of diagnostic fossils and its unknown relationship with younger strata, but lithologically similar ooid grainstones at Seaham and Blackhalls Rocks are probably of the same age as those at Hawthorn Quarry and are assigned to the Roker Dolomite Formation with reasonable confidence. The formation as a whole is interpreted as the shelf facies of the marginal carbonate wedge of Cycle EZ2 (Smith, 1971a, 1980a, b); its outcrop is restricted to north-east coastal districts from Whitburn southwards (Smith, 1980b, fig. 9), where its main exposures are in coastal cliffs at Whitburn (NZ 4161), Roker (the type locality, NZ 4059) and Seaham (NZ 4250), and in coastal rock platforms at Hartlepool (NZ 5234).

Structure

The geological structure requires no comment except for the narrow reverse shatter-belt formerly seen between the reef and younger ooid grainstones (?Roker Dolomite Formation) near the quarry entrance. The shatter-belt is roughly parallel with the strike of the reef crest and also with a normal NNW/SSE trending fault of 5-6 m displacement (downthrow to the east) in the underlying coal workings; it may be a surface expression of this fault, but it could also have resulted from differential compaction between the reef and the grainstones or from subsidence caused by dissolution of the Hartlepool Anhydrite that formerly lay against the steep reef-face. A combination of any of these causes is also possible, but the third suggested mechanism seems more likely than the others because it most readily accounts for the vertical displacement of 30 m+ in the Magnesian Limestone. Further evidence favouring this third mechanism comes from the partial (?collapse) brecciation of the ooid grainstone near the quarry entrance, and from the presence of fragments of red mudstone (from previously overlying strata) in some of the breccias there.

Future research

There are many unresolved geological problems in the rocks of Hawthorn Quarry, and correspondingly good opportunities for future research; some of these are currently being addressed. The ecology and biota of the reef, having been investigated by Hollingworth (1987), is now reasonably well understood, but the precise depositional conditions of the reef, its petrology and the nature and mode of origin of reef encrustations and laminar sheets still require further study. Other problems requiring further research include the nature, extent and origin of the erosion surface and overlying boulder conglomerate, the age, origin and diagenesis of the pisoids and algal laminites of the biostrome, and the age and diagenetic history of the ?Roker Dolomite Formation.

Conclusions

Hawthorn Quarry is an extremely important GCR site in that firstly, it is the largest exposure of late Permian (Ford Formation) reef-flat rocks in north-east England, and secondly, is the only exposure where their disconformable contact with the overlying Hesleden Dene Stromatolite Biostrome can be seen. The boulder conglomerate at the base of the Biostrome is seen elsewhere only in Crimdon Dene and at Blackhalls Rocks, whilst the contact between the biostrome and the overlying ?Roker Dolomite Formation is well-exposed only here. The site is ideal for further study and research into reef-rock characteristics, the age and diagenetic history of the Hesleden Dene Stromatolite Biostrome and the overlying ?Roker Dolomite Formation.

HORDEN QUARRY (NZ 435417)

Highlights

This small and very old quarry (box 14 in Figure 3.2) at Horden is now the best of the exposures in north-east England where a one-time crest of the shelf-edge reef of the Ford Formation may be seen and examined. The east side of the quarry also contains indifferent exposures of collapse-breccia (probably mainly of the Roker Dolomite Formation), showing that the Hartlepool Anhydrite, now dissolved, once lay against the steep reef-face here.

Introduction

This excavation, which formerly contained a concrete reservoir, lies near the foot of a steep slope on the west side of Horden township; it should not be confused with Yoden Quarry, which lies on the hill 400 m farther west. The slope into which the quarry is cut roughly marks the position of the seaward margin of the shelf-edge reef of the Ford Formation (Cycle EZ1b).

The quarry reveals two exposures of dolomite boundstone and bindstone at a one-time crest of the Ford Formation reef, and an adjoining collapse-breccia probably composed mainly of fragments of the Roker Dolomite Formation. The locality was probably that mentioned by Trechmann (1925) who reported *Epithyris (Dielasma)* and several bivalve and gastropod genera from his reef-limestone 'C' from a 'knoll behind Horden Colliery', and brief descriptions and illustrations were given by Smith (in Smith and Francis, 1967, p. 139 and Smith, 1973b, 1981a). The petrography of the reef dolomite at this exposure was considered by Aplin (1985) and the fauna of the rock here was used by Hollingworth (1987, fig. 6.18, reproduced in Hollingworth and Tucker, 1986, fig. 7) as the basis for his graphic reconstruction of a reef crest faunal community. There is some confusion regarding the local source of bivalves collected by Logan (1967), for he appears to have believed that they came from the same locality as those recorded by Trechmann (1925). He attributed them to 'Yoden Quarry (Horden Colliery)'.

Description

The location and outlines of the old quarry at Horden are shown in Figure 3.50, which also shows the position of the main points of geological interest. Figure 3.51 shows the section in the main surviving face in the west of the quarry, but similar features are also displayed in the south face.

The main geological interest at Horden is a one-time reef crest of the Ford Formation, exposed in two places on opposite sides of the quarry, and the presence of a collapse-breccia of later strata lying against the ultimate steeply-inclined reef-face in the north of the quarry.

Ford Formation, reef crest

The configuration of the reef crest at the old quarry at Horden, as it was towards the end of reef

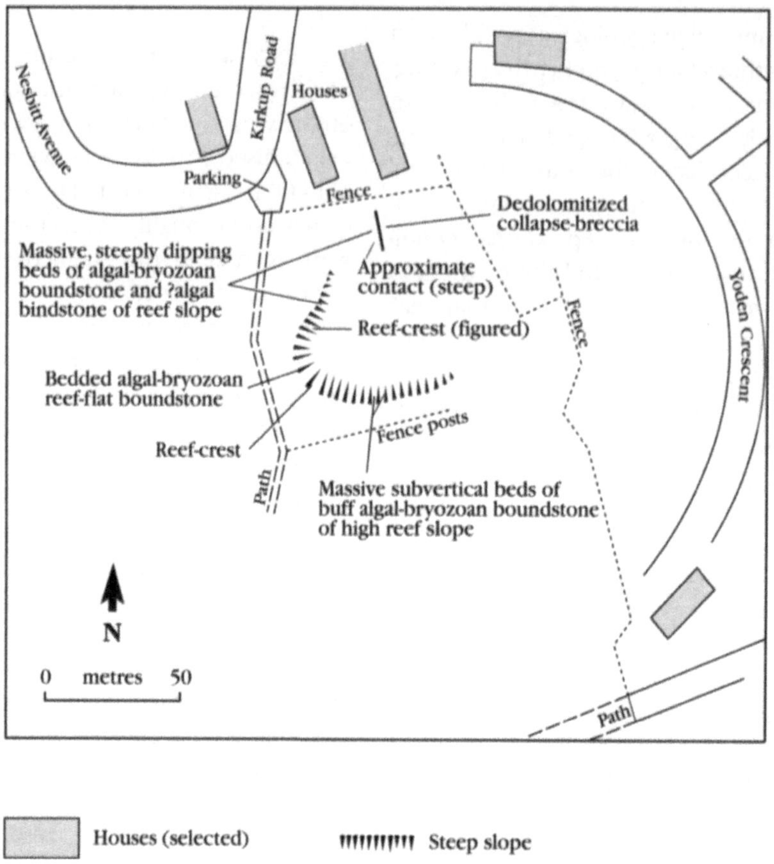

Figure 3.50 Location of Horden Quarry and the main features of geological interest.

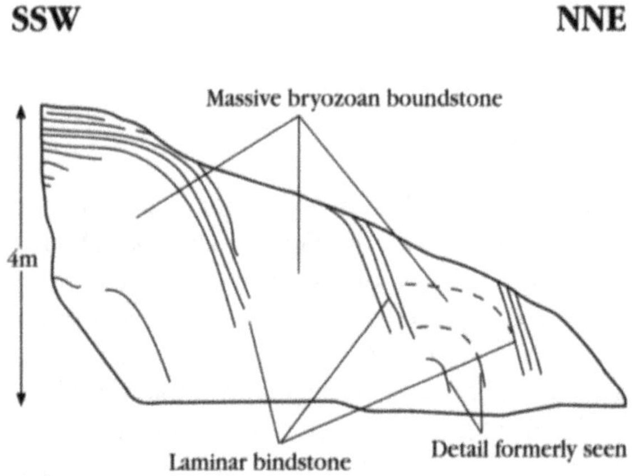

Figure 3.51 Sketch of the north-west face of Horden Quarry as seen in 1954 and later (parts of the face are now covered).

growth, is shown in Figure 3.52. This crest comprises a sharp transition from almost horizontally, fairly thin-bedded, dolomite boundstone of the reef-flat sub-facies to much thicker bedded, steeply dipping boundstone and bindstone of the uppermost reef slope sub-facies. Exposures near the floor of the western and southern part of the quarry show that the primary dip of the reef slope steepens there to 75° to 85°. The bindstone forms sinuous laminar sheets 0.1–0.3 m thick between exceptionally thick (1–3 m) boundstone beds, but appears to die out or become much thinner, at or just below the crest.

The dolomitized reef boundstone at Horden is a buff-coloured rock with an abundant fauna of low diversity. Lists by Trechmann (1925) and Pattison (in Smith and Francis, 1967) included ramose bryozoans (*Acanthocladia anceps*) and a small number of brachiopod, bivalve, gastropod, foraminifera and ostracod genera. These lists are confirmed by more detailed collecting by Hollingworth (1987), who quantified the relative proportions of the genera present; his observations showed that the bryozoans *Acanthocladia* (32%) and *Dyscritella* (18%) together make up half of the faunal elements present and that the remainder was dominated by *Dielasma* (18%), *Bakevellia* (16%) and

Figure 3.52 A reef crest in the north-west face of Horden Quarry (for position see Figures 3.50 and 3.51). Hammer: 0.33 m. (Photo: D.B. Smith.)

Pseudomonotis (12%). In his reconstruction of the reef crest community, Hollingworth also showed that most of the shelly organisms occupied (and presumably lived in) spaces in the tangled masses of *Acanthocladia zoaria*, and that most of the latter were heavily encrusted with ?algal laminae. Hollingworth (1987) found that almost all the *Dielasma* present in his sample died before reaching maturity, but only about 5% of the *Pseudomonotis speluncaria* collected by Logan (1967) were juveniles.

The laminar bindstone sheets between the massive boundstone beds are buff-cream in colour and are composed of finely crystalline, slightly calcitic dolomite. They are mainly lacking in skeletal fossils, but patches of densely crowded *Dielasma* and bivalves, and scattered fragments of bryozoans, occur parallel with the lamination of some of the sheets (Smith, 1981a). Presumably these organisms were all firmly attached to the seaward face of the reef. Bioclasts were also recorded by Aplin (1985, p. 92) in the algal-laminated lining of vertical tension fissures in reef-flat boundstone just landward (i.e. west) of the reef crest here, where fissures were bridged by reef-flat boundstone.

?Roker Dolomite Formation, collapse-breccia

Hard grey limestone (dedolomite) collapse-breccia, with some residual cream dolomite, is poorly exposed in the eastern part of a low rock eminence in the north of the quarry. The rock comprises angular fragments, up to a few centimetres across, of finely crystalline limestone in a grey to brown dense calcite matrix; few traces of its original lithology remain. Because of the poor quality of the exposure, the sub-vertical contact with the reef slope is somewhat obscure, but its approximate position and trend are shown in Figure 3.50.

Interpretation

Horden Quarry is of major importance, firstly in being the only good and readily accessible exposure of rocks formed at the crest of the shelf-edge reef of the Ford Formation, and secondly, because it illustrates the juxtaposition of the steep Cycle EZ1 reef slope and the Cycle EZ2 collapse-breccia. The exposures are thus vital links in the chain of evidence leading to present understanding of the reef profile, its communities and the mutual relationships of the reef and the Hartlepool Anhydrite.

Ford Formation, reef crest

The ultimate reef crest (i.e. that formed just before the reef ceased to grow) has been eroded off at the Horden exposure but, by comparison with former exposures at Hawthorn Quarry site (Figure 3.47) and at an old quarry (NZ 436437) at Easington Colliery (Figure 3.53), was probably only a few metres above present ground level. Elsewhere, the ultimate reef crest is exposed inaccessibly high in the north face of the working quarry (NZ 476344) near Hart, and earlier reef crests are indifferently exposed at Ford Quarry SSSI and in the Stony Cut site (NZ 418473), Cold Hesledon. As at Hawthorn Quarry, the relative thickness of the reef-flat and equivalent reef slope rocks at Horden and Easington Colliery show that the reef crest prograded basinward three to six times faster than it grew upward, perhaps because its upward growth was limited by an approach to sea level. There are, indeed, hints at all three quarries that, at times, the reef crest prograded without any corresponding

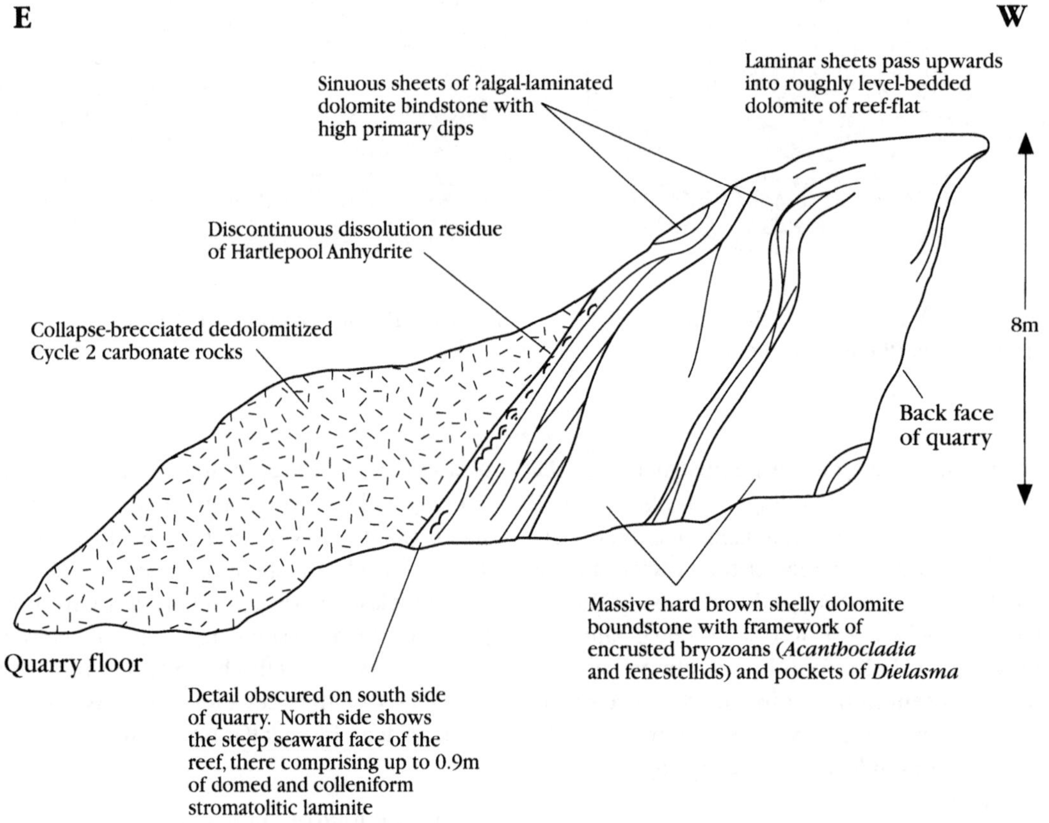

Figure 3.53 South side of old quarry (now filled) on the east side of Townfield Hill, Easington Colliery, showing the crest and stromatolitic seaward face of the shelf-edge reef, and succeeding residue and collapse-breccias. This quarry was recommended for SSSI designation but was filled before action could be taken; it is included here for the purposes of comparison.

build-up of its surface or even with reef-top erosion during phases of slightly lowered sea level.

The exposure of the ultimate reef-face in the old quarry (now filled) at Easington Colliery was uniquely important in that the youngest of the steeply-dipping laminar bindstone sheets was characterized by two short courses of unmistakable columnar stromatolites (Smith, 1981a, fig. 22). This showed that at least the last of the laminar sheets was a reef-face coating, perhaps implying a similar origin for the other steeply-dipping laminar sheets; probable stromatolitic laminite with domes up to 0.3 m across also coats the seaward face of the reef in the quarry at Hart, where the ultimate upper reef slope has a primary dip of 75–85° throughout its exposed height of about 13 m. The presence of a double layer of short columnar stromatolites on the outermost steep reef slope in the quarry at Easington Colliery invites comparison with parts of the Trow Point Bed (Smith, 1986; also this volume), which, in places, similarly comprises two short courses of columnar stromatolites and similarly lies immediately below the dissolution residue of the Hartlepool Anhydrite.

The faunal assemblage of the reef crest is of considerable interest in that it comprises only genera that could withstand high energy conditions and maintain their position on a near-vertical slope. The dominance of sessile flexible and robust bryozoans is predictable and, except for small gastropods, most of the shelly animals were adherent or encrusting forms. The smoothly even curvature of the crest, as compared with the sharply ragged angularity of some modern counterparts, presumably results from the smaller size of the late Permian reef frame builders and the abundance of laminar ?algal sheets.

From the known position of the reef crest and of two former quarries on top of the hill immediately west of the Horden exposure, the shelf-edge reef of the Ford Formation is shown to be at least 400 m wide between Peterlee and Horden. Its clear topographical expression here, at Easington Mill Hill and in Easington Colliery township, implies that a major reef re-entrant may be present between Horden and Easington Colliery, with the reef crest stepping back some 1.2–1.5 km to the west of its main position. It is not known whether the reef was continuous around this inferred re-entrant or whether, as is equally possible, it was discontinuous and present in sub-parallel stretches separated by open sea or large surge-channels. There is similar uncertainty over comparable re-entrants near Seaham and between Blackhalls Rocks and Hartlepool (see Figure 3.1).

?Roker Dolomite Formation, collapse-breccia

The juxtaposition of the reef slope and collapse-brecciated Roker Dolomite at Hawthorn Quarry, Easington Colliery and Horden indicates that here the Hartlepool Anhydrite must, before its dissolution, have lain against the steep reef slope and been at least as thick (?80–110 m) as the height of that slope. Foundering of the Roker Dolomite and higher strata in each place must have been by a similar amount, less a proportion resulting from a lower packing density, and was probably episodic. As at Hawthorn Quarry and in the nearby coastal cliffs, the foundering at Horden probably also involved the Hesleden Dene Stromatolite Biostrome that formerly overlay the reef, and must have affected all strata above the Roker Dolomite Formation. As elsewhere in the area to the south of Ryhope, reaction of the reef dolomite with brines rich in calcium sulphate from the dissolving anhydrite is assumed to have caused the dedolomitization in the collapse-breccia, but very little of the adjoining reef-rock at Horden has been dedolomitized.

Future research

The stratigraphy, palaeontology and petrology of this small but important exposure have all been investigated in considerable detail since 1980, and there is probably little immediate scope for further research on most aspects of the rocks exposed. Possible exceptions to this are a more detailed analysis of differences and similarities of faunal communities behind and in front of the reef crest and further research on the laminar sheets to determine if, like the youngest sheets at the Easington Colliery exposure, they are indeed stromatolitic in origin and at one time draped the seaward face of the reef.

Conclusions

This is the only GCR site where the one-time crest of the shelf-edge reef of the Ford Formation can still be seen in juxtaposition with collapse-breccias of the Roker Dolomite Formation. The reef crest is characterized by a sharp change from gently-dipping, mainly thin-bedded dolomite of the reef-flat to thicker bedded, steeply-dipping dolomite of the uppermost reef slope. The reef contains frame-building bryozoans and shelly fossils, the former encrusted with ?algal laminae.

The breccias comprise angular rock fragments with little evidence of the original lithology. The close juxtaposition of reef slope rocks and the breccias suggest that the Hartlepool Anhydrite must have lain against the reef slope before its dissolution. Both the exposure of the reef crest and the opportunity to relate the position of the reef to the anhydrite to the east, mark this site as being extremely important for the study of the stratigraphy and sedimentology of the late Permian marine rocks in Durham.

BLACKHALLS ROCKS (NZ 4683 3948 – 4763 3826)

Highlights

The coastal cliffs and shore platforms at Blackhalls Rocks (not shown in Figure 3.2) constitute the largest and best exposure of the Hesleden Dene Stromatolite Biostrome. The biostrome is almost entirely of dolomite rock and comprises a thick and highly varied boulder conglomerate overlain by a thicker unit of algal laminites ('stromatolites'). The conglomerate is formed mainly of rolled cobbles and boulders derived by erosion of the underlying (but unexposed) reef-flat rocks of the Ford Formation and the algal laminites include a strikingly complexly finely laminated basal layer and several generations of spectacular domes individually up to 1.5 m high and 18 m across. The sequence is capped by ooidal dolomite of the Roker Dolomite Formation and the overlying Seaham Residue.

Introduction

Blackhalls Rocks is a coastal site that exposes almost the full thickness of the Hesleden Dene Stromatolite Biostrome (?45 m) together with the whole of the overlying Cycle EZ2 Roker Dolomite Formation (?16 m) and much of the Seaham Formation. The sequence is gently anticlinal and a borehole near the core of the anticline is thought to have entered reef dolomite of the Cycle EZ1 Ford Formation almost immediately below the lowest beds currently exposed. The anticline is bounded to the north by the mineralized Blackhall Fault (Smith, 1964), which has a northwards downthrow of perhaps 12 m, and to the south by the steeply-dipping Seaham Residue of the Cycle EZ2 Fordon Evaporites. The age of the biostrome is uncertain, with delicately balanced arguments allowing either EZ1 or EZ2 affinities.

The boulder conglomerate at Blackhalls Rocks was first mentioned and scenically illustrated by Sedgwick (1829) and was termed a 'Shell-Limestone conglomerate' by Howse (1858) in recognition of its faunal similarity to the shelf-edge reef of what is now termed the Ford Formation. The section received little further attention until Trechmann (1913) published a brief summary of strata exposed there, and added a list of 24 invertebrate species from clasts in the conglomerate; later, Trechmann (1914) published chemical analyses of the conglomerate and of a thin 'large-grained pea-oolite' (= pisolite) from the top of it (also illustrated in thin-section), and subsequently (1925) gave an augmented fossil list of 29 species and a further five doubtfully identified forms. Woolacott (1918, 1919a) referred to the conglomerate at Blackhalls Rocks as a fossiliferous breccia composed of blocks that had rolled down the eastern edge of the reef, ie. a 'Vorreef', and illustrated it in 1919(a); Trechmann (1925) similarly referred to the conglomerate as a 'Vor-riff' of reef talus.

Apart from a brief mention by Trechmann (1931), the section at Blackhalls Rocks received no further attention until it was described and illustrated by Smith and Francis (1967). Pattison compiled a faunal list (in Smith, 1970a, repeated in Pattison *et al.*, 1973) comprising 28 species with an additional four doubtful identifications. Logan (1967) cited the locality as a host to seven species of bivalves, two of which were illustrated and designated as hypotypes. Further description by Smith (1981a) was within a proposed new lithostratigraphical framework in which most of the sequence at Blackhalls Rocks was ascribed to the newly-defined stromatolite biostrome. Finally a full investigation of the sedimentology of the whole of the sequence at Blackhalls Rocks was reported by Kitson (1982). The site also features in several field guides, excursion reports and popular articles (e.g. Smith, 1984).

Description

This site lies on the Durham coast about 8 km north-west of Hartlepool and comprises about 1.1 km of cliffs and shore-platforms (Figure 3.54); the cliffs are about 15–32 m high and comprise up to 24 m of southwards-thickening Quaternary (late Devensian) glacial drift deposits overlying up to 10 m of Magnesian Limestone. The drift deposits

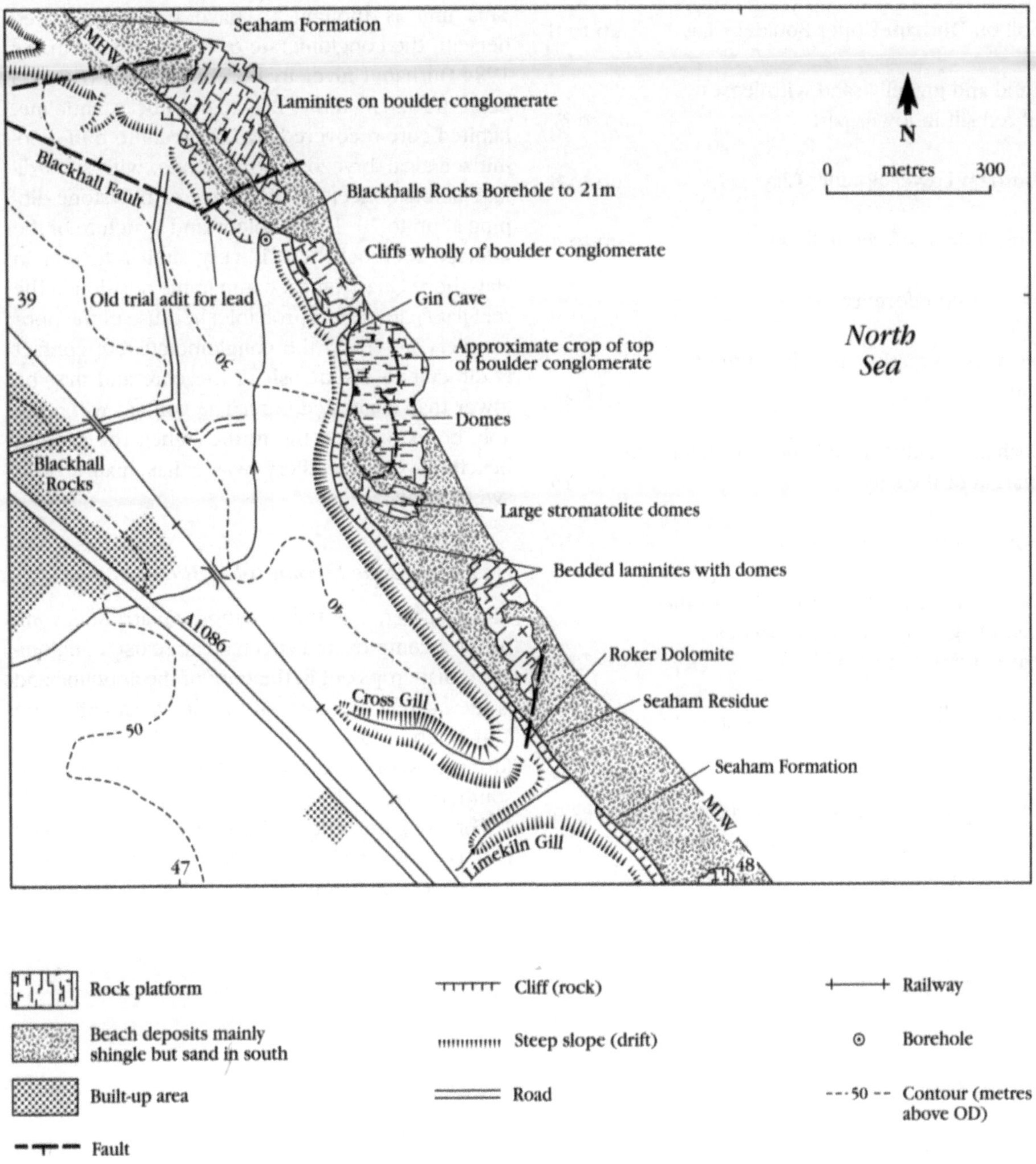

Figure 3.54 Blackhalls Rocks GCR site and its environs, showing the location of the main geological features.

form a layered sequence of two stony clays separated by a sand and gravel layer from which perennial seepages cause instability; this level of the cliffs is well known for its unusual plant communities and part of the cliffs is scheduled as a botanical SSSI. The northern section of the cliffs, extending to Blue House Gill, is managed as a local nature reserve by the Durham Wildlife Trust.

The general geological sequence in and adjoining the designated area is given on p. 116.

	Thickness (m)
Soil on Durham Upper Boulder Clay	up to 8
Sand and gravelly sand with lenses of red silt in lower part	up to ?7
Durham Lower Boulder Clay	up to 8
Gravel, in scattered hollows	up to 4
----- unconformity ------	
Seaham Formation, mainly south of the site	14.5+
Seaham Residue, mainly near the southern margin of the site	?5
Roker Dolomite Formation	?16
Hesleden Dene Stromatolite Biostrome, including conglomerate *c.* 18 m at base (1.5 m not seen)	?45
Ford Formation, reef-facies (doubtfully identified in borehole)	19.5+

The approximate positions of the main features of geological interest are shown in Figure 3.54 and their relationships are shown in Figure 3.55.

Ford Formation reef

This unit is thought to have been penetrated beneath the conglomerate of the biostrome in the 1984 Durham University borehole (NZ 4716 3911) near the axis of the Blackhalls Rocks anticline. Limited core recovered was of cream to buff dolomitized algal-bryozoan boundstone, with subordinate laminar sheets of bindstone or flowstone dipping at up to 70. The lithology and structure of the inferred reef-rock here is very similar to that in Hawthorn Quarry and a comparable position in the reef-flat facies seems probable; because of the poor recovery however, the conglomerate-reef contact is difficult to distinguish in the core and may be lower than suggested here. The contact will probably be exposed in the future, when the present beach cover of colliery waste has finally been swept away.

Hesleden Dene Stromatolite Biostrome

This member, as at the Hawthorn Quarry site, comprises a complex and spectacular coarse conglomerate that crops out in the core of the anticline and is the main component of the northern cliffs, and a thick upper unit of algal-laminated dolomite that forms most of the cliffs and shore platforms in the southern half of the site.

The conglomerate at the base of the biostrome is a complex and extraordinarily varied, poorly-sorted accumulation of cobbles and boulders of

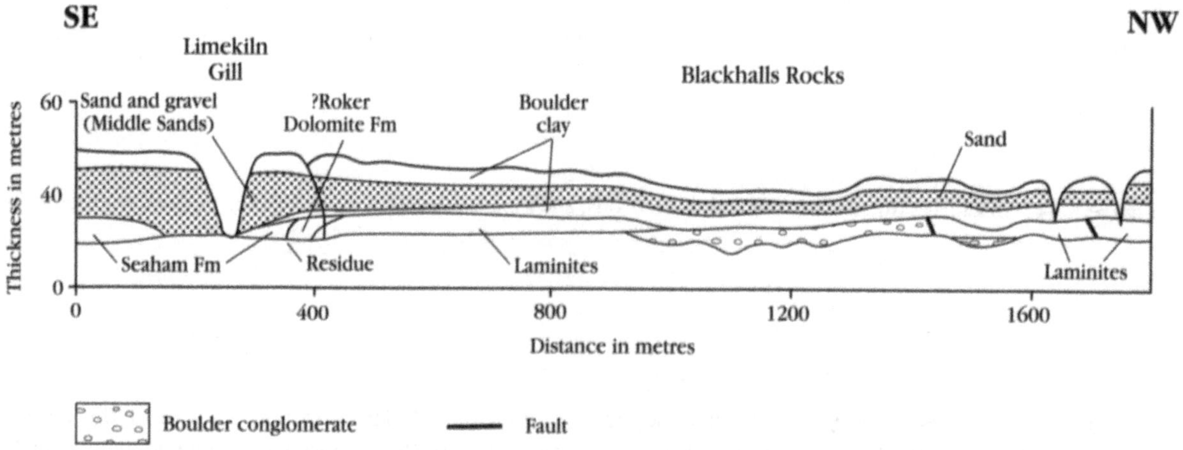

Figure 3.55 Geological strata in the cliffs of the Blackhalls Rocks GCR site (diagrammatic). The laminites and conglomerate together comprise the Hesleden Dene Stromatolite Biostrome. Slightly modified from Smith (1984, p. 24).

buff, algal-bryozoan boundstone (predominant) and algal bindstone in a 20–40% matrix of finely laminar clast-encrustations and fine debris (including bioclasts); it is divided almost chaotically into crude commonly eastwards-dipping beds, lenses and wedges by discontinuous, but locally extensive sinuous sheets up to 0.5 m thick of finely crystalline buff laminite and by local probable erosion surfaces. The clasts, matrix and laminar sheets are almost wholly of dolomite, but contain small amounts of pore-filling and cavity-lining late calcite.

The conglomerate is clast-supported and component clasts (commonly 0.15–0.4 m across, but exceptionally 0.9 m) are mainly subangular to rounded and roughly equidimensional (Figure 3.56); imbrication and cross-bedding are rare. Some clasts bear thin finely laminar coatings, but it remains to be established whether these coatings were formed before or after final deposition of the clasts. Non-laminar matrix generally post-dates any laminar coatings and is most abundant in middle and lower parts of the conglomerate; it comprises widely varied proportions of angular to subrounded sand- to pebble-grade debris of boundstone and bindstone (as in the major clasts) together with abraded (?derived) bioclasts, reworked fragments of laminar coatings and a few pisoids. Many of the smaller clasts have finely laminar concentric coatings and some bear evidence of several episodes of fracturing, recementation and recoating (Kitson, 1982). Similar debris forms scattered lenses up to 1 m thick, especially towards the northern end of the site, and probably accumulated in irregular hollows on the contemporary surface of the conglomerate.

The cobbles, boulders, matrix and debris lenses appear to have been the source of most or all of the fossils listed from Blackhalls Rocks by Trechmann (1913, 1925) and Pattison (in Smith and Francis, 1967, p. 143 and in Smith, 1970a, pp. 85–88). Smith (1981a), however, recorded *Peripetoceras* and Kitson (1982) reported *Bakevellia* and *Permophorus* from laminite sheets in the conglomerate, although it is not clear whether these fossils were part of an indigenous biota or are reworked bioclasts.

The laminar sheets that divide the conglomerate form 30–40% of its bulk in higher parts and generally somewhat less lower down; the thicker sheets are mainly low-dipping (5°–25°), but locally steepen to 50° or more. They divide and reunite grotesquely, and in places appear to surround trapezoidal to rectilinear conglomerate masses (Kitson, 1982) and to line both sides of subvertical fissures up to 1.5 m deep; some sheets are clearly bilateral, with a median void, void-fill or mutual contact. The laminae are fine and relatively even and, in the thicker sheets, are mainly gently undulate and flowing. Most of the thinner laminar sheets, however, clearly coat boulders forming the walls of primary cavities and display a wide range of simple or multistage 20–50 mm diameter botryoidal or laterally-linked hemispherical to columnar stromatolite-like structures with axes disposed at all angles. Kitson (1982, fig. 25b) illustrated probable algal relics in laminar sheets near the site of the 1984

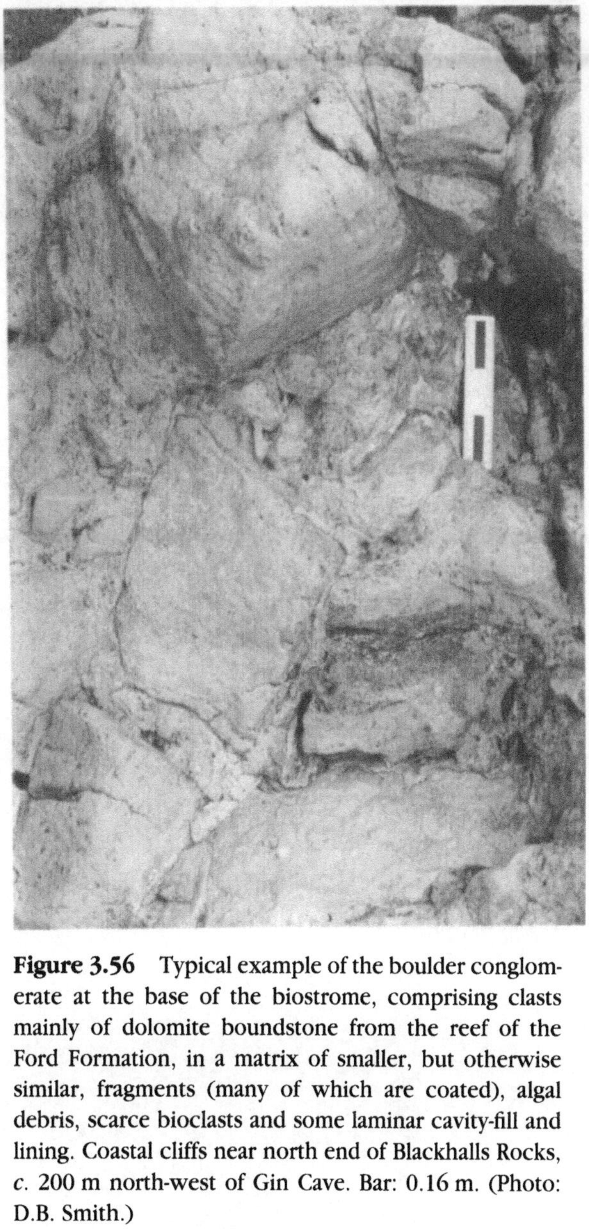

Figure 3.56 Typical example of the boulder conglomerate at the base of the biostrome, comprising clasts mainly of dolomite boundstone from the reef of the Ford Formation, in a matrix of smaller, but otherwise similar, fragments (many of which are coated), algal debris, scarce bioclasts and some laminar cavity-fill and lining. Coastal cliffs near north end of Blackhalls Rocks, *c.* 200 m north-west of Gin Cave. Bar: 0.16 m. (Photo: D.B. Smith.)

borehole. Late-stage partly botryoidal laminites also line cracks penetrating or crossing some of the larger clasts.

The uppermost bed (0.5–0.8 m) of the conglomerate is a highly complex buff laminar bindstone with relatively fewer boulders than most of the remainder. Lamination in this bed is strikingly tortuous and includes both obvious cavity-lining and a wide range of steep-sided stromatolite-like domes individually up to 0.5 m tall and 1.5 m across, but mostly much smaller; scattered inter-dome hollows in this bed contain much multi-coated laminar debris and also rare clusters of subspherical ?oncoids up to 40 mm across and with more than 30 concentric layers.

The algal-laminated unit is at least 28 m thick and comprises unfossiliferous cream to pale buff silt- to fine sand-grade saccharoidal dolomite with abundant empty or calcite-lined small cavities after dissolved secondary anhydrite; interstitial calcite is widespread. Much of the rock has been finely brecciated by diagenetic processes and Liesegang-type colour banding is widespread. The primary lamination is generally faint and in hand specimens is commonly almost invisible; it is, however, generally distinct on weathered surfaces and is particularly clear and spectacularly crinkly in a widespread 1.3–1.8 m bed (the 'Crinkly Bed', Figure 3.57) at the base where laminae average 45 per centimetre (Smith and Francis, 1967; see also Al-Rekabi, 1982, fig. 3.4). In at least one place this unique lowest bed passes into or includes ooid grainstone and it also contains steep-sided pockets up to 1.2 m deep filled with smoothly thinly multi-coated dolomite pisoids individually up to 50 mm across, but mainly 5–15 mm (Figure 3.58). First noted by Trechmann (1913), and later (1914) illustrated and analysed by the same author, these pisoids commonly have an abraded nucleus of fine dolomite laminite and many bear clear evidence of early lithification in the form of fractures, re-cementation and re-coating (Smith, 1981b, fig. 29; Al-Rekabi, 1982, fig. 3.2B; Kitson, 1982, fig. 30b).

Bedding in the algal laminites is generally even and parallel, but is widely diversified by minor domes and gentle undulations with a relief of up to 0.3 m. At several levels, however, continuous layers of large steep-sided rounded to flat-topped domes are present (McKay, in Trechmann, 1913; Smith and Francis, 1967; Kitson, 1982). These domes are formed entirely of crinkly or rippled finely laminated dolomite and exceptionally are up

Figure 3.57 Laminar algal bindstone of the 'Crinkly Bed' at the base of the Hesleden Dene Stromatolite Biostrome. Coastal cliffs *c.* 60 m south of Gin Cave (see Figure 3.54). Coin: 26 mm across. (Photo: D.B. Smith.)

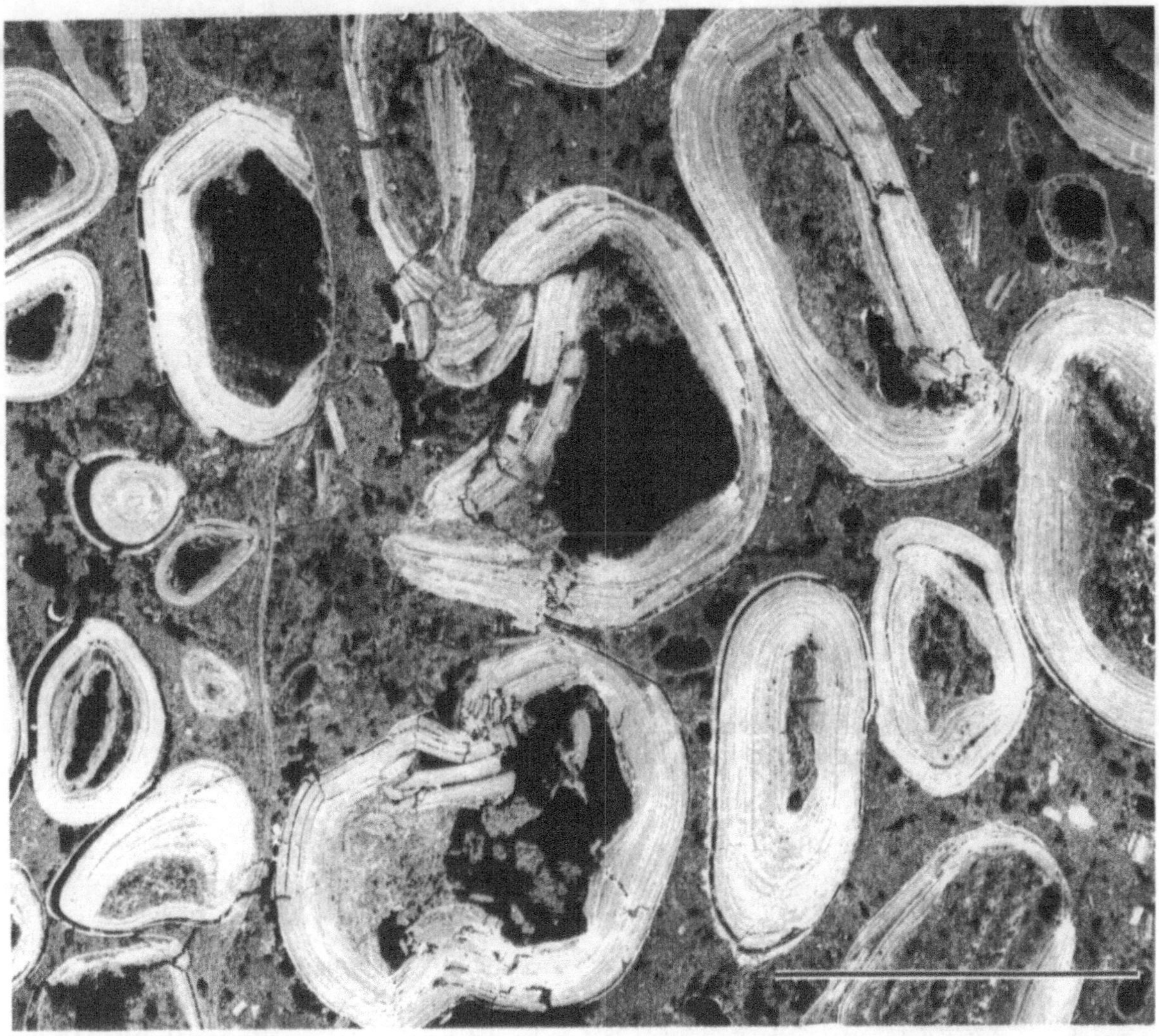

Figure 3.58 Thin section of multi-coated pisoids from steep-sided pockets in the upper part of the 'Crinkly Bed' in coastal cliffs near Gin Cave, Blackhalls Rocks. Interstices and the core of some formerly leached pisoids are occupied by equant dolomite microspar, but many of the pisoids are nucleated on to abraded dolomite clasts including portions of earlier fractured pisoids. Bar: 8 mm. (Photo: D. Kitson.)

to 1.5 m tall and 18 m across; they are best seen on the foreshore rock platform (Figure 3.59), but mutual relationships are clearer in the cliffs where the domes are locally seen to be separated by gently onlapping dolomite and limited fine contemporaneous debris.

Roker Dolomite Formation

This eastward-dipping formation forms much of the rock section in the cliff at the southern end of the site, its base rising north-westwards to reach rockhead near Cross Gill; it reappears patchily on the north side of the anticline, north of the Blackhall Fault. The uppermost 5-6 m of the formation comprise a markedly streaky and partly finely brecciated body of white to cream dolomite ooid grainstone and it is possible that this unit should be regarded as part of the Seaham Residue. Most of the formation, however, is of undisturbed cream dolomite ooid grainstone with strong tabular and trough cross-stratification at several levels; Kitson (1982) reported foresets dipping north-west and north-east at 10°-15°. Thin sections show that the ooids have been leached, and that some compound ooids and stromatolite flakes are present; Trechmann (1913, p. 201) reported cores of secondary fluorite in ooids of this formation in the north of the designated area. Trechmann also records *Schizodus* here, the only positive faunal

Figure 3.59 Broad dolomite stromatolite domes in about the middle of the Hesleden Dene Stromatolite Biostrome, on the foreshore near Green Stairs, Blackhalls Rocks. Hammer: 0.33 m. (Photo: D.B. Smith.)

identification in this formation at Blackhalls Rocks, though scattered bivalve-shaped cavities are present in the southern exposures.

The base of the Roker Dolomite Formation is generally difficult to distinguish in the cliff sections, where it is widely obscured by surface wash, but it is partly exposed in the lower ravine (NZ 4759 3825) of Cross Gill where it is interbedded with laminites and includes a 1.35 m ooidal dolomite bed that contains many coarse compound grains, pisoids and lumps. An interbedded basal passage was also recorded on the northern limb of the anticline (Smith and Francis, 1967, p. 150).

Seaham Residue

These insoluble clayey remains of the Fordon Evaporites lie in a gully near the southern boundary of the site; they are perhaps 5 m thick (depending on where the base is taken; see Interpretation). They comprise a lower 2.9–4 m layer mainly of cream ooid grainstone, but with many thin streaks, lenses and layers of brown and buff-brown argillaceous dolomite or dolomitic clay, a median 1.2–1.6 m layer of soft off-white dolomite ooid grainstone (partly brecciated) and an upper layer (maximum 1 m) of complexly interlayered multicoloured streaky clay or clayey dolomite. All the units of the residue are locally contorted and brecciated and the formation as a whole dips eastwards at 35°–50°.

Seaham Formation

Carbonate rocks of this formation occupy synclines both north and south of the designated area. They are best exposed in the south where they comprise a foundered and partly collapse-brecciated sequence similar to that at the type locality at the Seaham site; a detailed section here (Smith in Smith and Francis, 1967, pp. 157–160; see also plate XIIIC) shows thinly-layered basal shelly dolomite with *Calcinema* overlain by crystalline limestones patchily rich in spherulitic and globular calcite concretions. Red-stained and ochreous fine-grained carbonate or clay is a feature of the abundant infiltrated fill between disarticulated collapse-breccia fragments here.

Interpretation

The cliffs and rock-platforms at Blackhalls Rocks are doubly important in affording by far the best and most spectacular exposures of the Heselden Dene Stromatolite Biostrome, including most of the boulder conglomerate at its base, and also in furnishing a complete sequence from the reef of the Ford Formation up to (and including) the Seaham Formation. In this latter respect, the section at Blackhalls Rocks parallels and supplements that at the Hawthorn Quarry site, supporting the inferred identification of Roker Dolomite at the top of the sequence there.

The drift at Blackhalls Rocks epitomizes that of much of the Durham coast and requires no special comment; similarly, the Seaham Formation there has much in common with that at its type locality at the Seaham site, and the account (under 'Concretionary Limestone') by Smith and Francis (1967, pp. 157–160) will suffice. The Roker Dolomite Formation and the reef of the Ford Formation are normal for the area and also need no further discussion.

Hesleden Dene Stromatolite Biostrome

Blackhalls Rocks exposes almost the full thickness of this remarkable unit, a feature it shares with Hesleden Dene (and its downstream continuation) and the Hawthorn Quarry site; the boulder conglomerate at its base, however, is much thicker here than at the other localities.

The distribution of the conglomerate at the base of the biostrome is poorly known, but it is more variable in thickness and lithology and much less extensive than the overlying laminites, and in the Hawthorn Quarry site dies out sharply on its basinward side; the nature of its landward margin is unknown, though abutment against a basin-facing cliff notch may be speculated.

The origin of the conglomerate has been the subject of much discussion, but authors from Trechmann (1913) onwards agree that the major clasts were mainly derived from the shelf-edge reef of the Ford Formation. Woolacott (1918, 1919a) regarded it as an offshore (i.e. deep-water) equivalent of the reef and Trechmann (1925) interpreted it as reef talus with the clasts rounded by wave action on the eastern slope of the reef (ie. shallow water). Current research by the writer and Mr D. Kitson suggests that the conglomerate is a storm beach deposit that was formed on the reef-flat when a sea-level fall resulted in erosion of the Ford Formation reef, and that it was subject to phases of exposure, erosion, reworking and burial by conformable sheets of laminar carbonate. The angularity of some of the clasts indicates that the reef-rock was fully lithified before dislodgement and transport.

The origin of the laminar coatings on clasts in the conglomerate at Blackhalls Rocks and of the ramifying laminar sheets remains uncertain and is subject to continuing research. Field relationships and the abundance of laminar debris show that at least some of the laminar coatings and/or sheets must have been formed and lithified whilst the conglomerate was accumulating, and the localized fracturing of the conglomerate itself and of some of its large component clasts indicates that it too was cemented penecontemporaneously. The striking morphological similarity of many structures in the clast coatings and laminar sheets to algal stromatolites previously led the writer to infer deposition through the agency of blue-green (cyanophytic) algae, but the presence of (and lateral passage into) otherwise apparently identical linings of narrow fissures and of sheets of totally inverted laterally-linked hemispherical structures casts doubt on a wholly algal origin and deposition through other microbial or inorganic agencies may have been involved; in this connection the absence of the remains of a browsing fauna may be significant, possibly implying unusual salinity levels (either continental or marine). Final filling of fissures and inter-boulder voids by contemporaneous debris (including much fragmented laminite) distinguishes the Blackhalls Rocks conglomerate from that at Hawthorn Quarry in which much void space remained unfilled.

The general character, distribution and depositional environment of the main (upper) part of the biostrome are discussed in the account of the Hawthorn Quarry site, where it is noted that the biostrome is roughly co-extensive with the shelf-edge reef of the Ford Formation from Hawthorn Quarry southwards, but is not known farther north and only in fragmentary (brecciated) form at one exposure a short distance to the east. The main features of interest and importance in these upper strata at Blackhalls Rocks are exceptionally good exposures of large stromatolite domes on the foreshore rock platforms in the southern part of the designated area; although exposure varies greatly with the movement of sand and shingle, a selection of ripple-topped domes is almost always visible at about mid-tide levels. The basal 'Crinkly Bed' of this upper (stromatolitic) part of the biostrome is superbly exposed in the cliffs near the middle of the site (Figure 3.54), where it closely resembles its equivalent in Hawthorn Quarry site and in Hesleden Dene. This thick bed forms the roof of several of the main caves in the central headlands of the site.

The origin of the pisoids in the 'Crinkly Bed' at the base of the laminates is similarly uncertain. Originally described by Trechmann (1913) as 'pea-stone', Smith (1981b) reinterpreted them as oncoids (algal pisoids); provisional current thinking is that whilst derived algal debris may indeed form the core of many of these bodies, their smooth cortices have more in common with those of cave pearls and their travertine equivalents and they may, therefore, have formed at least partly inorganically in splash pools or other agitated areas. Other pisoids, especially those in pockets near the top of the conglomerate, have finely mammilar coats and resemble those at the Gilleylaw Plantation Quarry site from which F.W. Anderson (in Smith, 1958) recognized the algal growth-forms *Bevocastria conglobata* Garwood.

The considerable significance of the biostrome and the assumed underlying erosion surface in terms of the local and regional evolution of the Zechstein Sea is discussed in the account of the Hawthorn Quarry site.

Seaham Residue

This formation, where exposed near the southern boundary of the Blackhalls Rocks site, is somewhat thinner than at its type locality at Seaham, but this may be a result of attenuation on the dipping limb of a fold. Strict similarity between the two sections is not to be expected, but a broadly similar tripartite lithological subdivision is recognizable at both exposures, with ooid grainstone separating two more plastic clayey layers; a major difference lies in the apparent lack of dedolomite at the Blackhalls Rocks exposure, possibly indicating that the dissolved Fordon Evaporites here were halite-rich rather than sulphate-rich. The uncertainty regarding the thickness of the residue at Blackhalls Rocks arises from doubts on the determination of its base; a case could be argued for taking this at the top of the median dolomite unit and a less plausible case could be made for taking it 5–6 m below that chosen here, at the base of the brecciated ooid grainstone.

The relatively steep easterly dip of the Seaham Residue and adjacent strata at the southern end of Blackhalls Rocks is atypical of the local Magnesian Limestone sequence and may indicate that it overlies the crest of the shelf-edge reef of the underlying Ford Formation and its associated belt of differential dissolutional foundering (Figure 3.47).

Future research

The complexity of the different rock types and their mutual relationships at Blackhalls Rocks poses many problems, most of which have been addressed but few wholly solved. In particular, the mode and environment of origin of the laminar coatings of clasts in the boulder conglomerate and of related laminar sheets and void-fill needs to be established, and resolution of the doubt regarding the age of the sparse biota (derived or indigenous?) is crucial. Solution of these problems would throw much light on the local and regional depositional history including the key questions of the age of the biostrome and possible major sea-level fluctuations. Higher in the sequence, the uncertainty regarding the nature of the contact between the biostrome and the Roker Dolomite Formation prevents definition and full understanding of the Cycle EZ1–Cycle EZ2 boundary, and, near the top of the sequence, the base of the Seaham Residue needs to be defined satisfactorily.

Conclusions

This coastal site is the largest exposure of the Hesleden Dene Stromatolite Biostrome, including its distinctive coarse basal conglomerate of clasts of boundstone eroded from the Ford Formation reef. At this site, the algal laminates or stromatolites characteristically exhibit spectacular dome structures and a complexly finely-laminated basal bed. Above the biostrome are the Roker Dolomite Formation in the form of cream-coloured oolitic dolomite, and the Seaham Residue (the insoluble remnant of the Fordon Evaporites). The cliffs and shore platforms afford excellent exposures of the above sequence, and detailed evidence of the large variety of depositional and post-depositional structures can be observed along the shoreline. The site requires further study to understand the sedimentological nature and faunal character of the sequence, as well as to define the stratigraphical relationship of its upper part.

TRIMDON GRANGE QUARRY (NZ 361353)

Highlights

Trimdon Grange Quarry (not shown in Figure 3.2) is a secluded place where backreef to lagoonal limestone of the Ford Formation may be studied. Much of the rock has been severely altered by mineralogical changes linked to the dissolution or replacement of gypsum and/or anhydrite that were formerly abundant in it, but enough has escaped such alteration to show that the rock is mainly a cross-bedded, shallow-water ooid grainstone with traces of burrows and a sparse shelly fauna.

Introduction

This quarry, not to be confused with the Grange Quarry (NZ 362345) near Trimdon, lies about 1 km south-west of Trimdon Grange, County Durham; it is part of a nature reserve owned and managed by the Durham Wildlife Trust, mainly for its botanical interest, but with considerable faunal interest also.

The quarry is about 70 m across and 5–9 m deep, and is cut entirely into gently-dipping, shallow-water, backreef or lagoonal carbonate rocks of the Ford Formation (formerly the Middle Magnesian Limestone). Most of these rocks probably were primary ooidal limestone, but they have undergone extensive mineralogical changes and are now a complex mixture of secondary limestone and subordinate dolomite; there is much evidence of replaced gypsum and/or anhydrite. The formation exposed here may also be seen in a number of other quarries in the area, although many nearby quarries have been filled in recent years and others are under threat; details of strata in all these exposures are available in the field-notes files of the British Geological Survey.

Trimdon Grange Quarry was first described by Smith and Francis (1967, p. 129) and further details were given by Smith (1981d). It is a peaceful place, almost cut off from the outside world, and its fascinating geology and botany are best enjoyed in the summer when its abundant wild strawberries are in season.

Description

The position and outlines of the designated area are shown in Figure 3.60, together with the locations of the main features of geological interest. The main rock exposures are in the north-west and south-west faces, where exposures are up to 7 m high; smaller exposures are in the lower south face and behind undergrowth on both sides of the quarry entrance.

The rocks in the faces of Trimdon Grange Quarry are extremely varied, both vertically and laterally, and almost all display evidence of a complex diagenetic history; they dip gently eastwards, roughly parallel with the quarry floor, and about 9 m of strata are exposed. Strata in the north-west face and in some northern and upper parts of the south-west face are mainly of hard, off-white to pale buff-grey, finely saccharoidal secondary limestone, and most of the rock in the south face is of hard, coarsely saccharoidal, brown and grey-brown secondary limestone; rocks in the lower and southern parts of the south-west face are mainly of fairly soft, cream, finely crystalline dolomite which in places is soft and powdery. Traces of hollow ooids and of pisoids up to 4 mm across suggest that much of the rock was formerly an

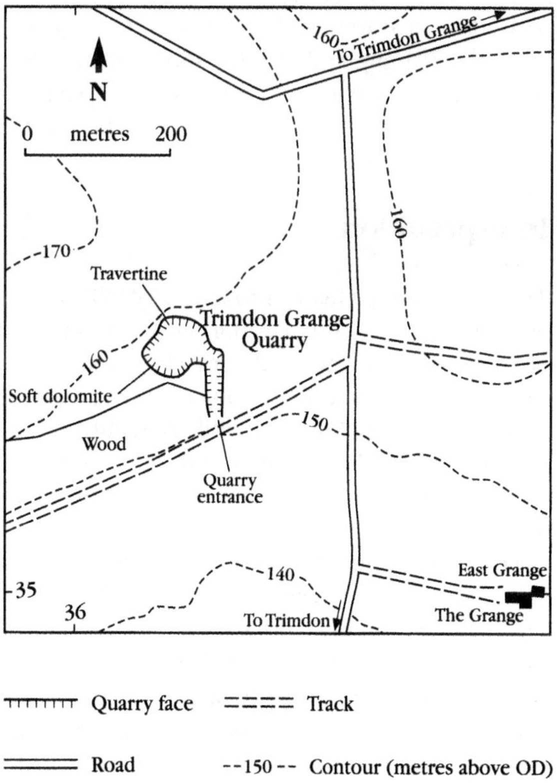

Figure 3.60 Trimdon Grange Quarry and its immediate surroundings, showing the main features of geological interest.

ooidal and pisoidal grainstone, though thin packstones and wackestones were probably also present. Much of the bedding has been blurred or obliterated by diagenetic changes, but enough has survived to suggest a mixture of beds, wedges and gentle lenses generally less than 1 m thick; abundant traces of low-angle, herringbone tabular cross-stratification (in sets up to 0.3 m thick) and hints of minor channels are present locally. Sub-vertical and U-shaped burrows up to 8 mm across are common in some beds and these have preferentially calcitized walls, but possible invertebrate remains are restricted to scattered bivalve-shaped casts.

Secondary features in the rocks of this quarry take the form of scattered, ovoid to irregular calcite-lined cavities up to 0.1 m across, of widespread, 'felted', platy calcite crystals after anhydrite or gypsum and of extremely numerous rectilinear and stellate intercrystalline dissolution voids; judging from the distribution and abundance of these features, the proportion of sulphate may once have ranged from 10 to as much as 90% of the rock. The 'felted' fabric is particularly eye-catching and is commonly associated with a boxwork of thin calcite veins; it was prevalent in cross-stratified ooidal dolomite exposed in a railway cutting (now filled) immediately east of the quarry. Other secondary features include widespread, patchy brick-red staining and a thick, sub-horizontal lens of coarsely-crystalline white calcite (recrystallized travertine?) in the north-east corner of the quarry.

Interpretation

Trimdon Grange Quarry provides a readily accessible and representative section in the backreef or lagoonal strata of the Ford Formation, and, where least altered, is typical of lagoonal ooid grainstones of all ages. Where highly altered, as in much of the quarry, it affords an unrivalled opportunity to study the diagenetic influence of former pervasive secondary calcium sulphate minerals.

Ford Formation strata landward of the shelf-edge reef occupy most of the outcrop of the Magnesian Limestone in County Durham, forming a triangular belt that exceeds 8 km in width in the Trimdon area; they are more than 100 m thick where adjoining the reef, but thin westwards (partly through erosion) and are probably mainly 30–60 m thick around Trimdon. Exposures are uncommon and mainly small in the narrow north of the outcrop, where they are near the reef and generally of shelly ooidal dolomite with scattered patch-reefs (as at Silksworth; see the account of Gilleylaw Plantation Quarry). In the wide outcrop to the south, by contrast, there are many quarries and natural exposures of only sparingly shelly ooidal strata lying several kilometres west of the reef (Smith and Francis, 1967, pp. 123–131) and it is these rocks that are epitomized by the beds exposed at Trimdon Grange Quarry. Ubiquitous low-angle cross-stratification, shallow channels and lenticular bedding all point to free grain movement under agitated (perhaps occasionally intertidal) shallow water, and the scarcity and low diversity of the shelly fauna may indicate slightly enhanced salinity. Lateral salinity gradients are common in many modern tropical lagoons and shallow marine shelves and may account for the westwards-diminishing faunal abundance and diversity in the Ford Formation. The general impression is of a broad, shallow marine shelf that evolved into a lagoon when the shelf-edge reef built up to sea level; near normal salinity probably characterized eastern parts, near the reef, but salinity increased gradually westwards (landwards) where the shelf/lagoon probably shelved imperceptibly into a marginal sabkha plain in which secondary evaporites may have been formed penecontemporaneously.

The predominantly dolomitic character of the rocks in the west face of Trimdon Grange Quarry is typical of the backreef/lagoonal peloid grainstones of the Ford Formation in a broad belt extending westwards to Bishop Middleham (NZ 3331) and northwards to South Hetton (NZ 3845), but most large exposures display some or all of the diagenetic changes noted in the rocks in the remainder of the quarry. In particular the 'felted' fabric is extremely common and dominates some exposures; it results from the volume-for-volume replacement by calcite of sheaths and aggregates of intersecting secondary calcium sulphate crystals (Jones, 1969). In addition, irregular layers and patches of secondary limestone (dedolomite) occur in many exposures, and fractured strata alongside faults are almost invariably of dense crystalline secondary limestone. As an exception, almost the whole of the formation exposed in Witch Hill Quarry (NZ 34439), Old Cassop, has been converted into massive and complex dedolomite. Field and petrographic evidence suggest that the proportion of former secondary sulphate and subsequent dedolomite in the rocks increases westward, perhaps in sympathy with the inferred westward salinity increase in the lagoon; it must be emphasized, though, that there is no evidence of the precipitation of primary evaporites on the lagoon floor.

The dolomitization of the backreef/lagoonal strata of the Ford Formation has been ascribed to refluxing dense brines relatively enriched in magnesium by the precipitation of thick calcium sulphate rocks in the succeeding Edlington Formation (Smith, 1981a, p. 179). Such brines might also have been generated in sabkhas marginal to the Ford Formation lagoon, which have now been eroded off. Harwood (1986) postulated a similar mechanism to account for the equally pervasive dolomitization of equivalent strata in the Yorkshire Province, and Lee and Harwood (1989) invoke a reflux mechanism for the dolomitization of the underlying Raisby Formation in Durham. The calcium sulphate ions that subsequently replaced much of the dolomite presumably were introduced in the brines that effected dolomitization, and much of the dedolomitization was probably accomplished when the anhydrite was hydrated and dissolved during the current cycle of uplift (?Tertiary to present).

Future research

There has been relatively little research into the diagenetic and depositional history of the backreef/lagoonal beds of the Ford Formation and both these aspects deserve attention; Trimdon Grange Quarry would form an excellent starting point.

Conclusions

Trimdon Grange is the only GCR site in which can be observed an accessible section of typical backreef or lagoonal carbonate sediments formed some kilometres west of the shelf-edge reef, the dominant feature of late Permian sedimentation in County Durham. Where unaltered, the oolitic limestones display bedding and channelling characteristic of lagoonal environments. However, in much of the exposure they are highly altered, the original limestone having been replaced by dolomite and this by calcium sulphate. The latter, in turn, was then either dissolved and its place taken by calcite, or replaced by calcite (together with some dolomite), producing a fascinating variety of rock fabrics and textures. The preservation of this site is important to safeguard the section of 'normal' lagoonal sediments, and the profound effects of later diagenetic processes.

RAISBY QUARRIES (NZ 3435)

Highlights

In addition to being the type locality of the Raisby Formation, the whole of which is exposed in the north face, Raisby Quarries (not shown in Figure 3.2) are amongst the largest man-made excavations in northern England. The formation mainly comprises well-bedded sparingly fossiliferous dolomite rock, but here includes an atypically thick and somewhat more fossiliferous limestone in its lower part. The underlying Marl Slate and uppermost part of the Yellow Sands are exposed in the quarry floor, below the Raisby Formation, and the basal beds of the overlying Ford Formation are present at the top of the main face. A varied suite of secondary minerals has been reported from these quarries.

Introduction

The enormous eye-catching north face of Raisby Quarries (formerly known as Raisby Hill Quarry) lies a short distance east of Coxhoe and just north of the Butterknowle Fault. The quarries have been worked for most of this century; they are the type locality of the Raisby Formation (Smith *et al.*, 1986), here perhaps 58 m thick, which is underlain by Marl Slate (0.2–0.9 m) and overlain by basal beds of the Ford Formation (6 m+). Basal Permian (Yellow) Sands of ?early Permian age are present in part of the quarry and Westphalian B Coal Measures have been temporarily exposed and worked in the quarry floor.

Raisby Quarries are well known because the Raisby Formation here contains a thick mass of limestone in place of the more usual dolomite rock (Trechmann, 1914; Woolacott, 1919a, b; Smith and Francis, 1967; Lee and Harwood, 1989; Lee, 1990, 1993). These authors and Jones and Hirst (1972) noted the presence of diagenetic breccias in part of the sequence, and, together with Fowler (1943, 1957), Jones and Hirst (1972) and Hirst and Smith (1974) reported a substantial range of secondary minerals including small amounts of native copper. Rocks from the quarry figure largely in Lee and Harwood's (1989) detailed investigation of the isotopic composition, sedimentology and diagenetic history of the formation.

The quarries are fully operational and permission to enter must be obtained; stout footwear and hard hats are essential and the bases of the higher faces should be avoided.

Description

The position of the site is shown in Figure 3.61, which also shows the location of the main points of geological interest. The principal face is more than 1.5 km long, though only part of this is to be preserved, and is worked in three main (17–29 m) benches with a subordinate basal 4-6 m bench. Beds dip gently north-northeastwards.

The general sequence in Raisby Quarries is given below.

	Thickness (m)
Soil on thin Durham Lower Boulder Clay	0–4
------ unconformity -------	
Ford Formation, at top of highest part of face	0–6
Raisby Formation, forming most of the worked faces	?58
Marl Slate, exposed only in deepest part of quarry	0.2–0.9
Basal Permian (Yellow) Sands, in floor of quarry, forming a gentle WSW-ENE ridge	?0–6.0+
------ unconformity -------	
Coal Measures (Westphalian B), temporarily exposed and worked	8.0+

Basal Permian (Yellow) Sands

Unfossiliferous, almost uncemented, aeolian yellow sand, ranging from parallel-laminated to coarsely trough cross-stratified, forms the lowest unit semi-permanently exposed in the floor of the quarry. It is overlain by a thinner unit of carbonate-cemented, buff-brown, redistributed sandstone in which winnowed *Lingula* are locally present at the top. Both units are medium- to coarse-grained.

Marl Slate

This comprises a basal unit of dark grey, finely-laminated, argillaceous and carbonaceous dolomite mudstone which thins against eminences on the surface of the underlying Basal Permian (Yellow) Sands, and a more uniform upper unit of buff, finely-laminated dolomite mudstone. Both units contain scattered fish scales.

Raisby Formation

Summaries of the Raisby Formation in Raisby Quarries have been given by Woolacott (1919b, p. 167), Smith and Francis (1967, pp. 111-112), Lee and Harwood (1989) and Lee (1990, 1993); they differ in detail on matters of thickness and dolomite/calcite content, but all are agreed that a thick, limestone unit separates a thin, basal dolomite from a thick upper dolomite unit.

The basal dolomite is about 5 m thick and is very finely crystalline; it comprises a brown-buff flaggy sequence overlain by a grey thin-bedded nodular sequence; the contact with the underlying Marl Slate is sharp and planar, without interdigitation.

The median limestone thins westwards (Francis, in Smith and Francis, 1967) and comprises uniformly bedded grey and blue-grey calcite microspar which, according to Lee (in Lee, 1993) is thinly layered in shades of grey in its lower two-thirds, and is of a more uniform dark blue-grey above; the base and top of the limestone is gradational, with limestone nodules in a dolomite matrix. Lee (1990, 1993) and Lee and Harwood (1989) recorded hemispherical concretionary masses averaging 0.1 m across of dolomite-replacive coarsely-crystalline calcite in upper parts of the limestone and Lee and Harwood (1989) investigated the isotopic composition of these masses. Patchy *in situ* brecciation in the transitional nodular beds between the limestone and the overlying dolomite has been recorded by several authors and was investigated in detail by Lee and Harwood (1989).

The thick uppermost unit of the Raisby Formation here is of relatively evenly-bedded cream and buff very finely crystalline dolomite with many calcite-lined irregular cavities (after secondary anhydrite); much of the unit appears to be unfossiliferous. Lee (1990) recorded patchy to pervasive calcite replacement of dolomite in this unit.

Fossils in the Raisby Formation at the type locality are concentrated in the lower beds of the limestone unit, in which a few beds are relatively rich in a varied assemblage of foraminifers, bryozoans, brachiopods, bivalves and ostracods (see list by Pattison in Smith and Francis, 1967, p. 111); amongst forms listed by Pattison and illustrated and further discussed by him in Smith and Francis (1967, plate V1 and p. 181) is an unnamed and previously undescribed cryptostome bryozoan resembling *Penniretopora waltheri*. Some beds low in the sequence bear abundant invertebrate burrows and trails, and complex grazing patterns are present at some levels (Figure 3.62); Trechmann

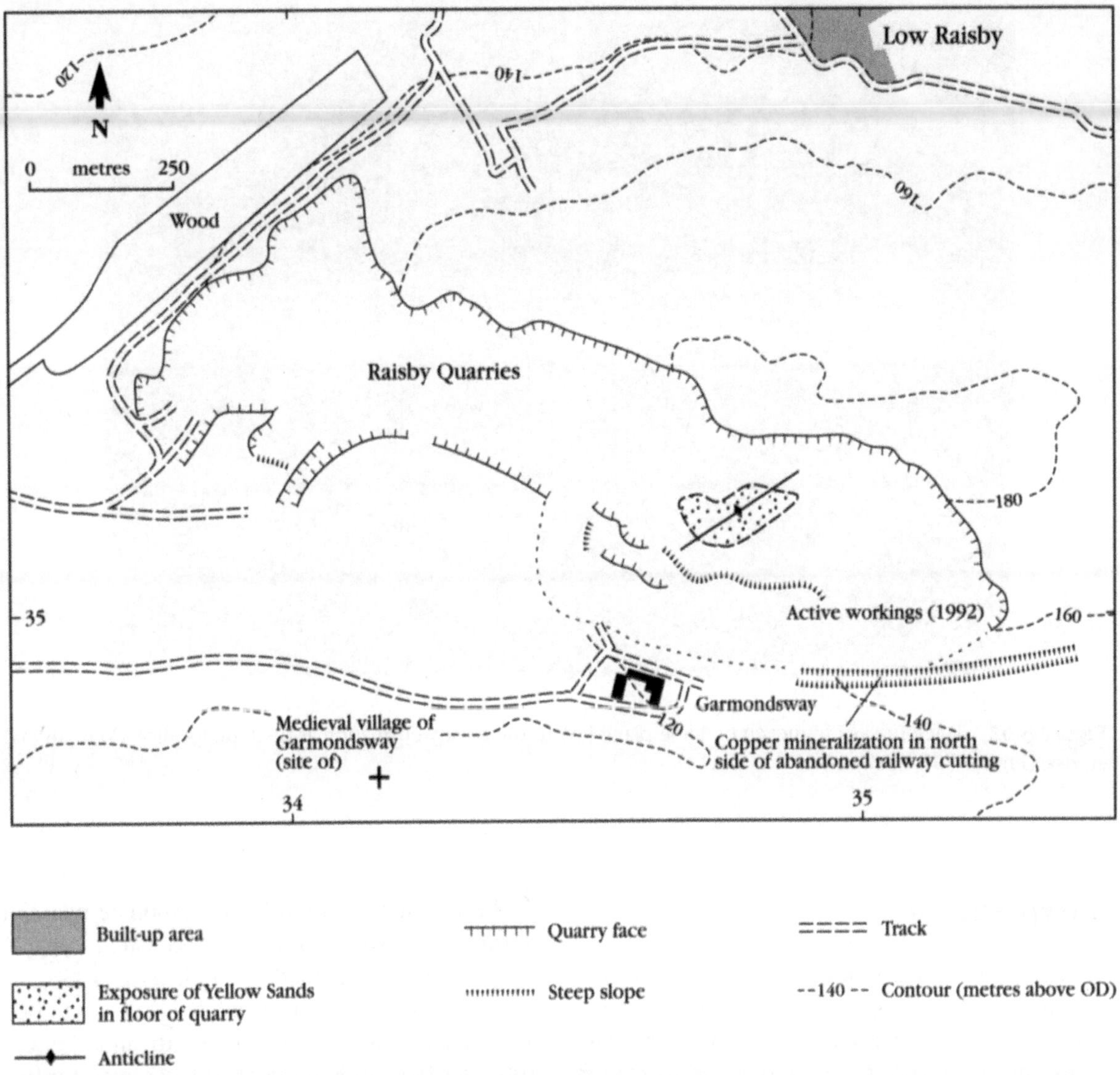

Figure 3.61 Raisby Quarries: the north face in the east is the type locality of the Raisby Formation (Smith *et al.*, 1986).

(1914) also recorded plants from these early Raisby Formation beds. The dolomite in the upper part of the sequence contains only a sparse fauna of poorly-preserved foraminifera and bivalves.

Ford Formation

Rocks of this formation lie at the top of the 26–29 m uppermost bench and are being progressively removed as the face recedes; they comprise up to 6 m of cream and pale-buff, cavernous, ooidal and pisoidal grainstones which, according to Lee (1990), are of calcitized dolomite with leached grain centres. The rocks are evenly bedded with traces of cross-lamination, and the base, though not readily accessible, appears to be conformable.

Secondary Minerals

The quarries have long been known for their suite of secondary minerals which occupy veins, geodes and replacive patches; authors reporting mineralization are listed in the introduction. Minerals recorded from here include azurite, baryte, calcite, chalcocite, fluorite, galena, malachite, selenite, pyrite and sphalerite. Of these, the copper minerals and much of the calcite and some dolomite occupy fractures, and the remainder (including much calcite) occur either in geodes or are replacive.

Figure 3.62 Trace fossils on uneven bedding plane low in the Raisby Formation at the type locality. Coin: 26 mm across. (Photo: T.H. Pettigrew.)

Interpretation

Raisby Quarries are unique in exposing the full thickness of the Raisby Formation, and the section also includes an unusually thick unit of primary limestone. Beds below and above the formation are typical of their respective stratigraphical units.

The thin Basal Permian (Yellow) Sands at Raisby Quarries lie at the north-eastern extremity of the Chilton sand ridge documented by Smith and Francis (1967, fig. 18) and the clear evidence of onlap of the lower part of the Marl Slate against the flanks of the ridge is typical of this formation. Similar relationships, in which the lower (dark grey) carbonaceous part of the Marl Slate is overlapped by the grey, less carbonaceous, upper part which then thins out over the highest ridges have also been seen at Houghton Quarry (NZ 340506), Sherburn Hill Quarry (NZ 345417) and Quarrington Quarry (NZ 327380). The Marl Slate is a principal feature of the fish- and plant-bearing GCR site at Middridge, some 15 km south-west of Raisby Quarries. Lower beds of the Raisby Formation are gently arched over the Yellow Sands ridge, without onlap, but the arch dies out upwards by slight differential thinning of the beds.

The Raisby Formation is the carbonate member of the first sub-cycle of English Zechstein Cycle 1, and was formally defined by Smith *et al.* (1986, pp. 13–14). It is thickest in a roughly north–south belt in eastern County Durham (Smith and Francis, 1967, fig.19) and is thought to have been formed on the outer part of the Zechstein marginal shelf and adjoining slope (Smith, 1989, fig. 6). Evidence of mass downslope sediment movement such as is seen at High Moorsley Quarry and several other localities (Smith, 1970c) has not been recognized at Raisby. Other GCR sites at which the Raisby Formation is present are Claxheugh Rock (NZ 362574) at Sunderland, Trow Point (NZ 3467) and Frenchman's Bay (NZ 389662) at South Shields, and Dawson's Plantation Quarry (NZ 326548) at Penshaw.

Primary Limestone in the Raisby Formation is relatively uncommon and is concentrated in (though not confined to) the lower half of the formation. In addition to Raisby Quarries, it was reported by Woolacott (1919a, b) in the Cotefold ('Cotefield') Close Borehole (NZ 4319 3276) where it was about 30 m thick, and in the Sheraton Borehole (NZ 4338 3466) where it was about 36 m thick; farther south, Trechmann (1914) recorded a

3 m lens of shelly limestone low in the formation at Thickley Quarry (NZ 240257) and Mills and Hull (1976) record patchy limestone at several places in the Middridge area. To the north, Lee (1990, 1993) reported limestone in Penshaw Quarry (NZ 334544), in the nearby Dawson's Plantation Quarry, at Houghton Quarry (NZ 341506) and at High Moorsley Quarry (NZ 334455). In surface exposures the limestone may generally be distinguished by its pale grey colour, but in boreholes it is commonly buff or brown. A primary origin for the limestone at Thickley and Raisby was suspected by Trechmann (1914, p. 259) who remarked that the rocks seemed to have escaped the otherwise ubiquitous dolomitization of the formation, and a similar conclusion was reached by Lee (1990, 1993) who speculated that early cementation of the limestone may have made it less susceptible to penetration by dolomitizing fluids. If this speculation is correct, the spectacularly complex patchiness and near-vertical margins of several of the limestone bodies in the Raisby Formation in the quarries near Middridge, point to equally complex patchiness of the cements and the involvement of yet other factors.

The patchy brecciation of the transitional top of the limestone was ascribed by Woolacott (1919b, p. 166) to post-depositional internal volume changes and by Jones and Hirst (1972) to collapse following the dissolution of interbedded sulphates. Lee and Harwood (1989) deduced that the brecciated rock had undergone a complex diagenetic history and that the brecciation occurred at about the time when abundant replacive anhydrite in the rock was being dissolved; they speculated that this dissolution might have been effected by fluids introduced during Tertiary re-activation of the nearby Butterknowle Fault. Farther east, away from known faults, apparently secondarily brecciated dolomite was reported by Magraw *et al.* (1963) at a comparable stratigraphic level in Elwick No. 1 Borehole (NZ 4531 3117) and Dalton Nook Plantation Borehole (NZ 4811 3144).

The mineralization of the rocks at Raisby Quarries has been considered by a number of authors (see Introduction). There is reasonable agreement that the copper mineralization is related to the proximity of the Butterknowle Fault, but somewhat more diverse views on the mode of emplacement of the remaining minerals. It is generally agreed, however, that secondary anhydrite was once widespread and abundant in these rocks and, on dissolution, was the source of marine sulphate ions involved in the formation of baryte. Jones and Hirst (1972) speculated that the anhydrite, perhaps with a bacterial and organic carbon involvement, was the source of sulphide ions incorporated into galena and sphalerite. Lee and Harwood (1989) envisage a complex sequence of events leading to the precipitation of late-stage anhydrite, baryte and fluorite following the dissolution of early diagenetic anhydrite by meteoric-derived fluids during the Tertiary to present cycle of uplift.

Future research

The geology of this quarry is reasonably well known and understood and there is relatively little immediate scope for further research; its main use is as a superb reference section.

Conclusions

This site is one of the largest quarries in northern England. Its importance lies in that it exposes a complete section of the Raisby Formation, comprising a thick sequence of well-bedded limestone (below) and dolomite, underlain by the Marl Slate and Yellow Sands in the quarry floor. The basal few metres of the overlying Ford Formation are present at the top of the main face. Much of the limestone of the Raisby Formation is original, particularly in the lower part of the formation, but higher beds are mainly of dolomite. This limestone and dolomite contains a well-documented suite of secondary minerals. Fossils are concentrated in the primary limestones in the lower part of the formation. The principal significance of Raisby Quarries is that they provide an excellent type section of the Raisby Formation.

Chapter 4

North-east England (Yorkshire Province)

INTRODUCTION

In this province, lying to the south of Cleveland High, the outcrop of the Magnesian Limestone generally lies closer to the western shoreline than in the Durham Province and the position of the shoreline or of offshore islands is clearly seen at a number of places including Knaresborough and North Deighton (near Wetherby); here a number of prominent rounded hills of resistant Carboniferous sandstone rose above the general plane of the early Permian land surface and were progressively onlapped by Cycle EZ1 shallow-water carbonate rocks and even, exceptionally, by the mixed sediments of the succeeding Edlington Formation. This shows that, unlike Durham, the original basin-filling transgression failed to reach the position of the present outcrop in the Yorkshire Province, which was inundated somewhat later as a result of world sea-level changes or relative subsidence.

Virtually all the outcropping carbonate rocks of Cycle EZ1a in the Yorkshire Province are shallow-water shelf deposits indicative of at least moderate energy deposition, but borehole evidence shows that they grade eastwards over 10–30 km into much-resedimented, finer-grained, slope carbonates indistinguishable from equivalent strata in the Durham Province. This implies that a shallow-water shelf prism similar to that in Yorkshire may have been built up in Durham, but has since been eroded off. In contrast, the carbonate rocks of Cycle EZ1b in the Yorkshire Province differ fundamentally and as yet inexplicably from those in the Durham Province and it is providential that rocks of this age are well exposed in both provinces.

The two provinces also differ in that the shelf and barrier carbonate rocks of Cycle EZ2 in the Yorkshire Province lie too far east to crop out, their place being taken by highly varied but poorly exposed rocks of the Edlington Formation (Figure 1.4). These varied strata extend across the Cleveland High into the southern parts of County Durham and adjoining areas and presumably at one time overlaid lagoonal ooidal grainstone of the Ford Formation in central and northern parts of the Durham Province.

The waning influence of the Cleveland High is well displayed by the Cycle EZ3 carbonate rocks, which are faunally and lithologically similar in the two provinces; the lack of a full GCR site in these rocks in the Yorkshire Province is thus not wholly critical because the rocks are readily available for research at Seaham and elsewhere in the Durham Province.

With the exception of Bilham Quarry and the Ure River Cliff, all the marine Permian GCR sites in the Yorkshire Province are in carbonate units of Cycle EZ1 (Figure 4.1); these are by far the most diverse and well-exposed parts of the Magnesian Limestone here, but the site network concentrates on the spectacular and the ordinary is poorly represented.

Bilham Quarry (near Doncaster) epitomizes the Basal Permian (or Yellow) Sands, traditionally classed as early Permian, but in the Yorkshire Province now reclassified as late Permian because the desert sands are believed to have been reworked completely during the Zechstein transgression (Versey, 1925; Pryor, 1971). Quarries in the Basal Sands are all shallow and are especially vulnerable to waste fill; the preservation of part of Bilham Quarry ensures that at least one face remains accessible for future study, together with the lowest few metres of the Wetherby Member of the Cadeby Formation. The Bilham exposure may be supplemented by the re-exposure of the red pebbly Basal Sands and breccias at Ashfield Brick-clay Pit (Conisbrough) if plans to re-excavate the lower part of this site are implemented.

Most of the GCR sites in Cycle EZ1 strata (the Cadeby Formation) in Yorkshire are in quarries and together they span the whole formation. All are in dolomitized shelf carbonates and include rocks inferred to have been formed in many of the shallow-water environments observed in modern tropical marine carbonate shelves and platforms. Five sites contain patch-reefs in the Wetherby Member and include the classic exposure at Newsome Bridge Quarry (North Deighton, near Wetherby) where an inferred bryozoan–algal patch-reef lies atop an eminence on the Carboniferous–Permian unconformity and is surrounded by shallow-water shelly grainstones. Such grainstones also surround the superbly-exposed algal-stromatolite patch-reef at South Elmsall Quarry and the several atypically tall patch-reefs in the vast working quarry at Cadeby, but are not seen (though probably are present) at the Wood Lee Common (Maltby) site, where saccolithic bryozoan patch-reefs form striking tor-like masses on a grassy slope; elsewhere the relationships of saccolithic bryozoan patch-reefs to surrounding grainstones is especially clear at Ashfield Brick-clay Pit (Conisbrough) and in the many small exposures in the picturesque village of Hooton Pagnell (SE 4808, near Doncaster), itself not a GCR site.

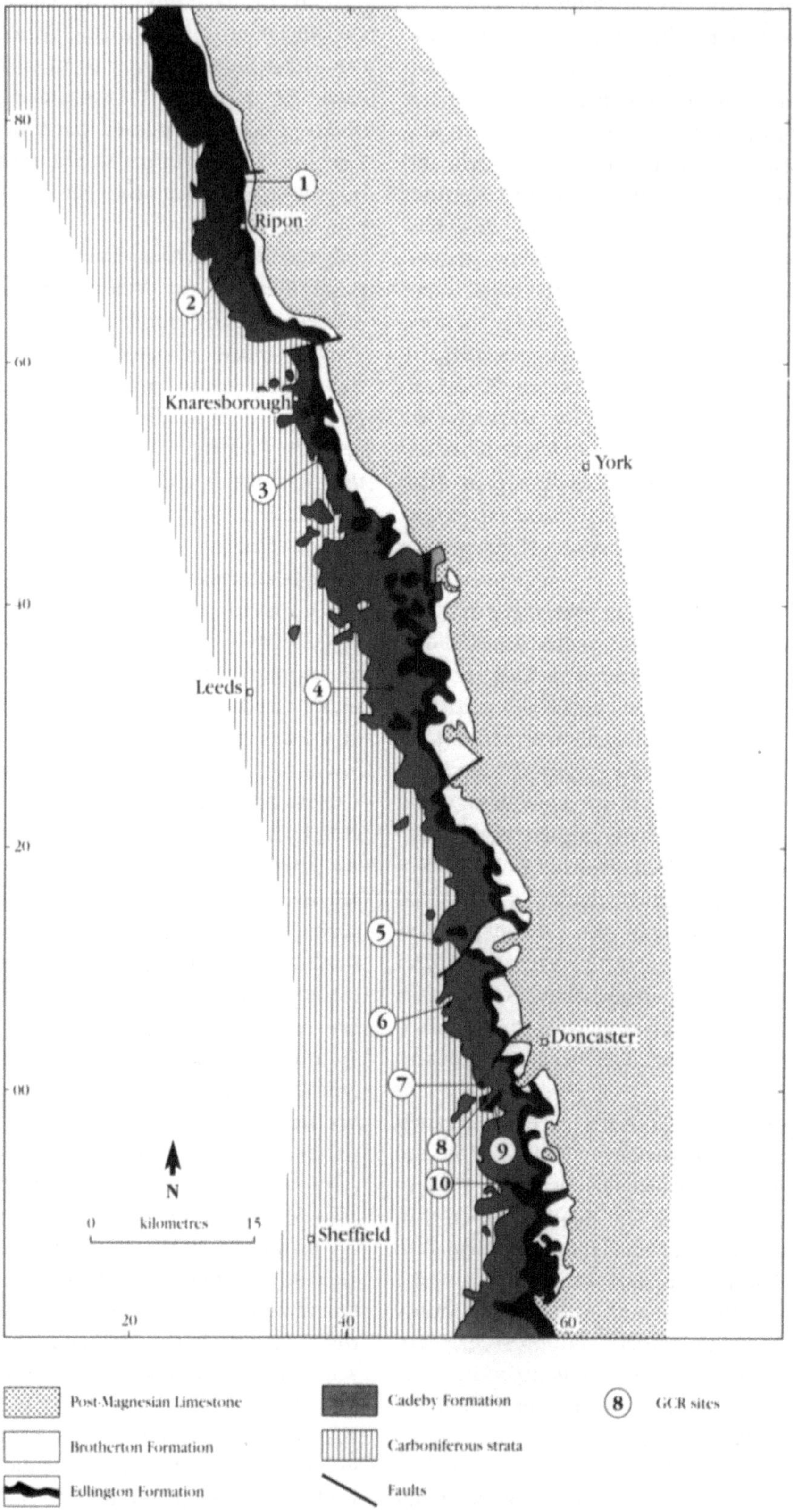

Figure 4.1 The distribution of Permian marine rocks in the Yorkshire Province, showing the location of Permian marine GCR sites: 1, River Ure Cliff; 2, Quarry Moor; 3, Newsome Bridge Quarry; 4, Micklefield Quarry; 5, South Elmsall Quarry; 6, Bilham Quarry; 7, Cadeby Quarry; 8, Ashfield Brick-clay Pit; 9, New Edlington Brick-clay Pits; 10, Wood Lee Common, Maltby.

Cadeby Quarry is the type locality of the Cadeby Formation (Smith *et al.*, 1986) and, in addition to patch-reefs and grainstones of the Wetherby Member, it exposes the typical ooidal grainstone sandwave facies of the Sprotbrough Member and unusually thick Hampole Beds; the latter span the contact between the two members and overlie an intraformational erosion surface, the Hampole Discontinuity, with a relief of up to 3 m. The Hampole Beds and the underlying discontinuity are also seen in Micklefield Quarry, together with the lower part of the sandwave facies of the Sprotbrough Member, but the largest and most impressive exposures of this facies include those at Knaresborough, Jackdaw Crag Quarry (SE 4641, near Tadcaster) and Warmsworth Quarry (SE 5300), which are not GCR sites. The remaining Cadeby Formation site in the Yorkshire Province is at Quarry Moor, Ripon, where ooidal dolomites and dedolomites at the top of the formation are partly algal-laminated and pass upwards by intercalation into interbedded ooidal grainstones and inferred evaporite dissolution residues (possibly of the basal Edlington Formation); a marine high-subtidal to intertidal shelf evolving to a marine sabkha is envisaged for this sequence, which is of a type unique amongst surface exposures in the Yorkshire Province, but has been seen in several cored boreholes east of the outcrop.

There are no thick Cycle EZ2 carbonate rocks at outcrop in the Yorkshire Province, where their approximate equivalent is partly represented by the interbedded siliciclastic and evaporite rocks of the Edlington Formation. These mixed strata, together with thin carbonate units, are exposed in the well-known River Ure Cliff section, near Ripon, where lower parts are only broadly folded, but upper parts are spectacularly tightly folded; the cause of the folding has been a matter of much controversy and remains uncertain, and recrystallization of the evaporites has obscured most primary fabrics. The River Ure Cliff is the only exposure of the Edlington Formation in the GCR site network, the type locality GCR site at New Edlington Clay Pits having recently been filled, and is one of only a very small number of exposures of this formation in the Yorkshire Province.

Cycle EZ3 carbonate rocks, the Brotherton Formation, are not the main interest at any Yorkshire Permian marine GCR site, but a few metres of rocks typical of this formation lie at the southern extremity of the Ure River Cliff site where they have foundered and been partly brecciated through dissolution of the gypsum of the Edlington Formation (Smith, 1974a; Cooper, 1987a). There are, however, many large exposures of these strata in and around Brotherton (SE 4825), Knottingley (SE 4923) and Womersley (SE 5319), although basal beds are generally poorly represented.

The main features of the GCR Marine Permian sites in the Yorkshire Province are summarized in Table 4.1 and their approximate stratigraphical positions are shown in Figure 4.2.

RIVER URE CLIFF, RIPON PARKS (SE 3073 7526–3083 7517)

Highlights

The low river cliff at Ripon Parks (box 1 in Figure 4.2) is one of the few exposures of the Edlington Formation in Yorkshire and is unique in containing thick beds of gypsum. The thickest gypsum lies at the base of the sequence and is only gently folded, but overlying thinner-bedded mudstone, siltstone, dolomite and gypsum is spectacularly tightly folded and fractured; these higher beds also contain many conspicuous veins of white fibrous gypsum.

Introduction

The river cliff lies on the west bank of the River Ure at Ripon Parks, about 3 km north of Ripon; it is partly concealed by vegetation and the base of the cliff is almost constantly washed (and undermined) by the fast-flowing river. Except for a mantle of red-brown boulder clay, all the strata seen in the cliff are thought to be part of the Edlington Formation, here in the floor of the narrow fault-bounded Coxwold–Gilling Trough. Strata in the cliff comprise several metres of gently folded gypsum overlain by a somewhat thinner, but partly strongly folded and faulted, sequence of thin-bedded mudstones, siltstones and dolomites. These upper beds contain many veins, lenses and sheets of secondary fibrous gypsum, and, at the southern end of the face, are separated from the underlying gypsum by a low-angle slip plane. Foundered beds of the Brotherton Formation lie in the bed of the river at the southern end of the main cliff.

The section has been known to geologists for well over a century, having been mentioned by Sedgwick (1829), Tute (1868a, b, 1870, 1884), Cameron (1881) and Fox-Strangways *et al.* (1885) in the last century. More recently the section has been described and illustrated in greater detail by

Table 4.1 Main geological features of the marine Permian GCR sites in the Yorkshire Province of the English Zechstein.

YORKSHIRE PROVINCE		
	Site	**Interest**
Cycle 1 / Cycle 2 Edlington Formation	River Ure Cliff, Ripon	The only permanent surface exposure of Permian evaporites in north-east England; much gypsum after anhydrite, partly strongly internally folded; many satin-spar veins; foundered limestones of Brotherton Formation (Cycle 3) with *Calcinema*
Cycle 1 Cadeby Formation (Sprotbrough Member), transitional to Edlington Formation	Quarry Moor, Ripon	Unevenly interbedded algal-laminated dedolomitized ooid grainstones and evaporite dissolution residues; expansion structures; algal-laminated dolomite ooid grainstones
Sprotbrough Member on Wetherby Member	Micklefield Quarry, New Micklefield	Typical dolomitized ooid grainstones of sandwave facies rests on full sequence of peritidal Hampole Beds; fenestral ('birds' eye') fabric; Hampole Discontinuity
	Cadeby Quarry, Cadeby	Typical dolomitized ooid grainstones of sandwave facies rests on atypically thick Hampole Beds; Hampole Discontinuity with relief of 3 m+; Wetherby Member with unusually tall patch-reefs and thick dolomite domed algal laminites
Wetherby Member	Wood Lee Common, Maltby	Selectively eroded dolomitized bryozoan patch-reefs form tors on grassy slope
	South Elmsall Quarry	Dolomitized bryozoan-algal patch-reef in peloidal and oncoidal shelf grainstones; stromatolite domes
	Ashfield Brick-clay Pit, Conisbrough	Dolomitized bryozoan patch-reef in dolomitized ooid grainstones, on bedded skeletal grainstones and rudstones (coquinas), on dolomitic siliciclastic mudstones
	Newsome Bridge Quarry, North Deighton	Dolomitized inferred patch-reef in peloidal and oncoidal shelf grainstones lies on eminence in Carboniferous - Permian unconformity; rock litter
Wetherby Member on Basal Permian Sands	Bilham Quarry	Basal shelf dolomite mudstones/wackestones of the Cadeby Formation on incoherent marine-redistributed aeolian sand-rock
	Ashfield Brick-clay Pit, Conisbrough	Basal dolomitic siliciclastic mudstones on atypically pebbly red friable sandstone

Kendall and Wroot (1924), Forbes (1958) and James *et al.* (1981). Sedgwick was unsure whether the gypsum at Ripon Parks formed part of the 'lower marl and gypsum' (now the Edlington Formation) or of the 'lower part of the upper red sandstone' (now the Roxby Formation), but Tute (1870, p. 5) favoured the former. Kendall and Wroot (1924) adopted the alternative (later) age and this view was accepted by visiting parties (e.g. Hudson *et al.*, 1938, p. 369) and by Forbes (1958). Smith (1974a, b, 1989), however, reverted to Tute's view on the evidence of increased knowledge of the local rock sequences and this interpretation has been accepted by James *et al.* (1981), Cooper (1986, 1987a, 1988) and Powell *et al.* (1992).

Description

The position of the River Ure Cliff is shown in Figure 4.3. The section is about 500 m long and

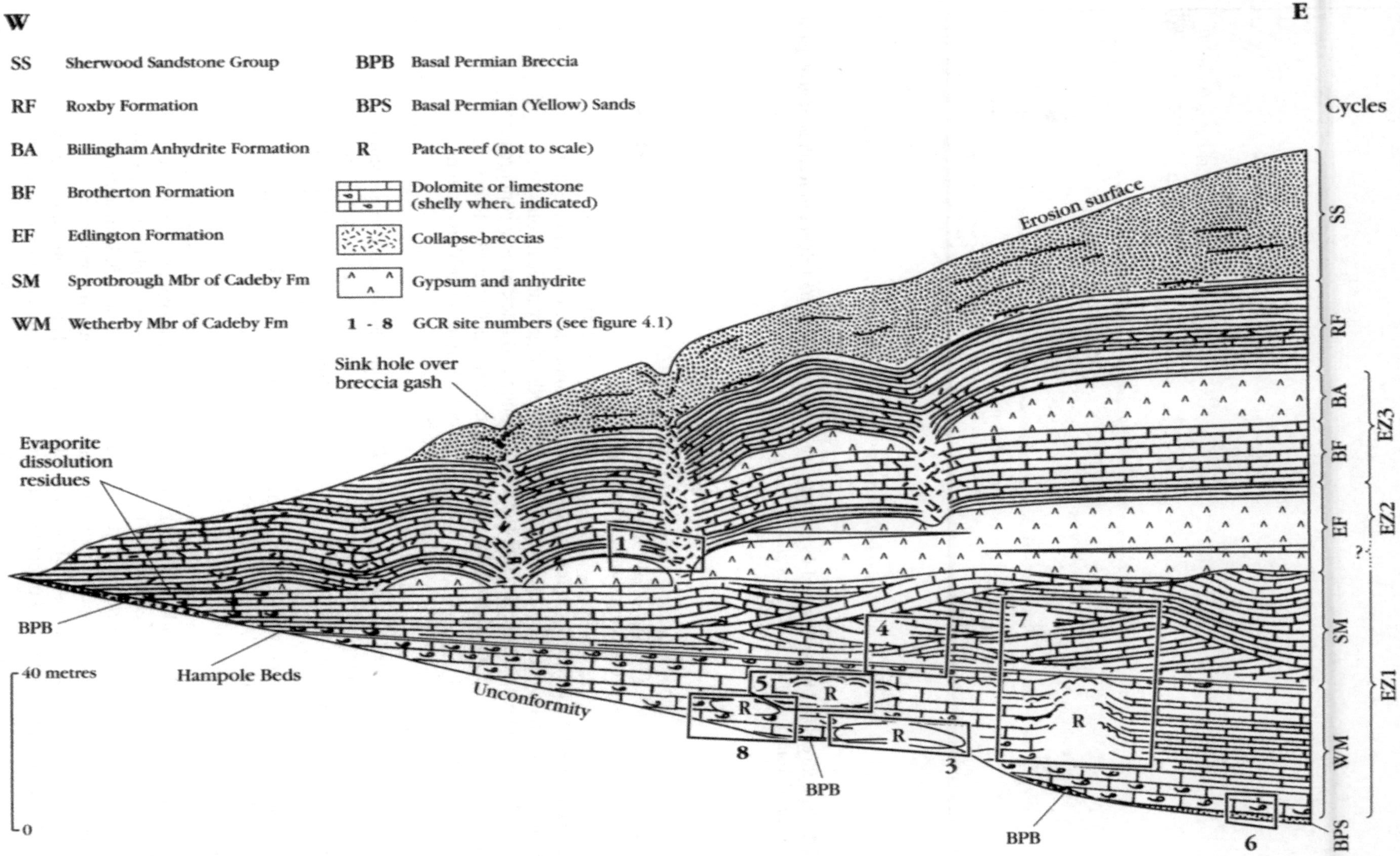

Length of section is approximately 10 kilometres

Figure 4.2 Approximate stratigraphical position of marine Permian GCR sites in the Yorkshire Province of north-east England (diagrammatic). Some sites cannot be shown on this line of section and have been omitted.

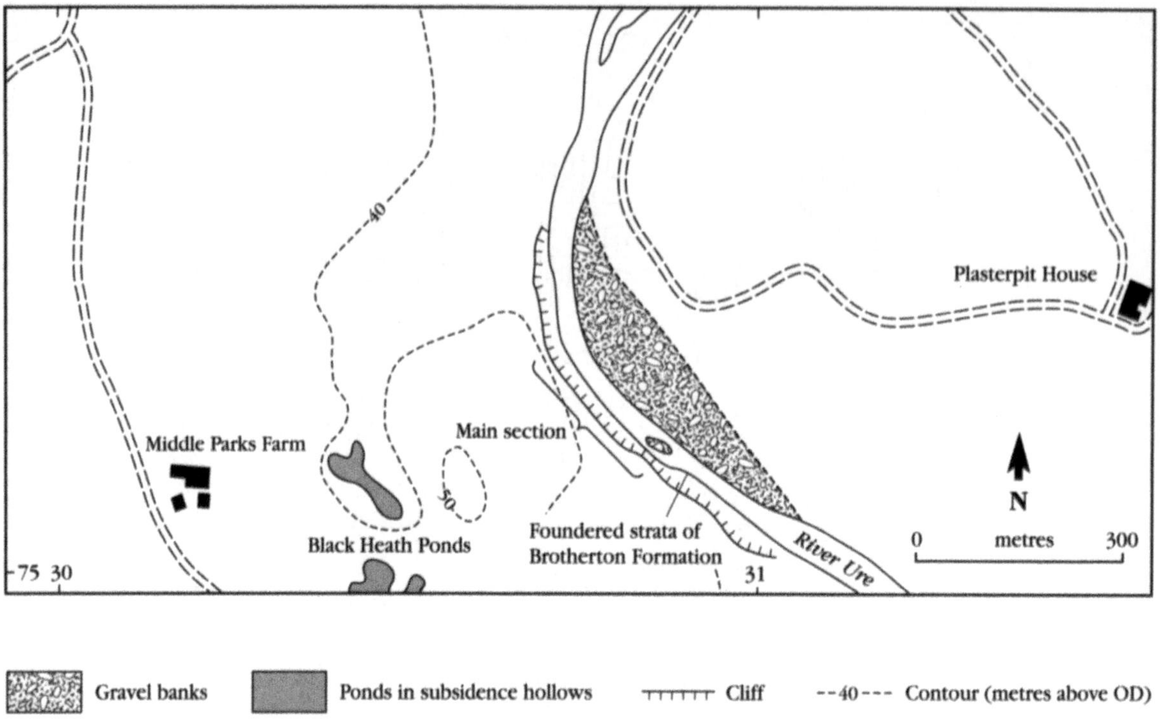

Figure 4.3 The River Ure cliff section and its environs, showing the position of the main features of geological interest. Modified from part of fig. 3 of James *et al.* (1981).

6-9 m high, but the main interest centres on the central 220 m where rock exposures are almost continuous.

The general appearance of the central part of the section is well documented in the literature and, despite indisputable evidence of rapid recession (James *et al.*, 1981), changed relatively little between 1923/24 (Kendall and Wroot, 1924), and 1956/57 (Forbes, 1958, fig.2, reproduced here as Figure 4.4) and 1980 (James *et al.*, 1981, plate 22).

Correlation of the various beds present is somewhat uncertain because of lateral variation and the presence of many folds and minor faults, but Forbes (1958, p. 353) tentatively reconstructed an 8.4 m sequence about 140 m north of the southern end of the main exposure and Cooper and Powell (in Cooper, 1987a, pp. 50-53, and Powell *et al.*, 1992, p. 14) measured a 13.7 m sequence in the central section as a whole; Forbes' measured sequence coincides with the upper part of that given by Cooper and overall agreement is good. The section measured by Cooper and Powell and Powell *et al.* and supplemented by observations by Forbes (1958) and Smith (1974b and unpublished notes) may be summarized thus:

	Thickness (m)
Mudstone, dull red, pink and grey, partly dolomitic, with laminae and thin beds of grey argillaceous dolomite and thick mainly concordant lenses and sheet-veins of white fibrous gypsum; some thin siltstone beds	*c.* 2.1+
Dolomite mudstone, grey to grey-buff, thinly interbedded with subordinate grey to pink mudstone and dolomitic mudstone and with concordant lenses and sheet-veins of white fibrous gypsum	*c.* 1.0
Mudstone, green-grey, grey and dull red, partly silty, blocky to laminated, with thin beds of argillaceous siltstone, silty sandstone, dolomitic mudstone and argillaceous to gypsiferous dolomite mudstone; abundant, mainly concordant lenses and sheet-veins of white fibrous gypsum	*c.* 1.2

Mudstone, grey to pink-grey and grey-pink, partly dolomitic, with subordinate grey argillaceous siltstone and scattered to abundant, concordantly-elongated, grey to pink and orange gypsum nodules and a few mainly concordant lenses and sheet-veins of white fibrous gypsum	*c.* 1.5
Gypsum, mainly grey, alabastrine to coarse grained, evenly to undulatedly thin-bedded to laminated at several levels, with a few thin grey and red mudstone beds and laminae; a few to abundant mainly concordant lenses and sheet-veins of white fibrous gypsum	7.6+

The thick gypsum at the base of this sequence forms most of the southern part of the cliff (Figure 4.5), but mixed strata dominate the remainder. Sedimentary structures in the siliciclastic beds include ripple lamination in some of the thin sandstones and desiccation cracks in some of the mudstones; casts of halite crystals also occur in some of the mudstones (Smith, 1974b). The basal layer of the gypsum is commonly porphyroblastic, especially adjoining gypsum veins and carbonate layers. Petrographic examination by Forbes (1958) showed that relic anhydrite is widespread and locally abundant in the bedded gypsum but there are few clues to the primary sulphate crystal fabric. Forbes identified dolomite as the dominant carbonate mineral in these rocks but noted that there is also a little widespread calcite, and he also recorded small amounts of celestite, some as a vein mineral.

The lenses and sheet-veins of white fibrous gypsum are up to 0.12 m thick and some may be traced for more than 15 m (Forbes, 1958); they form up to 40% of some of the higher beds in the section, but are less abundant below. Shorter 'feeder' veins connect the main sheets and locally contribute to a reticulate network; many veins and sheets are compound, with evidence of several phases of opening, movement and filling. Crystal fibres are sub-vertical in the extensive sheet-veins, and parallel with the axial planes of the folds elsewhere (Forbes, 1958); many are curved, in response to movement of the walls of the veins during crystal growth. Forbes concluded that much of the fibrous gypsum was emplaced after most of the folding and faulting but that some of the thicker fibrous veins and sheets must have been formed before faulting was completed.

Folds at Ripon Parks (Figure 4.6) occur on a wide range of scales and an element of overfolding towards the north is common; they are tightest and most common in the upper half of the sequence, and Forbes (1958) recorded a plane of accommodation (decollement) between the relatively competent gypsum and the less competent overlying beds. Forbes also noted that the axes of the folds in the strongly contorted part of the cliff lie between west to east and north-west to south-east but that those in the more northerly faces trend between north-west to south-east and NNE to SSW. Polished (slickensided) surfaces abound in the contorted sequence, and are a feature of the walls of many of the veins.

Interpretation

The River Ure Cliff is one of the few places in Britain where thick evaporite rocks are preserved in a surface exposure and is also by far the best and most instructive natural section in the Edlington Formation. Although not yet fully understood, the dislocation of strata in the upper part of the section is a superb example of a type of disturbance commonly associated with evaporite rocks that have been deeply buried and subsequently exhumed.

Abundant evidence in the Ripon area points to relatively rapid subsurface dissolution of gypsum in the Edlington Formation (e.g. Tute, 1870; Smith, 1972, 1974b; Cooper, 1986, 1987a, 1988) and estimates by James *et al.* (1981) suggest that surface dissolution of the gypsum has been the main cause of Ure cliff recession averaging about 1 m in every 10–20 years between 1853 and 1956. Much higher rates of dissolution were calculated and observed for detached gypsum blocks. Given these high rates of dissolution, the preservation of the Ure river cliff gypsum is remarkable and can probably best be accounted for by a combination of the unusually great primary thickness (up to 30 m) of the gypsum (and its anhydrite precursor) in the Ripon area, protection by its cover of relatively impermeable mudstone and siltstone, and by only fairly recent exposure to river attack, perhaps as a result of meander migration. Even so, the presence of steeply tilted foundered strata of the Brotherton Formation at river level at the southern end of the GCR site (Smith, 1974b; Cooper, 1987a) shows that dissolution rates have been capable of

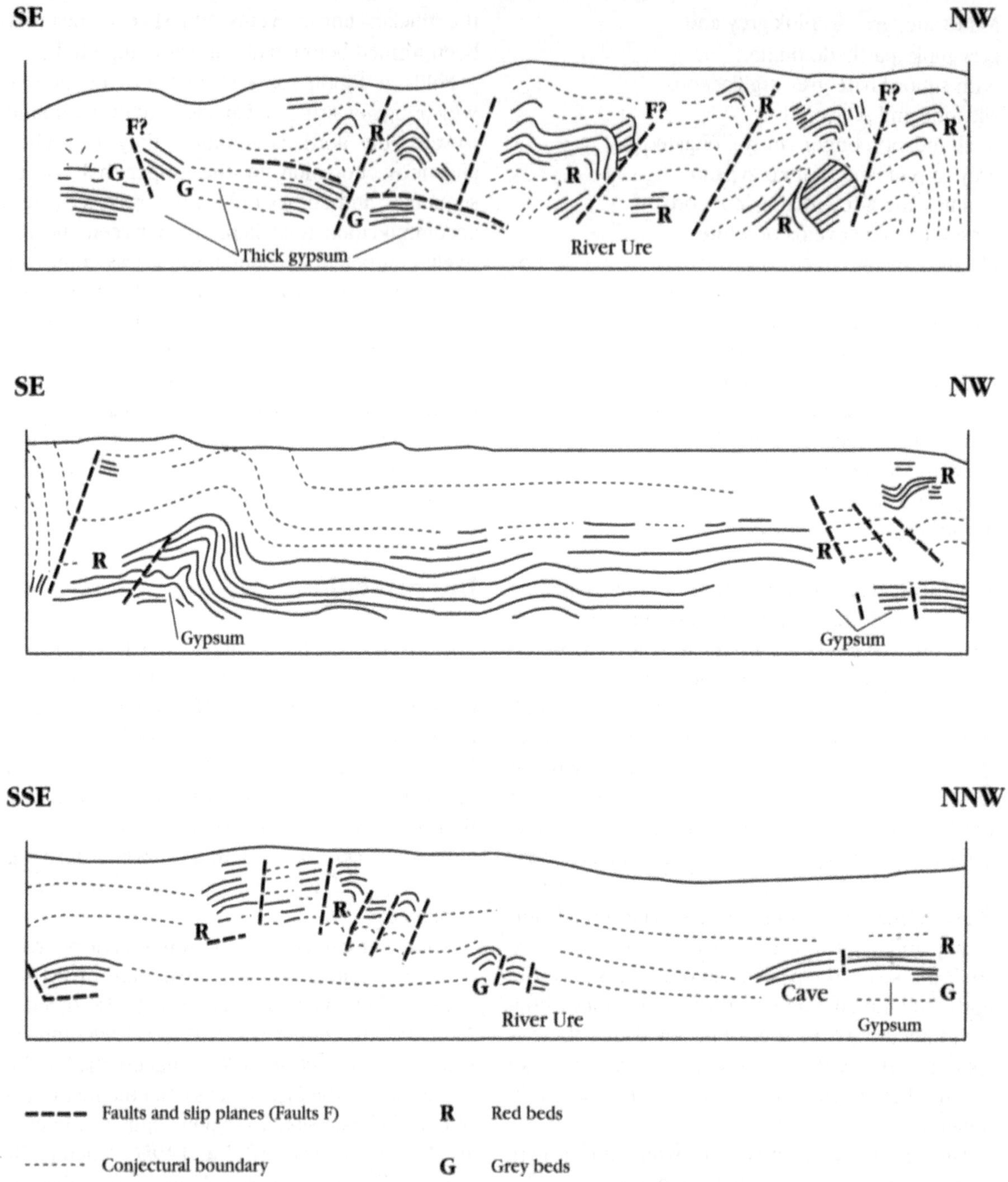

Figure 4.4 Sketch (top left to bottom right) of the main face of the River Ure Cliff, showing the principal geological features; slightly modified from Forbes (1958, fig. 2). Gypsum lies mainly at the base of the cliff except at the southern end. Total length of section shown is about 220 m, height about 7.5 m.

removing most or all of the gypsum there, perhaps indicating that other special factors accounted for the preservation of the gypsum cliff a few metres to the north.

The gypsum of the Ure cliffs is now thought likely to be the hydrated equivalent of anhydrite that is extensive at the base of the Edlington Formation (Smith, 1974a, b; James *et al*., 1981; Cooper, 1987a) and which locally makes up more than half of the formation; this unit is tentatively correlated with the Hayton Anhydrite of English Zechstein Cycle 1b age (Smith, 1974b, 1980a, 1989; James *et al*., 1981), but no direct connection can be demonstrated and it seems more likely to be an approximate age equivalent than part of a continuous rock body. Primary fabrics having been obliterated by hydration, the gypsum of the Ure river cliffs offers few clues to its original depositional environment and, even in the subsurface, the equivalent anhydrite is almost entirely of the mosaic type with a dolomite net; this

Figure 4.5 Gypsum, possibly equivalent to the Hayton Anhydrite, forming most of the river cliff at the southern end of the main rock section at Ripon Parks. Note the abundant sub-concordant sheet-veins of fibrous gypsum (white) in the upper part of the section. (Photo: A.H. Cooper.)

Figure 4.6 Sharp fold in mainly siliciclastic strata of the Edlington Formation with bedded gypsum (?=Hayton Anhydrite) at the base, and with many sub-concordant sheet-veins of fibrous gypsum (white). The fold is still recognizable at the southern end of the middle sector in Forbes' drawing of 1958 (Figure 4.4). The cliff is about 6 m high at this point. (Photo: A.H. Cooper.)

fabric, too, may be secondary and hence may throw little light on the rock's early history. Farther north, however, equivalent sulphate beds in the Edlington Formation on Tees-side were shown by Goodall (1987) to have been formed subaqueously in a stratified, hypersaline lagoon complex subject to oscillating brine levels and shorelines, and Smith (1989, fig. 8) envisages a generally comparable, but more extensive, lagoonal setting for the Ure river cliff sulphates.

The mainly siliciclastic rocks above the gypsum in the river cliff are typical of more marginal deposits, perhaps formed when the area lay near the shoreline and sedimentation was on an extensive, brine-soaked coastal plain or sabkha that was periodically extensively inundated and subaerially exposed as the lagoon expanded and contracted. The dolomite in the upper part of the measured section may be a feather-edge of the Kirkham Abbey Formation, but this supposition, like the correlation of the anhydrite, is difficult to prove and the rock may be of lagoonal rather than marine origin. Thin dolomite beds are widespread in much of the comparable inner shelf/lagoon area of the Edlington Formation (Smith, 1974a, b, 1989, fig. 9). Halite, too, is widespread in the Edlington Formation in a NNW to SSE belt through York and may formerly have extended into the Ripon area but has since been dissolved.

The cause and locus of the dislocation in the higher parts of the section remains controversial, and both deep and shallower settings have their proponents. The latter setting, favoured by James *et al.* (1981), Cooper (1987a) and Powell *et al.* (1992), ascribes the dislocation to pressures created during the hydration of the precursor anhydrite, theoretically involving a 63% increase in volume. Most gypsum beds formed from anhydrite elsewhere are not strongly folded, however, and the most strongly dislocated beds here apparently contained less anhydrite than the relatively undeformed massive beds at the base of the section. Forbes (1958), moreover, commented that his petrographic evidence from the Ure river section tended to show that the anhydrite–gypsum transition took place at depth on a volume for volume basis. Though not venturing a positive opinion on the cause of the folding, Forbes nevertheless remarked on the readiness of gypsum to flow under stress, implying a deep-seated cause. The writer sympathizes with this view, believing that the initial phases of deformation may have taken place at considerable depth (perhaps before hydration) and resulted from plastic flow of the evaporites (including halite) caused by differential loading; in this interpretation the pressure differential could have arisen during the Coxwold–Gilling faulting episode or early phases of evaporite dissolution, and the initial dislocation was probably augmented by foundering related to continuing evaporite dissolution during the current cycle of uplift.

Future research

As one of the few remaining surface exposures of the Edlington Formation, the main value of this spectacular section lies in its general appearance and as an excellent example of what evaporites look like in the field. The petrology of secondary gypsum rocks is now reasonably well understood so that research into this aspect might not be fruitful. The crucial problem of what caused the dislocation of higher strata in the section remains unsolved, however, and awaits satisfactory resolution.

Conclusions

This is the only GCR site in which the Edlington Formation is exposed, and one of the few exposures of this formation in Yorkshire. The site is unique in that the sequence contains gypsum interfolded with associated mudstones and dolomites. The gypsum of the River Ure Cliff is considered to be the hydrated equivalent of precursor anhydrite, and as such, affords little evidence on the original environment of formation. The deformation of the gypsum has resulted from the plastic flow of the evaporites under pressure, probably at depth and perhaps in association with halite (rock salt). The preservation of evaporite rocks is also unusual, because in most parts of England only relic texture, evidence of foundering, and insoluble residues survive as reminders of their former presence.

QUARRY MOOR (SE 308691)

Highlights

Quarry Moor, Ripon, uniquely exposes sea-marginal strata of the uppermost part of the late Permian Cadeby Formation, the Sprotbrough Member, and the transition to the overlying Edlington Formation. The rocks are mainly algal-laminated ooidal dolomites and limestone but higher beds in the section are contorted and include several clayey beds that may be evaporite-dissolution residues; all were formed at or very close to the Permian shoreline and tepee-like expansion structures, not reported elsewhere in British Permian marine strata, occur at some levels.

Introduction

The exposures of late Permian marine strata at Quarry Moor lie on the west side of the Harrogate–Ripon road just south of the City of Ripon. The quarry is now almost filled, but a section preserved along the western face reveals thin boulder clay overlying a gently northwards-dipping sequence of mainly algal-laminated ooidal carbonate rocks that span the transition between the Cadeby Formation (below) and the Edlington Formation. Together the Permian strata are about 9.5 m thick, but a further 3.7 m of underlying dolomite was visible in 1968, and parts of the quarry clearly cut into even lower strata. Neither shelly fossils nor signs of bioturbation have been found in any of the strata now exposed.

The quarry at Quarry Moor has existed for well over a century, and, judging from the distinctive lithology, provided much of the stone from which the earlier walls and buildings of Ripon were constructed. It was first mentioned in the literature by Sedgwick (1829) and later by Tute (1868a), who included a sketch of slightly disturbed strata in one unspecified face; Fox-Strangways recorded several minor faults in sketches of Quarry Moor in his field notebooks (now lodged at the British Geological Survey, Keyworth) and subsequently mentioned it in the second edition of the Harrogate (Sheet 62) Memoir (Fox-Strangways, 1908, p. 13). More recently the section in the west face was described in detail by Smith (1974b, 1976) and was also discussed by Kaldi (1980, pp. 20, 154–155; 1986b) and Harwood (1981, pp. 32–33, 109–111). The lower part of the sequence is interpreted as having been formed under shallow water on a broad tropical marine shelf and the upper part is interpreted as the product of a peritidal to supratidal marine sabkha or coastal plain subject to periodic marine inundation; some of the clayey layers in this upper part are interpreted as the residues of dissolved primary and/or secondary evaporites with perhaps some siliciclastic terrigenous input. The widespread contortion, 'tepee'-like structures and partial brecciation here are regarded as the result of plastic flow, secondary volume changes and contemporaneous lithification.

The preservation of part of the western face at Quarry Moor followed representations to the local authority in 1968 by the Yorkshire Geological Society and the Yorkshire Naturalists' Trust (now the Yorkshire Wildlife Trust); tipping of domestic waste was suspended, landscaping ensued and the site was ultimately (1993) scheduled as an SSSI. An information board provides full geological information for visitors.

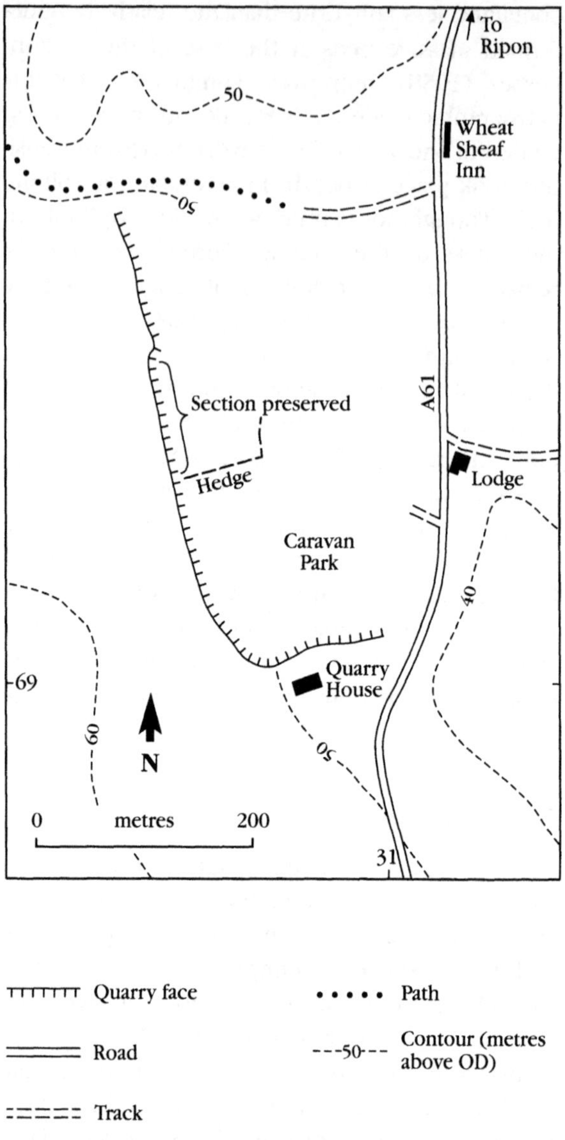

Figure 4.7 Quarry Moor, Ripon, showing the location of the preserved face.

Description

The preserved rock face at Quarry Moor is about 110 m long and mainly 2–3.5 m high; it lies just within the western margin of the site, most of which was scheduled on botanical and entomological grounds. The position of the face and its geological sequence are shown in Figures 4.7 and 4.8 and the general disposition of strata is shown in Figure 4.9. Lower parts of the sequence extend into private property to the south of the preserved face and should not be approached without prior authorization; higher parts of the section are repeated in a heavily overgrown 60 m exposure immediately to the north of the preserved section, where strata dip unevenly southwards.

All the strata in the preserved face at Quarry Moor are provisionally assigned to the Cadeby Formation, but strata above bed 2 were clearly influenced by an intermittent but generally progressive change in the depositional environment and it can be argued that some of the higher beds in the section, especially bed 12, should be assigned to the Edlington Formation. Petrographic details of the Quarry Moor rocks were given by Smith (1976) who showed that most (if not all) of the carbonate rocks there were originally ooidal but that many of the ooids had been diagenetically altered and are now obliterated or scarcely recognizable. Algal (stromatolitic or cyanophytic) lamination is a feature of most of the carbonate beds and thin dense cryptocrystalline layers (?crusts) occur at several levels in the upper part of the section; they also underlie erosion surfaces at the tops of beds 3 and 4. Some beds in the upper part of the sequence feature a dense network of narrow calcite veins. The petrographic evidence leaves little doubt that the limestone beds and patches at

Figure 4.8 The sequence of late Permian strata at Quarry Moor. Most of the section lies in the uppermost part of the Sprotbrough Member of the Cadeby Formation, but some of the higher beds may be part of the overlying Edlington Formation.

Quarry Moor

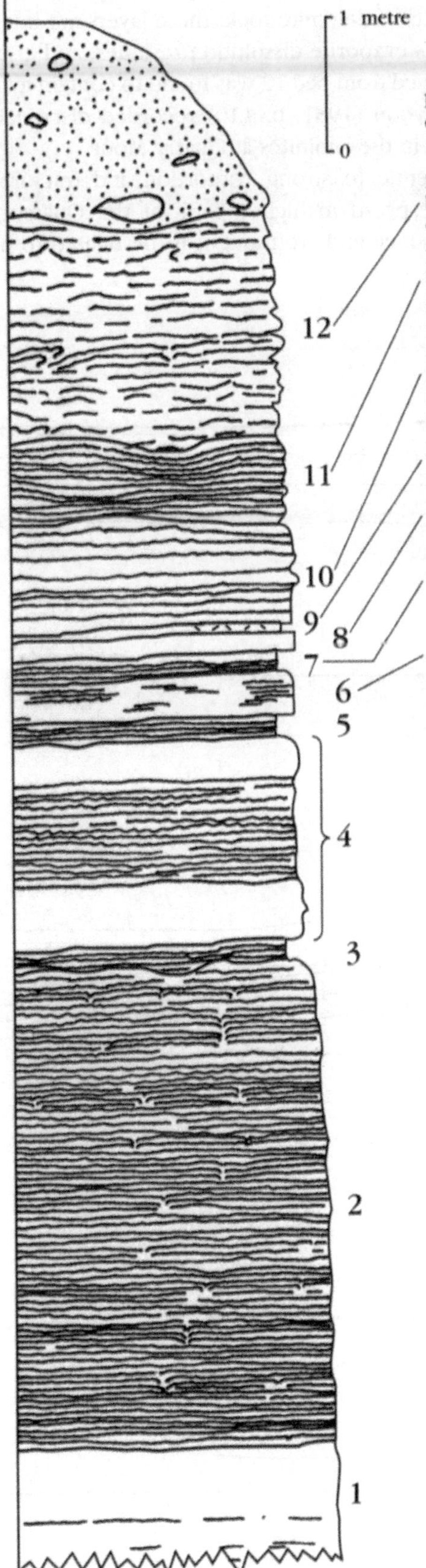

Description	Thickness (m)
Red-brown gritty stony clay (drift), with traces of cryoturbation in places. Largely overgrown	1-2.5
Dolomite, very calcitic and dolomitic limestone, grey and brown, crystalline (sand-grade), interbedded with thin layers of earthy and clayey dolomite. Locally strongly contorted	1.7+
Dolomite, calcitic, brown and grey-brown, irregularly thin-bedded, weakly laminated, with several thin earthy layers. Locally contorted, in places strongly	0.45-0.75
Dolomite, calcitic, pale grey-buff and dolomitic limestone, irregularly thin-bedded, with many traces of poorly-preserved wavy ?algal-stromatolic lamination and of fine hollow ooids. Abundant grey patches are of dedolomitized grey limestone with only vague traces of original ooids	0.90
Dolomite, grey and brown, laminated, earthy, partly passing into fine breccia	0.025-0.075
Limestone, grey, fine-grained, hard, dolomitic, with a sharp chunky fracture	0.15-0.20
Dolomite, grey and brown, thin-bedded, earthy, laminated, with several laminae and thin beds of grey and brown soft clayey dolomite	0.10-0.15
Dolomite, calcitic, and dolomitic limestone, cream, grey and buff, weakly algal-laminated, with ooids preserved in cream dolomitic patches	0.25-0.35
Dolomite, argillaceous, and dolomitic mudstone, grey and buff, semi-plastic when wet, with thin beds of brown mudstone	0.075-0.15
Dolomite, grey, cream and buff, calcitic, ooidal in single 0.30m bed at top and 0.45m bed at base, otherwise irregularly thinly bedded and algal-laminated and with several thin irregular grey, cream and brown earthy argillaceous beds. Parts of the bed are sharply contorted	1.5
Dolomite, yellow-buff and dolomitic limestone, soft, finely and unevenly laminated, finely saccharoidal, generally earthy and with brown clayey laminae and lenses	0.075-0.20
Dolomite, yellow and cream, porous, fairly soft, generally finely and slightly unevenly algal-laminated, ooidal. The surface of the bed is channelled and a layer of grey calcitic concretionary patches lies 0.05m to 0.15m below this surface. Algal lamination is best preserved 0.15m to 0.60m from the top of the bed and at several levels is arranged in flat-topped biscuit-like growth forms up to 0.15m high but generally less than 0.075m high. Ooids generally spherical and hollow, but at some levels include ovoid and fusiform ooids up to 3mm long and a few small pisoids and botryoidal grains	*c.*3.5
Dolomite, generally as in bed 2 but without algal lamination, in beds more than 0.90m thick. Cross-lamination in bed 1.5m from top	3.7+

Note: the lower part of bed 2 and the whole of bed 1 are not visible in the preserved section

Quarry Moor are secondary and it seems likely that most of the calcitization was accomplished by reaction of dolomite with calcium-rich waters late during uplift and erosion. Beds 3, 4, 5, 7 and 12 contain laminae and thin layers rich in siliciclastic quartz and clay minerals but are composed mainly of very fine-grained calcite and dolomite; these are interpreted as normal low-energy coastal plain or lagoonal sediments. In contrast, thin earthy layers in beds 3, 4, 5, 7, 9, 11 and 12 are streaky or irregularly laminated and comprise rubbly aggregates of dolomite rhombs, small quartz crystals and fragments of dolomite rock; these layers are interpreted as evaporite dissolution residues, and a sample X-rayed from bed 12 was found to contain gypsum. Harwood (1981, p. 110) recorded discoidal gypsum in the laminites at Quarry Moor.

Gentle to strong contortion and fractures are widespread in higher parts of the Quarry Moor sequence and are thought to be related to plastic

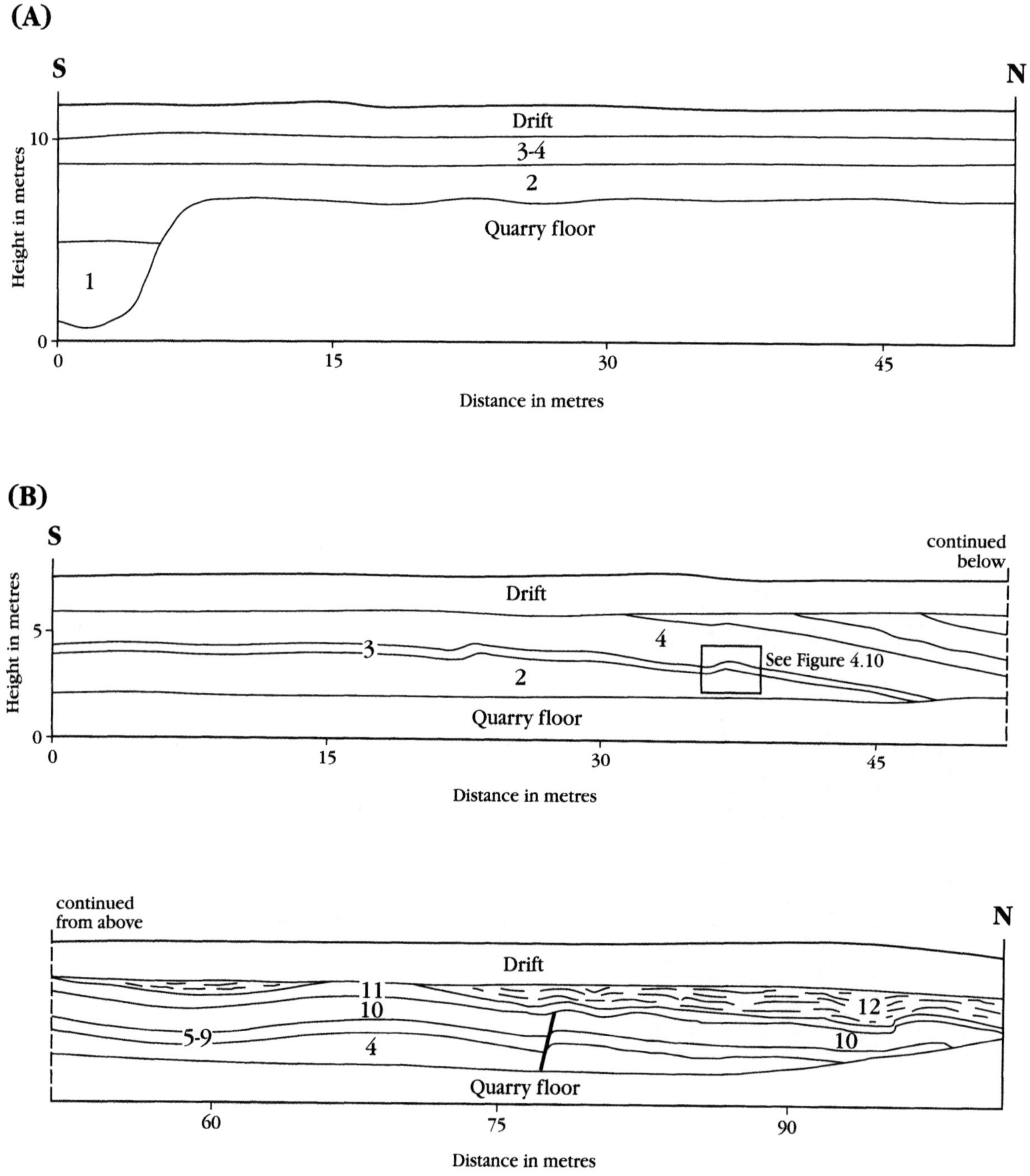

Figure 4.9 Sketches of late Permian strata at the west side of Quarry Moor. (A) In private property immediately south of the preserved face. (B) In the preserved face. Numbers refer to beds depicted in Figure 4.8.

flow, volume changes resulting from the formation, dehydration, hydration and dissolution of evaporites (mainly sulphates but possibly including halite) and, in some examples, to early lithification and expansion (Smith, 1976). The 'tepee'-like structures (Figure 4.10) comprise asymmetrical domes or anticlines about 1–1.5 m across and 0.5 m high in which competent (i.e. already lithified) carbonate beds have been fractured and thrust by lateral expansion; interbedded softer strata have been folded and squeezed-out by this process which, judging from onlap of overlying beds against the sides of the structures, and erosion at the top, must have been contemporaneous.

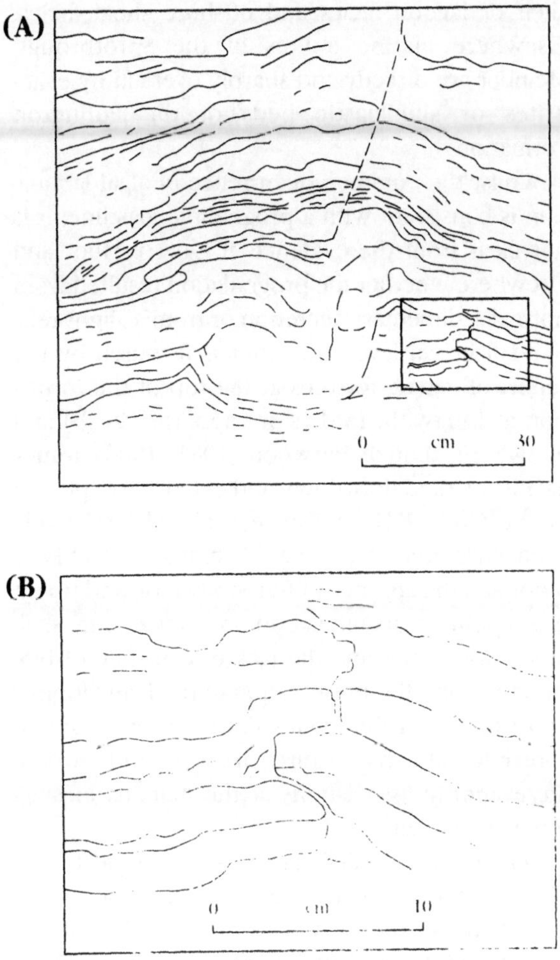

Figure 4.10 (A) Non-tectonic 'tepee'-like anticline in beds 2, 3 and 4 of the sequence at Quarry Moor, with a 0.3 m shortening shown by overthrust near the base and a minor fault near the top. Note the almost level bedding and apparent onlap in the uppermost strata, implying contemporaneous formation and burial. For position see Figure 4.9(B). (B) Detail of ?contemporaneous overthrust at bottom right of (A).

Interpretation

Quarry Moor is the only site where sea-marginal carbonate rocks at the top of the Sprotbrough Member of the Cadeby Formation may be studied in a large surface exposure and is also the only place where such rocks feature evaporite-related contortion and 'tepee'-like expansion structures. The exposed rocks are interpreted (Smith, 1974a, 1976; Kaldi, 1980; Harwood, 1981) as a shallowing-upward regressive sequence, evolving from high subtidal ooid grainstone sheets and shoals (bed 1) to extensive high subtidal algal (stromatolitic) flats (beds 2–4) and finally to a coastal plain (sabkha) environment subject to sporadic flooding and to the formation of primary and secondary evaporites. With a tidal range probably of less than 1 metre, slopes on the coastal plain must have been negligible.

Elsewhere, algal lamination at the top of the Sprotbrough Member has been noted in surface exposures at Wallingwells (SK 570843) (Kaldi, 1980, p. 202; Harwood, 1981, p. 32) and at Darrington (SE 494202 but now almost filled); few details of the Wallingwells exposure are available but Kaldi (1980, fig. 2.2e) illustrates columnar algal stromatolites about 0.10 m in diameter and height, that he describes as capping an ooid barrier shoal. Algal lamination has also been noted at the top of the Sprotbrough Member in a number of cored boreholes, including one a few kilometres north of Ripon where evaporite-bearing basal beds of the Edlington Formation overlie 1.25 m of algal-laminated ooidal dolomite (Smith, 1976); farther away, a similar sequence in the Bank End Bore (SK 706997) contains 2.5 m of algal-laminated sabkha dolomite at this stratigraphical level (Fuzezy, 1970, 1980; Smith, 1976; Kaldi, 1980, fig 2.6c; Harwood, 1981), and algal lamination has also been recorded at this level in the Camblesforth (SE 649358) (Fuzezy 1970), Milford Hagg (SE 533323) (Harwood, 1981), and Wistow Wood (SE 567358) (Harwood, 1981) bores, and in a number of boreholes seen by the writer in the Doncaster and Whitwell areas. The list is almost certainly not exhaustive and it is probable that other comparable sequences in boreholes have not been recorded adequately. The data do not allow full assessment of the overall distribution of Quarry Moor-type sequences but it seems likely that they cap the Sprotbrough Member over perhaps one-tenth to one-fifth of the crop and near-crop area and overlie, in different places, rocks of restricted shelf or lagoon facies and offshore shoal facies. Elsewhere marine oolites of the Sprotbrough Member are directly and sharply overlain by evaporites or siliciclastic beds of the Edlington Formation.

Whilst the common occurrence of algal lamination is consistent with a prograding coastline, it is not clear from the evidence at Quarry Moor and elsewhere whether the progradation resulted from normal sedimentary accretion or from a slight relative fall in sea level; the latter is favoured by the report of karstic features at the top of the formation at Langwith, Derbyshire, (Smith, 1974b) and elsewhere, though Harwood (1981, 1986) argues against marked drawdown and extensive exposure of the Cycle EZ1 carbonate shelf. The relatively even character of the algal lamination at Quarry Moor and the absence of fenestral fabric and rip-up clasts point to accumulation on a low-energy shallow marine shelf and the lack of evidence of bioturbation and the preservation of the lamination is consistent with abnormal salinity and/or an anoxic substrate; all these features are typical of carbonate/evaporite associations at the margins of shallow tropical seas.

Dolomitized bedded algal laminites (bindstones) also occur in the English Zechstein sequence at the top of the Wetherby Member (Cycle EZ1 Ca (a)) and the Seaham (Brotherton) Formations (Cycle EZ3 Ca) but have not been reported at the top of Cycle EZ2 carbonate rocks. Those at the top of the Wetherby Member mainly comprise the lower dolomite of the Hampole Beds (Smith, 1968) and are up to 4 m thick but generally 0.2–0.8 m; they differ from those at Quarry Moor in being widely fenestral, in their common content of small clasts derived from the fracturing of thin contemporaneous crusts, and, at Bramham (SE 4242) and Wetherby (SE 4049), by the presence of small volcano-like structures (see account of Micklefield Quarry). The algal laminites at the top of the Cycle EZ3 carbonate rocks are generally less than 1 m thick and are dolomite mudstones; in County Cleveland and eastern parts of North Yorkshire they are interbedded with nodular anhydrite at the transition to the overlying Billingham Anhydrite (Smith, 1974a).

Future research

Whilst there is general agreement on the overall environmental and diagenetic interpretation of the section at Quarry Moor and in the boreholes cited, much of the petrographic and geochemical detail is virtually unknown. Future research should aim to address these aspects, with the aim of further elucidating the environmental and diagenetic history of the rocks, and, in particular, of determining the role played by probable former evaporites and early lithification. Extra information on the distribution of Quarry Moor-type strata at the top of the Sprotbrough Member must await a thorough search of confidential records and the drilling and careful logging of additional cored boreholes.

Conclusions

The site is unique in that it exposes the transitional relationship of the Sprotbrough Member of the Cadeby Formation and the overlying Edlington Formation. The lower part of the sequence comprises algal-laminated oolitic carbonates and thin clay laminae, which were formed on a shallow marine shelf at a time of increasing salinity. The shelf later shallowed further and environments evolved to intertidal and then coastal plain. In the upper part of the sequence, some clay layers are probably the residues of former evaporites, and widespread contortion, expansion structures and partial brecciation are probably the result of flowage under pressure and secondary volume changes. The site is important because of the unique features displayed, and for the exposure of a rarely seen part of the late Permian sequence in Yorkshire.

NEWSOME BRIDGE QUARRY

Highlights

Newsome Bridge Quarry (box 3 in Figure 4.2) is a superb exposure of a late Permian marine bryozoan-algal inferred patch-reef; it rests in a textbook position on a low eminence on the Carboniferous-Permian unconformity, and is flanked and partly overlain by dolomitic oolites and pisolites.

Introduction

The quarry lies on the south side of West Lane about 100 m SSE of Newsome Bridge and 1.2 km west of North Deighton village, near Wetherby. The floor has been patchily filled with farm waste, but this does not impinge on the main rock faces in the east and south sides of the quarry. Working ceased long ago and the main faces look much the same today as when they were sketched by J. C. Ward for the Harrogate Memoir (Fox-Strangways, 1874, fig. 1; 1908, fig. 1).

The section comprises a lower unit of slightly reddened, coarse-grained Namurian sandstone, the Upper Plompton Grit (4.5 m+), unconformably overlain by dolomite (7.5 m+) of the Wetherby Member of the Cadeby Formation (formerly the Lower Magnesian Limestone). The dolomite mainly comprises a massive, inferred bryozoan-algal reef, but this passes sharply into bedded variably shelly ooidal and pisoidal grainstone at the northern end of the east face; basal parts of the Magnesian Limestone contain scattered fragments of Upper Plompton Grit (Fox-Strangways, 1874, fig. 1), some of which are exceptionally large.

In addition to the sketch by Ward in Fox-Strangways (1874, 1908), the main features of Newsome Bridge Quarry - the uneven nature of the unconformity, the presence of large sandstone clasts in the overlying Magnesian Limestone, the large inferred patch-reef and adjoining bedded dolomite - were mentioned by Smith (1969b, 1974a), who noted that lowest parts of the Magnesian Limestone were absent through onlap against hills on the buried Permian land surface. A more recent account is by Cooper (1987b, p. 30) in which reef-rocks are not recognized and the Magnesian Limestone is regarded as being mainly ooidal.

Description

The position and shape of Newsome Bridge Quarry are shown in Figure 4.11, together with the main features of geological interest. The preserved faces of the quarry have a total length of about 120 m and a maximum height of about 10 m.

The general geological sequence is shown below.

	Thickness (m)
Cadeby Formation, Wetherby Member; patch-reef (inferred) and peloid grainstones	up to 7.5
---unconformity, relief at least 2.5 m--	
Upper Plompton Grit	up to 4.5

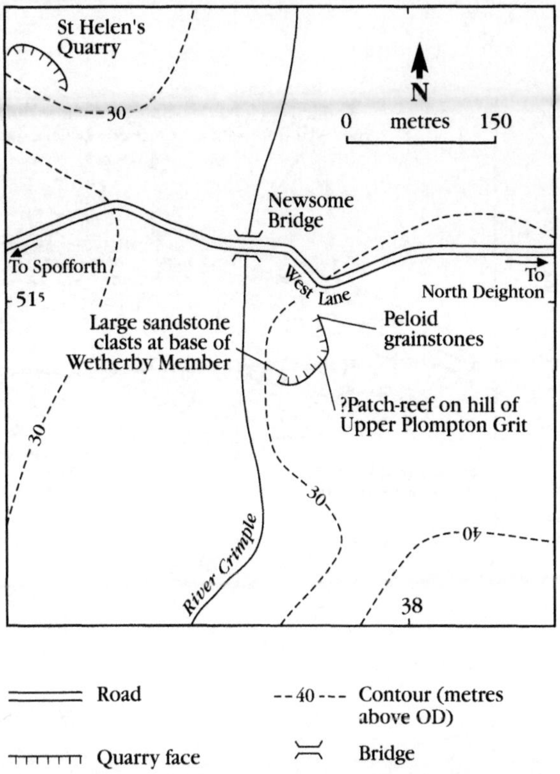

Figure 4.11 Newsome Bridge Quarry and its environs, showing the position of the main features of geological interest.

The disposition of the lithological units exposed in the east and south faces of the quarry is shown in Figure 4.12. All the Permian rocks are of buff and cream dolomite and contain scattered to abundant cavities up to 0.10 m across after secondary anhydrite.

Upper Plompton Grit

The lowest unit exposed in Newsome Bridge Quarry is a coarse-grained sparingly pebbly feldspathic sandstone assigned by Cooper (1987b) to the Upper Plompton Grit (Namurian). It is a resistant, thick-bedded (partly cross-bedded), pale grey to pale yellow-brown rock, with a faint purple discoloration consistent with its position immediately underlying the unconformity. Cooper records a patchy dolomite cement in the uppermost 1-2 m, where the rock is generally paler than below.

The Carboniferous–Permian unconformity

This is a smooth, sharp erosion surface with a total relief in the quarry of more than 2.5 m; it is

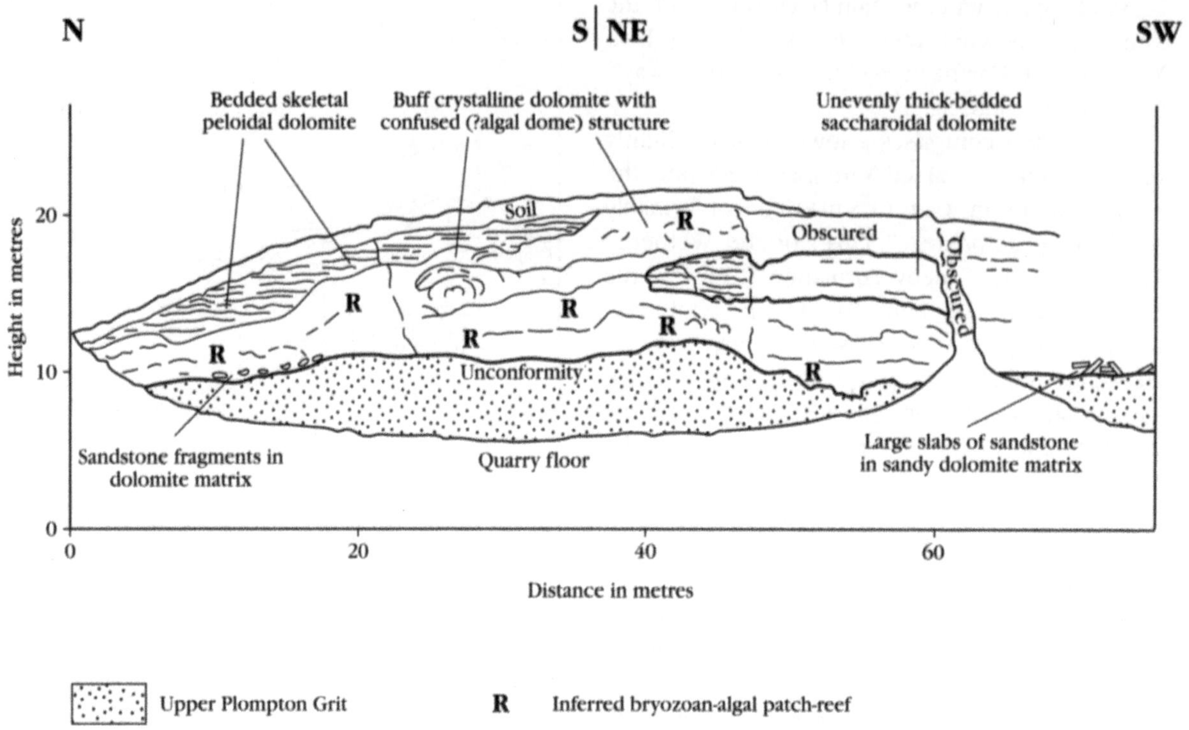

Figure 4.12 Sketch of the main faces of Newsome Bridge Quarry showing an inferred bryozoan–algal patch-reef centred on an eminence in the Carboniferous–Permian unconformity here cut onto Upper Plompton Grit.

highest in the south-eastern corner of the quarry and forms a low hill beneath the middle of the inferred patch-reef. Slopes on the surface are generally low, but in places they increase to 30° and some small near-vertical steps are present. The unconformity is generally clean-swept except where coarse debris lies in the basal 0.5 m of the overlying ?reef, and represents a hiatus of at least 75 million years.

Cadeby Formation, Wetherby Member, patch-reef (inferred) and peloid grainstones

The ?patch-reef at Newsome Bridge Quarry (Figure 4.13) extends across the full width of the preserved faces, but thins northwards towards an inferred margin just north of the quarry; it is largely obscured by vegetation on the south-west side of the quarry, where it includes a thick wedge of crudely bedded dolomite and may grade laterally into non-reefoid dolomite. In the south-east corner of the face the inferred reef forms all of the Magnesian Limestone exposed sequence, but there are hints that originally the reef was probably no more than 8 or 9 m thick.

Marked diagenetic changes have obscured most details of the primary fabric of the reef-rock and its interpretation as reefoid is based mainly on gross structure and overall relationships. In particular, the identification rests on the massive character of much of the rock, the apparent lack of ooids or pisoids, the sharp upper contact in the east face, the suggestion of saccoliths (pillow-shaped masses, Smith, 1981b) at the north end of the east face and, especially, the strong impression of presumed algal-stromatolitic doming in the upper part of the east face. Together these characteristics are so similar to those of the undoubted bryozoan stromatolite patch-reef at South Elmsall Quarry that interpretation as a reef is reasonably firm. Small bivalves (mainly *Bakevellia binneyi*) are present locally in the inferred reef but the distinctive and unmistakable framework of straggling *Acanthocladia* colonies that characterizes most reefs in this member has yet to be identified. J. Pattison (by letter, January 1990), however, records bryozoa at the south end of the quarry where the rock is believed by the writer to be reefoid; the presence of bryozoans strongly favours a reef interpretation, since in most of the shelf facies at outcrop, bryozoa are common only in and near the patch-reefs.

Figure 4.13 East face of the Newsome Bridge Quarry, showing the inferred patch-reef of the Wetherby Member of the Cadeby Formation and equivalent bedded strata, resting unconformably on Upper Carboniferous strata. For scale and interpretation see Figure 4.12. (Photo: Total Oil Company.)

Angular fragments of gritty sandstone are present in the basal part of the ?reef where it overlies low parts of the underlying unconformity (J.C. Ward in Fox-Strangways, 1874), and in places are crudely imbricated or vertical (Smith, 1974a; Cooper, 1987b); they are mainly flaggy and some fragments near the south-western end of the face are almost 1 m long.

The peloid dolomite that abuts against and laps onto the reef in the east face comprises a varied mixture of mainly thin-bedded, soft, skeletal ooidal and pisoidal grainstones and subordinate packstones. As in the ?reef, many fabric details have been obscured by advanced diagenesis, but ooids generally predominate. Most beds, however, contain a poorly sorted mixture of ooids and pisoids, and some beds near the top of the face are composed almost entirely of pisoids and shell remains; the pisoids include lumps, grapestones and coated compound grains, some of which bear evidence of contemporaneous fracturing, erosion and re-cementation. Some of the pisoids may be oncoids but undoubted algal filaments or structures have not been recognized. Shells in the grainstones range from thinly scattered to very abundant and most are only slightly abraded; most are small bivalves (chiefly *Bakevellia binneyi* and *Schizodus obscurus*) and gastropods, but J. Pattison (by letter, January 1990) also noted many reworked bryozoans in peloidal brash in the field above the east face.

Interpretation

Newsome Bridge Quarry is unique in combining superb exposures of both the Carboniferous–Permian unconformity and an inferred bryozoan–algal patch-reef with its enclosing grainstones; it is unmatched in the Permian of England and is rivalled in significance only by the famous quarry exposure at Bartolfelde, Germany (Richter-Bernburg,

1952, fig.1), where the sequence is generally similar but which, in addition, is richly fossiliferous.

Although the Carboniferous-Permian land surface has a low relief where it is cut onto Westphalian rocks in much of Yorkshire, there is more morphological diversity where the erosion surface is composed of resistant Namurian and earlier strata (Smith, 1974a, b). This diversity is particularly clearly seen in the Knaresborough area where local relief of up to 10 m and slopes of 30° are clearly visible; here also Aveline *et al.* (1874) demonstrated marked onlap indicative of a regional palaeorelief of at least 50 m (confirmed during the recent re-survey; see, for example, Summary of Progress of the Geological Survey for 1977, 1978, p. 37 and Cooper, 1987b). The unconformity at Newsome Bridge Quarry is probably more typical of these North Yorkshire areas, and is comparable with that exposed in St Helen's Quarry (SE 376517) nearby and in old quarries (SE 394456) near Collingham. The quarry is atypical, however, in that reddening of the underlying Namurian Sandstone is unusually faint compared with its more normal intensity as displayed, for example, at St Helen's Quarry and in some of the Knaresborough Gorge exposures.

The onlap of the late Permian marine Magnesian Limestone in the Knaresborough area locally resulted in the overlap of both the Cadeby Formation and the Edlington Formation, and Tute (1884) noted that earliest marine Permian strata are missing in Knaresborough Gorge. At Newsome Bridge Quarry, however, perhaps slightly farther from the basin margin, only the lowest few metres (?5–8) of the Wetherby Member sequence appear to be missing, though onlap is convincingly demonstrated both here and at St Helen's Quarry.

It follows therefore, that the inferred patch-reef at Newsome Bridge Quarry, although lying on the unconformity, is stratigraphically some distance above the base of the formation; this position is consistent with the overall inferred composition of the reef, with its massive core and its stromatolitic mantle (see also the account of South Elmsall Quarry). It is speculated that the reef exposed in Newsome Bridge Quarry was nucleated there because of the elevated firm substrate furnished by the Upper Plompton Grit.

As noted in the account of South Elmsall Quarry, patch-reefs in the Wetherby Member of the Cadeby Formation lie in a roughly north–south belt that is a few kilometres wide and extends from north of Harrogate to south-east of Sheffield (Smith, 1981b). The inferred reef at Newsome Bridge Quarry is thus towards the northern end of the known range, only those exposed at Brearton (Smith, 1974a, p. 375) and South Stainley (Cooper, 1987b, pp. 4 and 39) being farther to the north; it is slightly above the average size of about 20 m diameter and may not have projected much more than half a metre above the surrounding sea floor.

For further discussion of the character and biota of late Permian patch-reefs in the Wetherby Member see Smith (1981b) and the accounts (this volume) of the South Elmsall and Wood Lee Common sites.

The dolomitized peloid grainstones surrounding and onlapping the inferred patch-reef at Newsome Bridge Quarry are, like those at South Elmsall, typical of the belt of patch-reefs, though rip-up clasts have not been noted at Newsome Bridge. The rarity of carbonate muds and the large size of some of the sandstone clasts at the base of the sequence point to phases of moderate to high energy, and the general impression is of a broad unevenly shelving shallow tropical sea floored by shelly peloid sands and fine gravels and unevenly dotted with both subaqueous marine patch-reefs and irregular rounded islands, formed by incompletely submerged sandstone hills; the shoreline lay perhaps 1–5 km to the west.

Future research

As with South Elmsall Quarry, the section at Newsome Bridge Quarry derives most of its impact from spatial relationships; for this reason, it is best viewed from a distance and the main (east) face ought not to be scarred by intensive sampling. Nevertheless, it is important to try to establish whether the inferred reef has an *Acanthocladia*-rich framework and is therefore a true patch-reef or if, as thought probable by Cooper (1987b and in conversation, 1990), the reef-like body is an expression of differential diagenesis of the peloid grainstones.

Conclusions

The site is highlighted by the exposure of a probable patch-reef resting on older sandstones marking the Carboniferous–Permian unconformity. The character of the ?patch-reef is considered to be similar to the bryozoan–stromatolite patch-reef at South Elmsall Quarry, but has undergone greater diagenetic change which has obscured much of the original internal detail. Bedded oolitic and pisolitic dolomite lap against the ?reef, and still display their original texture. The Carboniferous-Permian unconformity is an undulating erosion surface, formed on resistant felspathic sandstones of the Upper Plompton Grit of Namurian age. The exposure eloquently illustrates the variability of the nearshore carbonate sedimentation and reef building on the newly-inundated erosion surface near the western margin of the Zechstein Sea.

MICKLEFIELD QUARRY (SE 445325)

Highlights

The abandoned face at Micklefield Quarry (box 4 in Figure 4.2) is the best and most readily accessible exposure of the Hampole Beds, which span the contact between the Wetherby and Sprotbrough members of the Cadeby Formation (Lower Magnesian Limestone), and also affords excellent views of the Hampole Discontinuity and large-scale cross-bedding in the ooidal dolomite of the Sprotbrough Member.

Introduction

Micklefield Quarry lies behind houses near the south end of New Micklefield village and about 100 m west of the Great North Road; most of the floor of the former quarry has been filled with builders' and domestic waste. Strata now exposed comprise the lower part of the Sprotbrough Member (7.5 m+) and the uppermost part of the underlying Wetherby Member (2 m+), but the main features of interest and importance are the Hampole Beds (about 1 m) and the underlying Hampole Discontinuity. The discontinuity is regarded (Smith, 1968) as an erosion surface cut during a minor sea-level fall and the Hampole Beds are thought to be a product of an oscillating tropical shoreline on an extensive carbonate shelf at the edge of the English Zechstein Sea. The distinctive lower dolomite of the Hampole Beds was a much valued building stone and has been extensively used in the construction of the nearby walls and houses.

The quarry was first mentioned in the literature by Edwards *et al.* (1950), who recorded 60 ft (18 m) of 'limestone' (consisting mainly of the Wetherby Member), and summaries of the strata exposed now were given by Smith (1969b); a sketch of part of the quarry face was given by Kaldi (1980, fig. 3.36, 1986a, fig. 3b) who also included (fig. 6b) a photograph of a complex burrow system there.

Preservation of the face at Micklefield Quarry followed representations to the local council in 1969. Since then vital parts of the face have twice been covered and, following protests, twice re-excavated; nevertheless they remain vulnerable to illegal tipping. An information board provides a geological interpretation for visitors.

Description

The position of Micklefield Quarry is shown in Figure 4.14; the preserved face is about 90 m long and up to about 9.5 m high. The Hampole Discontinuity is well exposed for only a short distance towards the south end of the face.

Strata present in Micklefield Quarry belong entirely to the Cadeby Formation (the Cycle EZ1

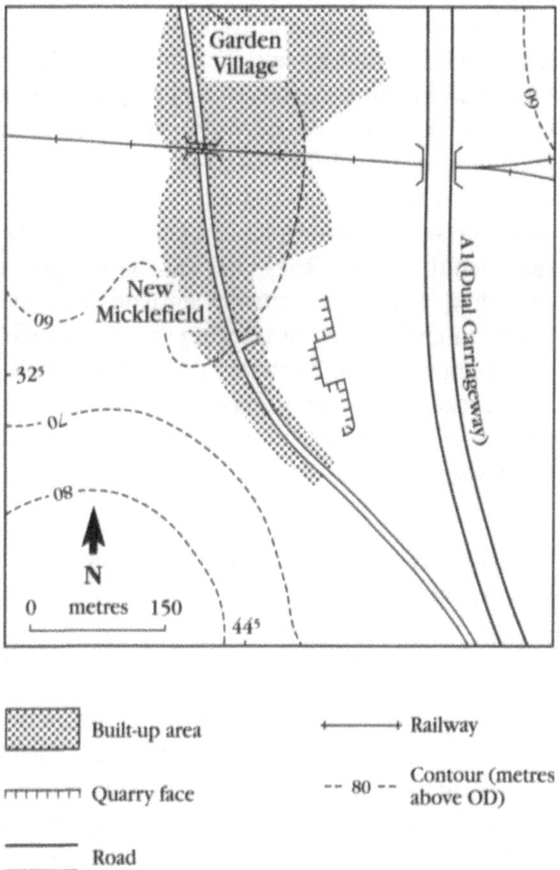

Figure 4.14 Position of Micklefield Quarry GCR site.

carbonate unit of the marginal English Zechstein sequence) and comprise parts of the Wetherby Member (below) and the Sprotbrough Member. The general geological sequence in the quarry is summarized in Figure 4.15, which emphasizes the Hampole Beds that span the contact between the two members. Cavities after secondary anhydrite are present at all levels and many of the ooids have leached centres; some of the cavities are calcite-lined.

The sequence depicted is uniform throughout the quarry and dips gently eastwards. The overall apppearance of the southern part of the face is shown in Figure 4.16.

Interpretation

Micklefield Quarry is the most accessible and best exposure of the Hampole Discontinuity and Hampole Beds in central Yorkshire, supplanting the type locality of Hampole Limeworks Quarry (SE 515097), most of which is now covered. The discontinuity (Figure 4.17) is important in that it is evidence of a phase of erosion and probably subaerial exposure near the end of deposition of the Wetherby Member, and the Hampole Beds are important in that they furnish unambiguous evidence of a phase of peritidal sedimentation between the end of the main phase of Wetherby Member sedimentation and the beginning of the main phase of Sprotbrough Member sedimentation (Smith, 1968). They are also important in providing a readily mappable horizon between the two members and in showing that the depositional surface of the Wetherby member had widely approached contemporary sea level and was thus particularly sensitive to minor sea-level changes.

Wetherby Member below the Hampole Discontinuity

Less than 1 m of this member is now exposed below the Hampole Discontinuity, but up to 3.65 m of strata were seen by the writer before the site was landscaped; they were of typical open-shelf shallow water to intertidal peloid shoal or marginal facies, lithologically similar to equivalent strata at the type locality at the former site of Wetherby Station (SE 397484) and at Hampole Limeworks Quarry (SE 515097).

Hampole Discontinuity (HD in Figure 4.15)

The gently rolling low relief of this crusted surface at Micklefield Quarry is typical of its configuration throughout the province; as elsewhere, it bears only a few minor eminences and hollows (?channels) and only slightly truncates bedding in underlying strata (Figure 4.17). This widespread low relief is known to be exceeded only in Cadeby Quarry and its immediate surroundings, but greater relief may also be inferred in places where the lower dolomite is abnormally thick (e.g. Bramham and Wetherby) and in places such as Kirk Smeaton (SE 5116) where the lower dolomite is underlain by interbedded, multicoloured dolomite and siliciclastic mudstones like those at Cadeby Quarry.

Hampole Beds

Early workers from Sedgwick (1829) onwards recognized that the Cadeby Formation (formerly the Lower Magnesian Limestone or equivalents) could be readily divided into two main units on the basis of lithology and sedimentary characteristics, but the two units were nowhere fully defined and their mutual contact was commonly regarded as diachronous. Mitchell (in Mitchell *et al.*, 1947, p. 122), however, described a distinctive bed about 0.6 m thick at the junction between the two units in the Don Valley near Sprotbrough and this bed was informally defined in his memory as the lower dolomite of the Hampole Beds by Smith (1968); the full normal sequence of the Hampole Beds at Micklefield Quarry (Figure 4.15) differs only slightly from that at the type locality. Within the Hampole Beds, the contact between the two members is taken at the base of the middle mudstone, though Moss (1986) advocated taking the contact at the top of an inferred palaeosol in the middle of the middle mudstone in parts of the Don Valley area where this bed is thicker than its usual 10–30 mm.

Figure 4.15 Section of the Hampole Beds and other strata at Micklefield Quarry. Abbreviations signify parts of the typical Hampole Beds sequence: HD, Hampole Discontinuity; LD, lower dolomite; MM, middle mudstone; UD, upper dolomite; UM, upper mudstone. The lower mudstone is absent. The Wetherby Member-Sprotbrough Member contact is taken at the top of the lower dolomite.

Micklefield Quarry

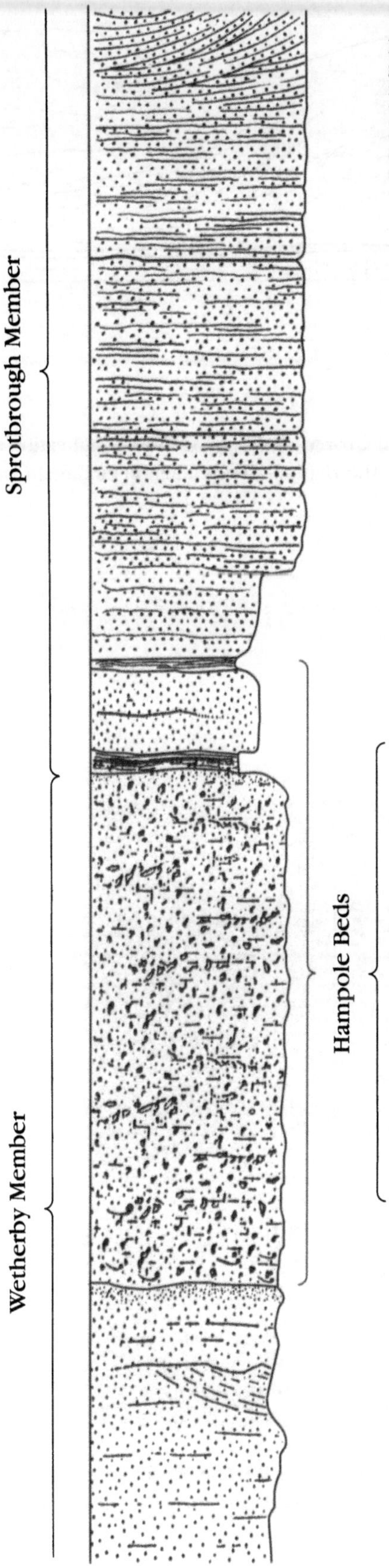

Description	Thickness (m)
Dolomite ooid grainstone, pale cream, fine-grained, well-sorted, fairly evenly bedded in lowest 0.3-1.3m where locally weakly unevenly laminated to trough cross-laminated in sets up to 0.15m thick, increasingly coarsely cross-stratified above in sets up to 5m thick; 1-5mm layer of grey-green plastic argillaceous dolomite or dolomitic clay at base	7.00+
Dolomite ooid grainstone, pale cream, fine-grained, partly weakly unevenly laminated, with roughly concordant minor stylolites	0.25-0.30
Dolomite ooid grainstone, pale cream, fine-grained, thin-bedded to flaggy in lower 0.15-0.18m where soft, unevenly weakly laminated above; some small-scale cross-lamination	0.35-0.40
Dolomite-illite rock, grey-green, with subordinate quartz and chlorite; shaly, plastic when damp (UM)	0.01-0.04
Dolomite ooid grainstone, pale cream, fine-grained, partly faintly finely (?algal) laminated, slightly argillaceous in lower part; a few small pisoids and ?rip-up clasts (UD)	0.10-0.15
Dolomite-illite rock, grey-green, with subordinate quartz and chlorite; shaly, plastic when damp (MM)	0.01-0.03
Dolomite ooid grainstone, pale cream-buff, mainly fine-grained, unevenly (?algal) laminated, with a few traces of small-scale cross-lamination and shallow channels; generally strongly fenestral; dense layers at some levels may be crusts and scattered (partly imbricated) rip-up clasts may be crust fragments; casts of small bivalves and gastropods locally common near base (LD)	0.75-0.90
Erosion surface (HD), gently rounded to slightly uneven, local relief 5-10cm; some minor steep-sided hollows (?channels)	
Dolomite peloid grainstone, pale cream-buff, slightly lenticularly unevenly thick-bedded with some low-angle tabular cross-lamination, cut-and-fill structures and ripple-laminaton; mainly ooidal with some compound grains; traces of fenestral fabric towards top; no fossils seen. Most of this bed now covered but formerly seen to...	3.65+

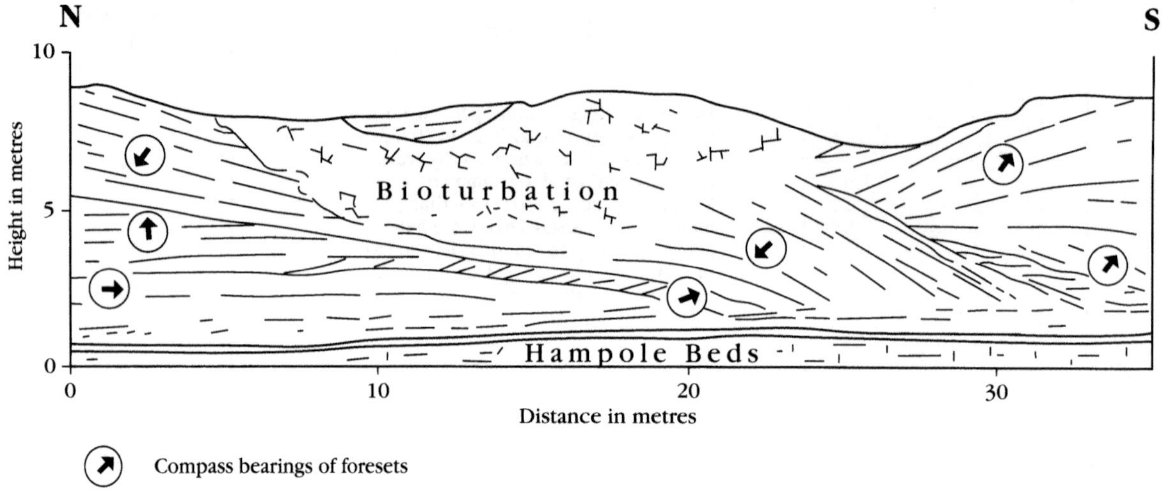

Figure 4.16 Sketch of the southern part of the main face at Micklefield Quarry, showing the cross-stratification in the Sprotbrough Member of the Cadeby Formation and the position of the Hampole Beds. Slightly modified from Kaldi (1980, fig. 3.3).

Figure 4.17 The Hampole Discontinuity (arrowed) and adjoining strata, as seen in 1967 before filling of the lower part of Micklefield Quarry. The white layer is the lower dolomite of the Hampole Beds and the grassy cleft conceals the less resistant upper parts of the Hampole Beds. (Photo: D.B. Smith.)

The most striking feature of the Hampole Beds is the unusual and environmentally significant lithology of the lower dolomite. The pronounced fenestral fabric of this algal-laminated ooidal bindstone is exceptionally clearly seen, and, in the Magnesian Limestone, is confined to this youngest bed of the Wetherby Member and a few thin lenses in grainstones a short distance below the Hampole Discontinuity. The combination of algal lamination, inferred crusts, imbricated rip-up clasts, minor channels and a convincing fenestral fabric suggests with reasonable confidence that this bed can be interpreted as the deposit of a high intertidal to predominantly low supratidal tropical marine-marginal sabkha or coastal flat (Smith, 1968); the presence at Mansfield of amphibian footprints on the surface of the underlying discontinuity (Hickling, 1906) is consistent with this interpretation.

The Hampole Beds have been traced along the depositional strike for more than 150 km from near Ripon to near Nottingham and, for a peritidal shoreline sequence, are extraordinarily uniform. All the component beds are lithologically consistent, the greatest variations being in relative and absolute thicknesses. In these respects the lower dolomite is the most variable, in keeping with its inferred beach or sabkha origin, ranging to more than 2.5 m thick at Wetherby (SE 49945) (but see Harwood, 1981, p. 82 for an alternative view) and Bramham (SE 429422), but generally being 0.2–0.8 m thick; at Wetherby and Bramham, this bed features abundant small contemporaneous volcano-like structures (?gas- or fluid-escape vents) in addition to its other distinguishing features. The lower dolomite is, of course, absent or unrecognizable above and below the tidal range (i.e. west and east of the intertidal belt). A local variant, in which a sequence of bedded dolomite and subordinate siliciclastic mudstones lies between the Hampole Discontinuity and the lower dolomite of the Hampole Beds, is discussed in the account of Cadeby Quarry.

The Sprotbrough Member

Micklefield Quarry furnishes a representative (though relatively small) cross-section through the lower part of the grainstone shoal facies of this member, but displays most of the main distinguishing features; these are a parallel-laminated to thin-bedded basal unit, a coarsely cross-stratified main body, a scarcity of shelly fossils but local abundance of burrows, and an overall makeup of well-graded small (0.10–0.15 m) dolomite ooids (Kaldi, 1980). The member is also exposed high in the face at Cadeby Quarry and in many quarry and natural sections scattered along the outcrop between Ripon (where it is in a different facies) and Nottingham. The best and most spectacular exposures of the grainstone shoal (or sandwave) facies include those at Knaresborough (SE 348571-359559), Jackdaw Crag Quarry, Tadcaster (SE 4641), Warmsworth Quarry (SE 5300), Cresswell Crags (SK 5374) and Pleasley Vale (SK 517649-525651).

The grainstone shoal facies of the Sprotbrough Member is up to 60 m thick and forms most of the outcrop from Knaresborough southwards (Smith, 1989, fig. 7). It is interpreted as marking the site of an offshore field of high-energy subaqueous dunes (Smith, 1968, 1970b), sandwaves (Kaldi, 1980, 1986a) or grainstone barriers (Harwood, 1981, 1989), and separates a shallow protected shelf or lagoon to the west from a deeper-water open marine outer shelf to the east. The sedimentology and diagenesis of these rocks was investigated in detail by Kaldi (1980, 1986a, b) and their mineralization by Harwood (1981, 1986). By analysis of the trends of the prevalent large-scale cross-stratification, Kaldi showed that the sandwaves were constructed mainly by currents from the north-east (i.e. oblique to roughly normal to the inferred contemporary shoreline) with occasional storm currents from the south-east. The change of style from parallel-bedded at the base to coarsely cross-stratified above, points to a rapid deepening of the sea following the inferred peritidal phase of the Hampole Beds (Smith, 1974a, b, 1979; Kaldi, 1980, 1986a; Harwood, 1981, 1989).

Future research

There is scope here for further detailed research on the petrography and sedimentology of the lower dolomite of the Hampole Beds and on the mineralogy and origin of the various argillaceous beds present both here and at other exposures of these strata.

Conclusions

This site provides an excellent exposure of the Hampole Beds, a remarkably persistent and uniform sequence of passage beds spanning the contact between the Wetherby and overlying Sprotbrough Members of the Cadeby Formation, and of the Hampole Discontinuity. This marine carbonate sequence evolved from shallow water to intertidal and coastal plain sediments, before the Zechstein Sea again transgressed westwards and shallow marine deposition was resumed. The site is important in recording these changing phases of deposition, and in providing evidence in the form of the Hampole Discontinuity of a phase of emergence and erosion near the top of the Wetherby Member.

SOUTH ELMSALL QUARRY (SE 483116)

Highlights

South Elmsall Quarry (box 5 in Figure 4.2) is of national importance because it provides an unusually complete and readily accessible section through a typical patch-reef in the Wetherby Member of the Cycle EZ1 Cadeby Formation. The reef has a core of massive bryozoan dolomite and a broader spectacularly domed algal mantle; it passes laterally into well-exposed shallow-water ooidal and pisoidal dolomite of types that typify a wide north to south belt that extends along much of the outcrop in Yorkshire.

Introduction

The quarry lies on the south side of Field Lane, a few hundred metres east of South Elmsall village and was cut into about 15 m of dolomitized peloid grainstones of the Wetherby Member of the Cadeby Formation; the basal unconformity was probably a few metres below the quarry floor. Most of the quarry is now filled, but the main feature of interest, a bryozoan-algal patch-reef in the upper part of the sequence in the north-east corner, has been preserved in a 9 m high vertical face. The reef has the shape of a broad inverted cone surmounted by a complex gentle dome, and is at least 8 m thick; it was described and illustrated by Smith (1981b).

The reef was discovered and brought to the Nature Conservancy Council's attention in 1966, and its conservation involved ownership disputes, complete filling and re-excavation; resolution of these problems was followed by landscaping and enclosure of the site as one of the last acts of the West Yorkshire Metropolitan County Council before it was abolished in 1986. The official opening of the site, now known as the South Elmsall Interpretative and Study Centre, was on 14 February, 1986. An information board provides a geological interpretation for visitors.

Description

The position and shape of the GCR site at South Elmsall Quarry are depicted in Figure 4.18, which also shows the location of the main feature of geological interest. The preserved faces are about 170 m long and up to 9 m high.

The entire quarry was cut into the lower half of the Wetherby Member of the Cadeby Formation (formerly the lower division of the Lower Magnesian Limestone), here composed mainly of a varied mixture of dolomitized, partly skeletal, peloid grainstones. Although well within the belt of abundant patch-reefs (Smith, 1981b, 1989), the only reef exposed when working was ceased is in the north-east corner of the quarry; it is about 105 m across and at least 8 m high (the top is not

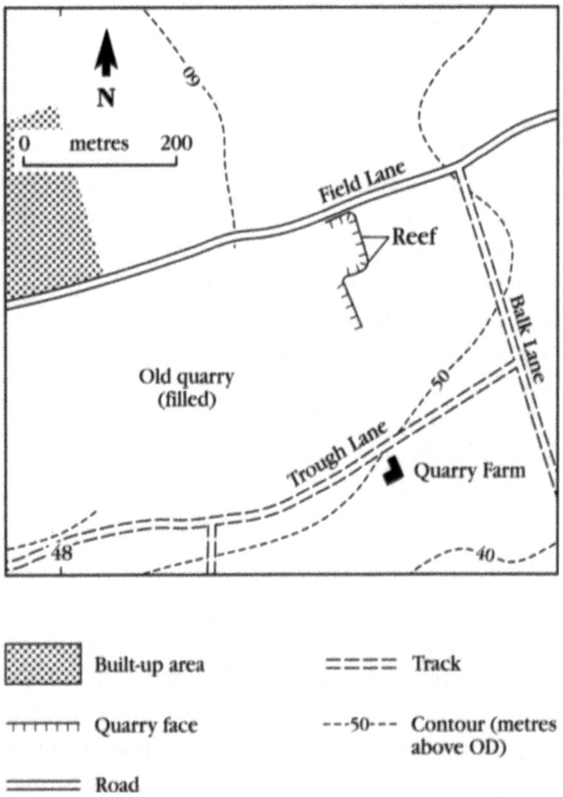

Figure 4.18 South Elmsall Quarry, showing the position of the GCR site.

exposed). This reef and its relationship to enclosing grainstones is shown in Figure 4.19.

As is common at this level in the Wetherby Member, the patch-reef in South Elmsall Quarry is in two main parts (Figure 4.19). The lower part, about 55 m across and up to 3.5 m thick, comprises buff, massive bryozoan boundstone (framestone/bafflestone) formed of a sparse framework of slender arborescent *Acanthocladia* colonies, and a predominant matrix of slightly turbid dolomite microspar and micrite, and, although complex diagenesis has obliterated much of the primary reef fabric, a few bivalve casts (mostly of *Bakevellia binneyi*) may still be found. The upper part of the reef (0–5 m) is composed of complexly domed, buff, laminated saccharoidal dolomites (Figure 4.20), interpreted on their morphology (Smith, 1981b, fig. 13) as algal stromatolitic (cyanophytic) bindstones; here, too, most of the delicate structures have been almost obscured by diagenetic changes and no undoubted algal remains have been detected. The base of the massive part of the reef is sharp and apparently slightly discordant, but the base of the stromatolitic mantle, where it oversteps the massive core, is less sharp, and at the northern margin the stromatolites grade almost imperceptibly into the surrounding grainstones.

Grainstones and subordinate packstones (11 m, including about 2 m now covered) exposed in the north-eastern corner of the site mainly comprise level-bedded, buff and cream-buff, peloidal dolomite. Ooidal rocks form most of the uppermost 6 m of the section and also occur in parts of the lowest 4 m, and some beds feature low-angle cross-stratification in sets up to 0.3 m thick; casts of bivalve and gastropod shells occur at several levels and are scattered abundantly, and most beds contain a few stromatolite flakes, pellets, compound coated grains and other pisoids. Pisoids up to 8 mm across are abundant, however, in a 0.9 m bed 6–7 m below the top of the section, and are accompanied by reworked flaky clasts of ooidal and pisoidal grainstone exceptionally up to 0.1 m across, but generally less than 0.03 m. Such clasts were first noted at this quarry by Mitchell *et al.* (1947, p. 122), who referred to them as 'pebbles'. The pisoids may be oncoidal (algal) in origin, but no algal filaments have been recognized. Leaching has removed the cores of many peloids and parts of the grainstone sequence also contain cavities up to 0.1 m across after leached secondary anhydrite. The biota of the grainstones in the Wetherby Member at South Elmsall Quarry has not been investigated in detail, but Mitchell *et al.* (1947, pp. 118 and 121) recorded *Bakevellia antiqua* (*binneyi*), *Liebea squamosa*, *Pleurophorus*, *Permophorus costatus*, *Schizodus truncatus* (= *S. obscurus*) and several species of small

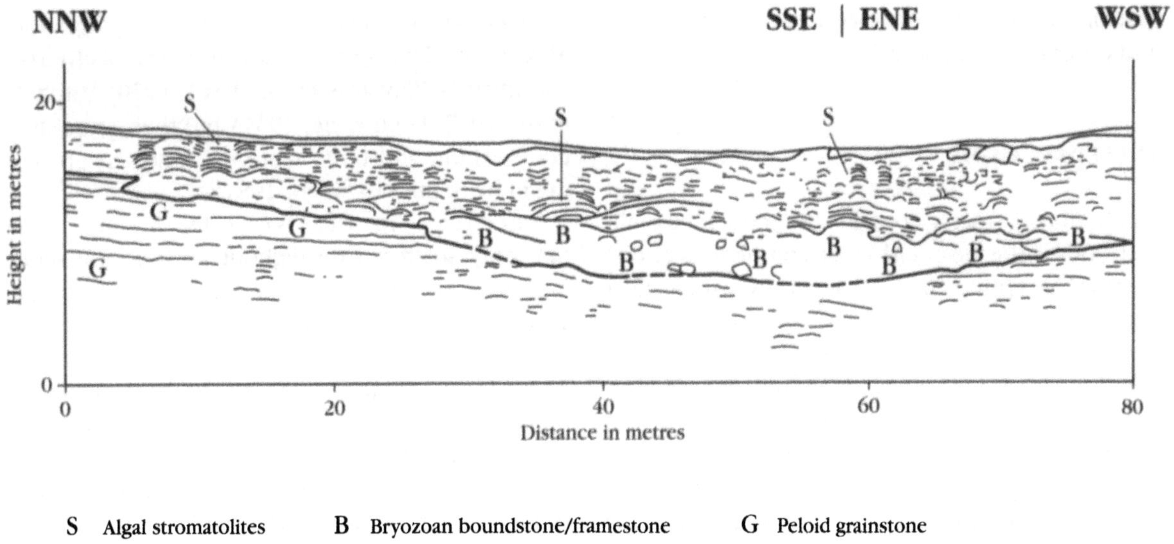

Figure 4.19 Cross-section of an algal-stromatolite reef in peloid grainstones of the Wetherby Member of the Cadeby Formation, South Elmsall Quarry. The core of the reef is of massive bryozoan boundstone and is overlain by more extensive stromatolites that pass laterally NNW into sparingly skeletal peloid grainstone. The stromatolites extend at least 30 m to the right of the area depicted. Note: lowest strata depicted are now covered. Slightly modified from Smith (1981b, fig. 12).

Figure 4.20 Complexly domed dolomitized algal-stromatolites discordantly overlying thin-bedded dolomite peloid grainstones. Central part of east face of South Elmsall Quarry. Bar: 1 m. (Photo: D.B. Smith.)

gastropod from the north-west part of the quarry and other exposures nearby.

Interpretation

South Elmsall Quarry contains the most accessible and one of the best and most complete sections through a typical late Permian bryozoan-algal patch-reef in the Magnesian Limestone of the Yorkshire Province. Its complexly domed stromatolitic mantle is instantly impressive (Figure 4.20), and the relationships of the reef to enclosing peloidal grainstones, and the nature of the latter, is particularly clear.

Although the presence of unbedded masses of dolomite in the Cadeby Formation was mentioned by several authors, including Kirkby (1861) who recognized that they contained a sessile fauna that had probably grown *in situ*, these were first described as reefs by Mitchell (1932a). These reefs were subsequently documented briefly in a series of Geological Survey memoirs covering the Magnesian Limestone outcrop from Wetherby southwards (Edwards *et al.*, 1940, 1950; Mitchell *et al.*, 1947; Eden *et al.*, 1957) and their structure, composition and biota were discussed in greater detail by Smith (1974a, b; 1981b).

Patch-reefs in the Wetherby Member are unevenly scattered throughout an 8-12 km wide belt that generally follows the present outcrop and extends from near Harrogate to Barlborough (SK 4777), south-east of Sheffield (Smith, 1989, fig. 6). Most of these appear to have projected no more than 2 m above the surrounding sea floor. The reefs lie at all levels in the Wetherby Member between the top of the widespread Bakevellia Bed (which may have provided a stable substrate) and the Hampole Discontinuity, and range from simple hemispherical bodies less than 1 m across and 0.4 m thick to complex bodies more than 100 m across and up to 30 m thick; most are 10 to 25 m across and 3 to 8 m thick, and many of the largest bodies were formed by the coalescence of a number of smaller reefs. In places such as Hooton Pagnell village (SE 4808), patch-reefs make up at least half of the Wetherby Member, and more than 20 reefs have been partly to wholly quarried away during the excavation of the 1 km^2 site at Cadeby Quarry; elsewhere, as at South Elmsall Quarry, reefs are relatively uncommon.

The character and shape of the reefs varies according to the stratigraphical level at which they occur within the member (Smith, 1974b, 1981b), and all the main types are represented in one or more of the reef GCR sites in the Yorkshire Province. Those formed near the base of the formation, as exemplified by the reefs of the Wood Lee Common site, Maltby, comprise an untidy aggregate of bryozoan saccoliths, and those near the top of the member, such as the youngest of those at Cadeby Quarry, are mainly of domed algal stromatolites. Those at stratigraphically intermediate levels, such as the reefs at the South Elmsall and Newsome Bridge sites, have a core of bryozoan saccoliths and a mantle of algal stromatolites. It is possible, of course, that some apparently wholly stromatolitic reefs near the top of the member may be founded on saccolithic cores outside the plane of section. Other stromatolite-mantled reefs were formerly exposed in a road cutting at Collingham (SE 398460), near Wetherby, and in Alverley Grange Quarry (SK 554992) near Doncaster, and are poorly exposed behind houses in the village of Bramham (SE 435428); those at Cadeby Quarry differ in some respects from the reef at South Elmsall and these differences are described in the relevant account.

The open-shelf patch-reefs in the Wetherby Member of the Cadeby Formation are all older than those in the lagoonal beds of the Ford Formation of the Durham Province and differ from them greatly in their structure and biota (see the account of Gilleylaw Plantation Quarry in Chapter 3); in particular, the reefs in the Wetherby Member (1) have a much less diverse range of frame-builders and other indigenous organisms than those in the Durham Province, (2) contain virtually none of the lamellar encrustations that characterize much of their Durham counterparts, (3) are not associated with contemporaneous talus and (4) many have evolved into stromatolite bodies that have no parallel in Durham. No patch-reefs like those in the Yorkshire Province have been reported from the contemporaneous Raisby Formation of the Durham Province, but the Durham rocks belong mainly to a deeper-water facies found east of the reef belt in Yorkshire, and reefs could have lain west of the present Durham outcrop, but since been removed by erosion.

The dolomitized grainstones surrounding the South Elmsall reef are typical of much of the outcropping Wetherby Member wherever patch-reefs are present and are also well-exposed in several neighbouring quarries. The generally good grading and the cross-lamination of the ooidal rocks, and the comparative rarity of carbonate muds, point to accumulation and winnowing under at least moderately agitated conditions though large bedforms are uncommon, and the local abundance of compound grains and rip-up clasts suggest phases of sea-floor cementation and perhaps of relative quiescence punctuated by occasional storms. The general impression is of a tropical open-shelf sea no more than a few metres deep and widely dotted with generally small patch-reefs; there is no firm evidence of subaerial exposure either of the reefs or surrounding grainstones. The grainstone floor clearly supported an abundant, but low-diversity bivalve-gastropod biota, but the growth of small bush-like bryozoan colonies led to the creation of a more varied suite of reefy subenvironments in which a rather more diverse and different fauna flourished.

Future research

The main strength of the exposures of reef and surrounding strata at South Elmsall Quarry lies in their visual impact and their scale and mutual relationships; because of the profound alteration of much of the reef-rock, the exposure is probably better suited to a study of the advanced diagenesis rather than to research into reef fabric and ecology. The main face must remain, however, one best viewed from a distance and there is much to be said for a 'no hammering' policy and the preservation of the impressively photogenic faces unscarred by heavy sampling.

Conclusions

This GCR site is of national importance in that it contains one of the best sections through a patch-reef and its surrounding shallow-water carbonate rocks. The sequence is within the Wetherby Member which forms the lower part of the Cadeby Formation in Yorkshire. The morphology of the patch-reef is uniquely displayed as an inverted shallow cone with a gently dome-shaped top; the lower part contains a bryozoan framework and the dome-shaped top is composed of algal stromatolites. The surrounding oolites contain a restricted suite of fossils, chiefly bivalves and gastropods. The site illustrates the structure and spatial relationships of the reef to the surrounding sediments, and, for this reason, South Elmsall Quarry has been preserved for further study and research.

BILHAM QUARRY (SE 487066)

Highlights

Bilham Quarry is important as a representative section of the basal few metres of the Permian sequence in the central part of the Yorkshire Province. The sequence comprises most of the Basal Permian ('Yellow') Sand and lowest beds of the Cadeby Formation of the Magnesian Limestone.

Introduction

The section at Bilham Quarry (box 6 in Figure 4.2) lies amongst trees on the east side of Bilham Lane, about 1100 m south of Hooton Pagnell village church and 300 m north-east of Bilham House; it is all that remains of a large shallow quarry that has otherwise been filled and landscaped. Exposed strata comprise bedded dolomite of the Wetherby Member of the Cadeby Formation (*c.* 3.45 m+) on Basal Permian ('Yellow') Sands (*c.* 3.3 m+).

There are no published records of the section now preserved at Bilham Quarry, but Mitchell (1932a, plate 8) illustrated a similar sequence 'at Bilham House' (about 300 m to the south-west) and Mitchell *et al.* (1947, p. 117) recorded (possibly repeated) this section, then specified as 800 yards (730 m), north-east of Bilham House and at a site shown on the six-inch Geological Survey map as a small sand pit.

The face at Bilham Quarry was preserved and landscaped by the South Yorkshire County Council specifically to facilitate geology teaching and research.

Description

The position of the face at Bilham Quarry is shown in Figure 4.21; it is about 60 m long and up to 6.8 m high. The face is commonly partly obscured by vegetation and soil-wash and may be difficult to locate in high summer. The general geological sequence in Bilham Quarry is shown below.

	Thickness (m)
Cadeby Formation, Wetherby Member (shelf facies)	up to 3.45
Basal Permian ('Yellow') Sands, base not seen	up to 3.3

The strata are roughly horizontal and there are no faults; the Carboniferous–Permian Unconformity is not now exposed but was revealed temporarily during landscaping and lies an estimated 1–1.5 m below the preserved section.

Basal Permian Sands

The Basal Permian (or 'Yellow') Sands in the preserved face at Bilham Quarry comprise yellow-brown and yellow-grey, weakly-cemented to

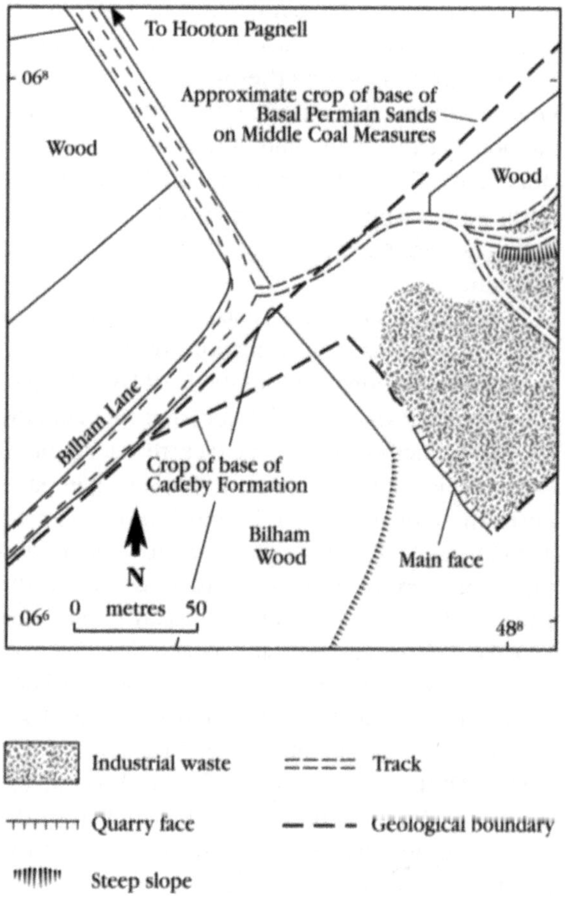

Figure 4.21 The preserved area of Bilham Quarry and its environs.

almost uncemented, medium-grained sand in generally ill-defined horizontally bedded units; a pale grey-brown bed about 1.35–1.55 m below the top has a dolomite cement and may be a sandy dolomite, and the lowest 1.8 m of the exposed sequence features rhythmic secondary banding similar to that produced by Liesegang rings. Grains in the sand are mainly of quartz and most are subangular; many beds contain a few coarse rounded to well-rounded grains, some frosted, and green and red coarse grains are scattered sparingly throughout.

The Cadeby Formation

The Wetherby Member of the Cadeby Formation at Bilham Quarry mainly comprises flaggy and thin-bedded (beds 0.04–0.20 m thick) buff dolomite wackestones and mudstones with several laminae and thin beds of soft dolomitic clay or clayey dolomite in the uppermost 2.65 m; the lowest 0.8 m of dolomite present is superficially similar to the overlying beds but may be an altered ooid grainstone or packestone. Scattered casts of bivalves (mainly species of *Bakevellia* and *Schizodus*) occur at some levels throughout the section, but the well-known and extensive Bakevellia Bed, though present in parts of the former quarry, is absent at this southern extremity. The base of the formation is flat and basal beds are only sparingly sandy.

Interpretation

Bilham Quarry is an exposure of the Basal Permian (or 'Yellow') Sands, a formation that is generally poorly and only temporarily exposed; though small, the exposure exhibits most of the features typical of this formation at outcrop in Yorkshire. Although most Permian sandstones in Britain are continental in origin and of early Permian age, the Basal Permian Sands in Yorkshire qualify for inclusion in the Marine Permian Review because they are considered (Versey, 1925; Pryor, 1971) to have been completely redistributed during or soon after the late Permian Zechstein transgression.

The Basal Permian Sands crop-out patchily along the escarpment and this patchiness is also apparent from borehole and shaft provings farther east (see Versey, 1925, and the Geological Survey memoirs for Sheets 70, 78, 87 and 100); at outcrop the sands are generally only a few metres thick, but reach a reported 8.2 m at crop at Laughton en le Morthen (SK 5288) (Eden *et al.*, 1957). The general distribution of the formation, and its geographical relationship to associated basal breccias, was summarized by Smith (1989, fig. 3); it should be noted, though, that the breccia and sand facies are not mutually exclusive and that a thin breccia not uncommonly underlies the sands (Versey, 1925; Smith, 1974b).

The petrography and sedimentology of the Basal Permian Sands at outcrop in Yorkshire were investigated by Versey (1925) and Pryor (1971), and were summarized by Smith (1974b). The deposit is mainly of uncemented to weakly-cemented, fine- to medium-grained, yellow subarkose, which is grey and blue-grey in the subsurface farther east. It owes its outcrop colour mainly to grain-surface pellicles of hydrated iron oxides and, though authigenic kaolinite is relatively abundant, the main cement (where present) is calcite. Grains are mainly of quartz, but up to 10% of potash feldspars are ubiquitous and up to 20% of rock fragments are widespread; they are predominantly rounded to subrounded in the area investigated by Pryor (from Glass Houghton northwards), with less than 10% of the coarse well-rounded grains for which the formation is noted. Level, thick bedding predominates in most exposures, including Bilham Quarry, but is commonly masked by coarse rhythmic Liesegang-type colour banding. The dominant sedimentary structure in the Leeds area is said by Pryor (1971, p. 244) to be tangential trough cross-lamination in sets typically 0.10–0.65 m thick. A suite of abraded heavy minerals dominated by garnet, tourmaline and zircon was thought by Versey (1925, p. 209) to have survived long transport or be multicyclic.

The Basal Permian Sands in Yorkshire were once widely worked for moulding sand, and in a number of places were followed underground by bord and pillar workings in faces up to 3 m high. In a number of places the roofs of such workings have proved to be unstable, leading to severe subsidence problems. All the underground workings are now closed and most of the quarries are filled; there are no present workings and the few remaining exposures are in road and rail cuttings, which are commonly bricked-over, and in small quarries that have so far escaped filling.

Future research

The Basal Permian Sands in Yorkshire, including those at Bilham Quarry, are now indifferently exposed and no longer readily susceptible to regional petrographic and sedimentological investigation. There is, however, scope for further investigation of these aspects in borehole cores from farther east, and for refining knowledge of the distribution of the formation as new borehole data become available.

Conclusions

Bilham Quarry is one of only two listed GCR sites in the Yorkshire Province that contain a section in the basal Permian deposits and the lowest beds of the overlying Cadeby Formation, the other being Ashfield Brick-clay Pit. The sands are uncemented or weakly cemented, yellow and thickly bedded. They have been extensively worked for moulding sand, but the underground workings have now been closed and most of the surface exposures have been filled in. The lower beds of the Cadeby Formation are of fine-grained, thin-bedded dolomite with thin clayey layers and contain a scattered fauna chiefly of bivalves. The sands are considered to owe their origin to reworking of former aeolian sands during the late Permian marine transgression. The site's principal claim for preservation is that it is one of the last remaining exposures of the basal part of the late Permian sequence in Yorkshire.

CADEBY QUARRY (SE 5200)

Highlights

Cadeby Quarry (box 7 in Figure 4.2) is the type locality of the Cadeby Formation and provides by far the largest and most comprehensive exposure in Yorkshire of both the Wetherby and Sprotbrough members and their mutual contact; the Wetherby Member here is of open shelf facies and contains more than 20 patch-reefs (at least some of a type found only at Cadeby) and the Sprotbrough Member is of offshore sandwave (shoal) facies with exceptionally large-scale cross-bedding. The intervening Hampole Beds are thicker here than anywhere else and atypically have yielded plant remains, and the erosion surface of the Hampole Discontinuity has a uniquely high relief of up to 3 m.

Introduction

The great working quarry at Cadeby lies on the north side of the River Don, south and east of the hamlet of Cadeby; most of the quarry is scheduled for filling, but the 18–23 m high north face is to be preserved when working ceases. Strata worked in Cadeby Quarry make up most of the Cadeby Formation, and comprise the Wetherby Member (13 m+) and the overlying Sprotbrough Member (10 m+); the latter is typical of this unit in the offshore sandwave or shoal belt of the English Zechstein marginal shelf, but the Wetherby Member is a uniquely spectacular mosaic of large byozoan-algal patch-reefs and surrounding skeletal grainstones. The Hampole Beds (?1–4 m) are atypical in containing abundant siliciclastic mudstones.

There are no comprehensive published accounts of Cadeby Quarry though summaries of strata have been given (Smith, 1969b, 1981b; Smith *et al.*, 1986). Reefs and associated strata exposed in nearby railway cuttings and on the opposite bank of the Don were mentioned by Mitchell (1932a) and Mitchell *et al.* (1947) and a typical Cadeby Quarry reef was discussed and illustrated by Smith (1981b, fig. 5). The presence of plants in the mudstones of the Hampole Beds was mentioned by Downie (1967) and a comprehensive account of the geochemistry of the Hampole Beds exposed in an abandoned railway cutting a short distance east of the quarry was given by Moss (1986).

Description

The position of Cadeby Quarry and its extent in 1990 is shown in Figure 4.22, which also shows the position of the retreating north face where the GCR site is concentrated. This face is about 18–23 m high (according to position) and 300–400 m long.

The quarry is cut through most of the Cadeby Formation, an estimated 7 m of which (proved by boreholes) underlies the quarry floor and a small thickness (probably less than 5 m) has been eroded from the top. The general geological sequence in the quarry is shown diagrammatically in Figure 4.23.

Wetherby Member, including most of the Hampole Beds

The Wetherby Member at Cadeby Quarry comprises three main rock types, all dolomite; peloid grainstones (with some packstones), bryozoan-rich boundstone (in patch-reefs) and domed stromatolitic laminites. In general the reefs are scattered

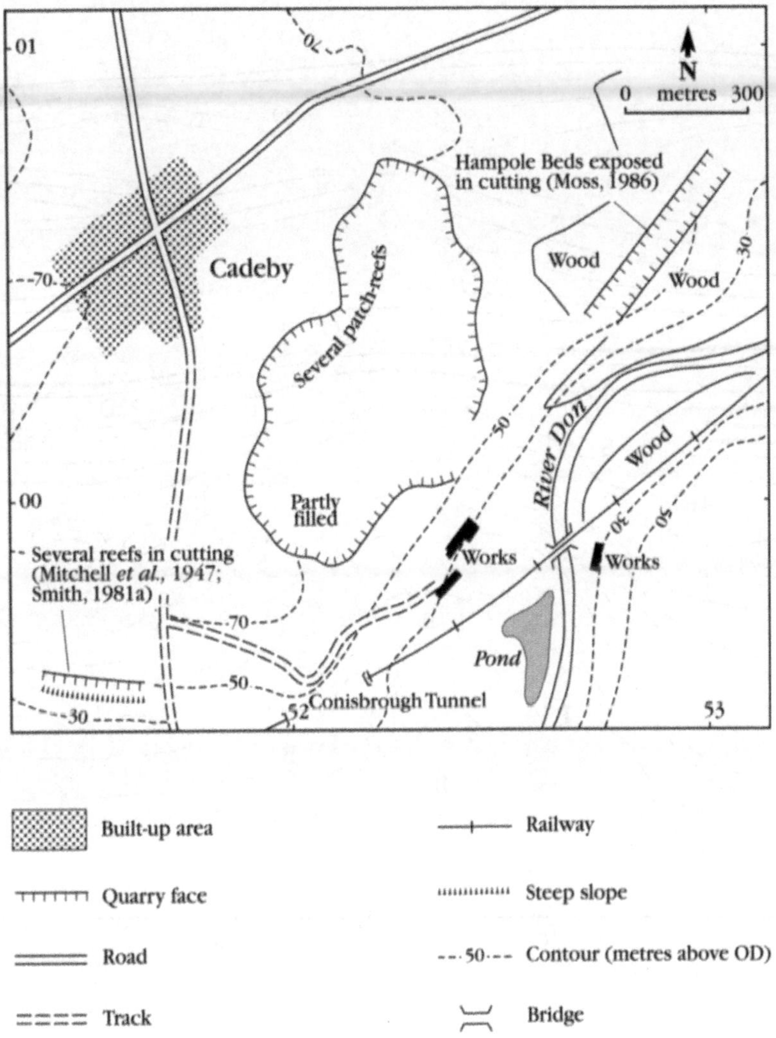

Figure 4.22 Cadeby Quarry and its environs, showing the location of the main features of geological interest.

unevenly in an otherwise continuous sheet of grainstones and grade upwards into stromatolite mantles that progressively extend to form much of the upper part of the member.

Grainstones in the Wetherby Member at Cadeby Quarry are similar to those surrounding the patch-reefs at South Elmsall Quarry and Newsome Bridge Quarry and comprise a varied mixture of pale buff, poorly-sorted ooids (predominant), pellets, compound grains, botryoidal grains, lumps, pisoids, stromatolite flakes, scattered tabular clasts up to 0.1 m across of ooid grainstone, and bioclasts. Some or all of the pisoids may be oncoidal (algal) in origin but no undoubted algal filaments have been recognized. The bioclasts are mainly represented by casts of largely unworn bivalves (*Bakevellia, Liebea, Permophorus* and *Schizodus*) and small gastropods, and they occur in profusion at some levels. Cavities after secondary anhydrite abound.

Patch-reefs in the Wetherby Member at Cadeby Quarry occur at intervals of 100–200 m in the north and west faces and similarly-spaced low eminences on the quarry floor show where others have been almost quarried away; more than 20 reefs are, or have been, present. Most of the reefs are roughly circular in plan and 15–50 m across at the level of the quarry floor; they have a buff and grey-buff, hard bryozoan boundstone (framestone) core at least 8 m high (bases not seen) and most are steep-sided mounds (Figure 4.23). In detail, the reef-cores are dense masses of dolomite mudstone and siltstone with a varied 5–50% framework of straggling and pinnate bryozoans, including species

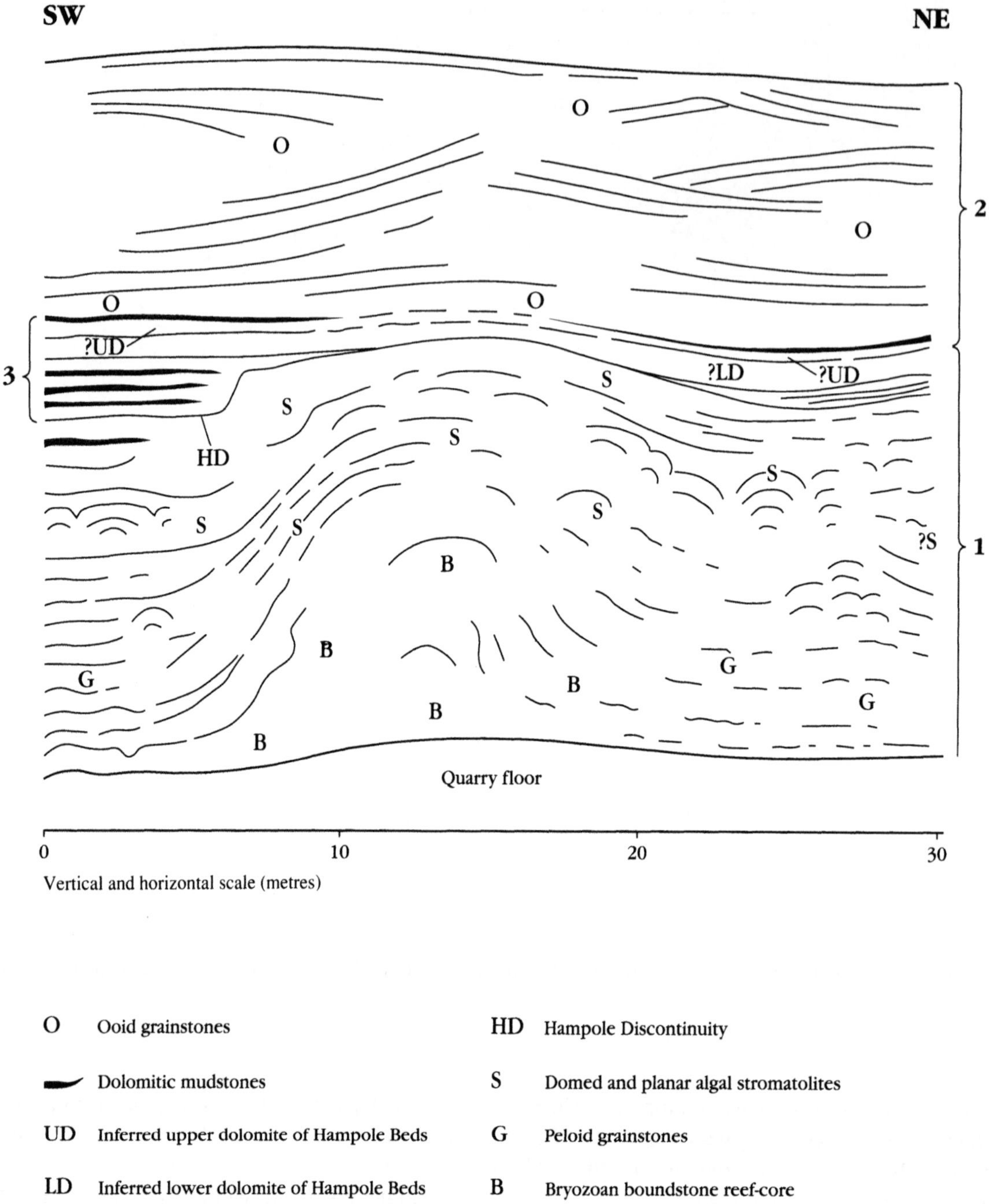

Figure 4.23 Diagrammatic sketch showing the relationships of the stratigraphical units and main rock types in the north-west face of Cadeby Quarry. The face is about 22 m high. 1, Wetherby Member; 2, Sprotbrough Member; 3, Hampole Beds, resting on the Hampole Discontinuity.

of *Fenestella* which have not been recorded from other patch-reefs in this member; a photomicrograph of typical reef dolomite from a patch-reef near to Cadeby Quarry was published by Smith (1981b, fig. 11B). At least one of the reefs features striking arrays of finger-sized stromatolite columns, also not reported from other patch-reefs in the Wetherby Member.

Upper parts of the bryozoan patch-reefs at Cadeby Quarry are thickly draped with complexly-domed, cream to pale buff, finely saccharoidal laminites which extend part of the way down the reef sides and either merge with mantles draping adjacent reefs or pass into peloid grainstones. Details are difficult to determine because of the inaccessibility of most of the exposures, but it appears that much or most of the upper part of the Wetherby Member, up to the Hampole Discontinuity, is composed of complexes of stromatolite domes that mainly overlie bryozoan reef-cores and are separated by relatively flat-lying stromatolites and/or grainstones in which domes are present only locally. Fallen blocks from these strata show that lamination is faint, fine and commonly mammilar, and that at least some of the stromatolite domes are composed of stacked minor domes each 0.10–0.30 m across and a few centimetres high. It is possible that the draping effect of stromatolitic dolomite over the reefs is partly compactional, but some primary relief and early cementation is indicated by the evident differential resistance of the reef mantles to erosion when the Hampole Discontinuity was cut.

The Hampole Discontinuity in Cadeby Quarry is unusually uneven, having a local relief of 2.5–3 m and, in places, sub-vertical cliffs up to about 1.5 m high (Smith, 1981b, fig. 5). High points on the discontinuity coincide with upstanding reefs in the main part of the Wetherby Member, against which lower parts of the Hampole Beds display marked onlap; higher parts of the Hampole Beds appear to thin only slightly over these eminences.

The Hampole Beds at Cadeby Quarry are not readily accessible but their position, thickness and relationship to the main part of the Wetherby Member are relatively clear (Figure 4.23) when viewed from the quarry floor. From this vantage it is difficult to recognize the typical Hampole Beds sequence (Smith, 1968, fig. 2) but it seems likely that the usual members of this sequence form the upper (more uniform, Sprotbrough Member) part of the Hampole Beds at Cadeby Quarry and are augmented by a discontinuous lowest unit (up to 1.8 m thick, part of the Wetherby Formation) which occupies hollows in the discontinuity surface. Fallen blocks from this lowest unit reveal it to be mainly of thin- to thick-bedded, buff to grey-green, slightly argillaceous dolomite mudstone with thin beds and lenses of grey, green and red siliciclastic mudstone from which Downie (1967) recorded remains of a land conifer.

Sprotbrough Member, including the upper part of the Hampole Beds

This forms the upper part of the faces all round the quarry and is about 7–10 m thick in the north face; it is not readily accessible, but can be seen from the quarry floor to be mainly coarsely cross-stratified in wedge-shaped sets individually up to about 8 m high. Fallen blocks from this unit show that the rock is typical of the Sprotbrough Member throughout the belt of sandwaves, being composed of pale cream, well-graded, fine ooid grainstones with only a few compound grains and less than 1% of siliciclastic grains (mainly subangular quartz); no fossils have been found in this member at Cadeby Quarry. The rock is of almost pure dolomite, but a little calcite thinly lines some of the many cavities that mark the sites of former patches of secondary anhydrite; the ooids average about 0.10–0.15 mm across (Kaldi, 1980) and many have leached cores. Large-scale cross-stratification is less prevalent in lower parts of the member than in middle and higher parts, and the lowest beds are mainly thick and sub-parallel with the slight rolling relief (?1 m) of the base of the member. The upper dolomite of the Hampole Beds at the base of the member cannot be distinguished with certainty, but is presumed to be present and 0.1–0.3 m thick.

Interpretation

Cadeby Quarry is unique in Yorkshire in providing a large and almost complete section through the Cadeby Formation (hence its choice by Smith *et al.*, 1986, as the type locality) and also in furnishing exceptionally clear information on the distribution pattern of patch-reefs in the lower (Wetherby) member of the formation. The patch-reefs themselves differ in a number of respects from those found elsewhere in the Wetherby Member, and the Hampole Beds are atypically thick and more diverse than in most other places.

Wetherby Member

The main interest in Cadeby Quarry, and its main asset as a GCR site, lies in (a) the many spectacular patch-reefs and enclosing strata and (b) the unusually high relief of the Hampole Discontinuity and the atypically great thickness of the Hampole Beds.

Patch-reefs and enclosing strata

The distribution and character of patch-reefs in the Wetherby Member in Yorkshire were described by Smith (1974b, 1981b, 1989) and are summarized here in the accounts of the Newsome Bridge, South Elmsall and Wood Lee Common sites; they occur in a roughly north to south belt a few kilometres wide that is parallel with the depositional strike and present outcrop, and lie at a range of levels between the Bakevellia Bed and the Hampole Discontinuity. Despite the relatively large number of exposures of patch-reefs, their spacing and relationship to enclosing strata is nowhere as clearly seen as in Cadeby Quarry.

Reefs are particularly abundant in parts of the area west of Doncaster and were first identified as such by Mitchell (1932a); the presence here of beds packed with bryozoans was however, recognized by Kirkby (1861) who inferred that they were the remains of prolific sessile benthic communities that would probably now be termed reefs. Mitchell *et al.* (1947) noted the presence of reefs in a railway cutting (SK 513996 - 519995) (Figure 4.22) and Smith (1981b, figs 10 and 11B) illustrated bryozoan boundstone from there. In general, however, few of the reefs in the Don Valley between Cadeby and Conisbrough are fully exposed and their detailed composition and relationships are less clear than at Cadeby Quarry.

Most large patch-reefs in the Wetherby Member in the Yorkshire Province have a bryozoan boundstone (framestone) core and a mantle of coarsely domed algal stromatolites, and in these characteristics the Cadeby Quarry reefs are no exception. They differ from most reefs in the member, however, in being narrower in relation to their height, in not being constructed of obvious saccoliths based on ramose bryozoans such as *Acanthocladia* (though this fossil is very abundant), in containing fenestrate bryozoans such as *Fenestella,* and also in containing narrow ('finger') columnar stromatolites and sinuous stromatolite sheets. The significance of these differences from the general range of patch-reefs in the Yorkshire Province is unclear and merits further study; slightly greater initial water depth and slightly lower salinity are two possible causes.

Grainstones and packstones enclosing the Cadeby Quarry patch-reefs are normal for the reef belt, and require little special comment. They are well seen in the railway cutting a short distance to the south where many weathered faces show the constituent grains more clearly than in the fresh quarry sections. The grainstones there and nearby are exceptional only in their local content of scattered reworked grainstone clasts up to 0.13 m across and of even larger clasts of rock described by Mitchell *et al.* (1947, p. 120) as 'close-grained dolomite'; the presence of these implies contemporaneous (?submarine) cementation and occasional considerable energy levels.

Algal-stromatolite (cyanophyte) mantles widely characterize patch-reefs that extend into the upper part of the Wetherby Member, and were discussed in the accounts of Newsome Bridge Quarry and South Elmsall Quarry; those at Cadeby Quarry are, however, exceptionally thick and extensive and it appears (but is difficult to prove in the available vertical faces) that most of the uppermost 6 m of the member here may be stromatolitic. Only the section at the former Alverley Grange Quarry (SK 554992, near Doncaster) was comparable in this respect. As with the differences between the patch-reefs at Cadeby Quarry and those elsewhere, the reasons for the atypical thickness and extent of the stromatolites is unclear.

The Hampole Discontinuity and the Hampole Beds

Almost throughout its range from near Ripon to near Nottingham, the Hampole Discontinuity has a gentle rolling relief of only a few centimetres, that at the exposure at Micklefield Quarry being typical (see Figure 4.17); the relief of up to 3 m at Cadeby Quarry and of 2.5 m at a nearby abandoned railway cutting (Moss, 1986) is therefore exceptional, as are the buried cliff notches surrounding reefs high in the quarry faces at Cadeby. The thickness of strata removed during the erosion of the discontinuity and, by implication, the sea-level decline that caused the erosion, cannot be estimated elsewhere in the Yorkshire Province but is shown at Cadeby to have been at least 3 m. The more normal relief was seen in former exposures of the Hampole Discontinuity at Boat Lane Quarry (SE 533013) at Sprotbrough, a little over 1 km to the north-east.

The atypically large thickness of much of the Hampole Beds at Cadeby Quarry is clearly a response to the abnormal relief of the discontinuity, the additional beds filling a local hollow and surrounding the more resistant reef-top mounds. Smith (1969b, p. 177) and Moss (1986) suggest that the beds filling these hollows may be estuarine. For further discussion of the Hampole Beds near Cadeby Quarry see Moss (1986) and for discussion of the Hampole Beds in general see Smith (1968) and the account here of Micklefield Quarry.

Sprotbrough Member

Difficulty of safe access to the high faces in which the Sprotbrough Member is seen in Cadeby Quarry limits the value of these for teaching and research purposes, but they nevertheless provide an excellent section through an assemblage of large ooid sandwaves and show their spatial relationship to the underlying Hampole Beds. Other, thicker, sections of these strata may be seen and are more readily accessible at many other places, however, including the nearby Warmsworth Cliff (SE 538009). The distribution and environmental significance of the sandwave facies of the Sprotbrough Member of the Cadeby Formation are discussed in more detail in the account of Micklefield Quarry.

Future research

The present inaccessibility of much of the main face at Cadeby Quarry has resulted in much uncertainty and has left ample scope for future research. In particular the biota, ecology and petrography of the patch-reefs is in need of careful study in view of the apparent differences between the Cadeby Quarry reefs and those elsewhere, and the character and flora of the multi-coloured beds between the Hampole Discontinuity and the lower dolomite of the Hampole Beds would amply repay further research.

Conclusions

This exceptionally large quarry is the type locality of the Cadeby Formation, although the lowest beds lie below the floor of the quarry and the highest beds have been eroded off. The most complete sequence is in the north face of the quarry, where the Wetherby (lower) Member comprises cross-bedded oolitic dolomite and algal-laminated dolomite, and contains a number of patch-reefs that are taller and steeper sided than patch-reefs found in the Wetherby Member elsewhere in Yorkshire. The overlying Sprotbrough Member mainly comprises finely-oolitic dolomite and features spectacularly large-scale cross-bedding of a type thought to indicate deposition in offshore oolite shoals. The Hampole Beds at the contact of the two members are unusually thick here, and the underlying Hampole Discontinuity has an erosional relief of 2.5–3 m. This unusually steep relief is thought to have been caused when sea level fell by a few metres, exposing the Wetherby Member to a phase of weathering and intertidal conditions before the sea returned and the Sprotbrough Member was formed.

The site is extremely important for the study of reef development, and is more extensive than others in Yorkshire. This should afford opportunities for the closer study of fauna, palaeoecology and petrography of the reefs and the surrounding strata, as well as the changes in depositional environment indicated by the character of the overlying Hampole Beds.

ASHFIELD BRICK-CLAY PIT, CONISBROUGH (SK 515981)

Highlights

This section, preserved for geological study since 1955 but nevertheless partly filled and soon to be re-excavated, vividly illustrates how much geology can be packed into one quite small rock face. In a vertical range of less than 20 m of strata, it contains Carboniferous coal measures formed in an equatorial coastal setting, an exhumed desert land surface formed by the tropical erosion and removal of more than 500 m of Carboniferous strata, a suite of water-laid desert litter and sand trapped in minor hollows in the desert surface and, in the main part of the face, evidence that the desert was then flooded by the tropical Zechstein Sea in which was formed, successively, lagoonal muds and open-sea shallow-water oolites full of the remains of a teeming marine life; scattered small reefs complete the range.

Introduction

This exposure, in the south-east outskirts of Conisbrough (Figure 4.24), spans strata high in the local Upper Coal Measures unconformably overlain by thin basal Permian deposits and the lower part of the Wetherby Member of the Cadeby Formation. The section was first described by Gilligan (1918), who recorded an unusually pebbly facies of the basal Permian deposits and an atypical 3.5 m sequence of multi-coloured 'marls' and 'limestones' at the base of the Lower Magnesian Limestone (now the Cadeby Formation). Mitchell *et al.* (1947) gave some details of the section as it was in 1930 and Downie (1967) noted that unspecified marine microfossils were present in the lower (argillaceous) beds of the Cadeby Formation. Downie also recorded reefs in the higher Magnesian Limestone beds present.

Description

The mainly late Permian sequence exposed at Ashfield Brick-clay Pit in late 1993 is shown on page 170.

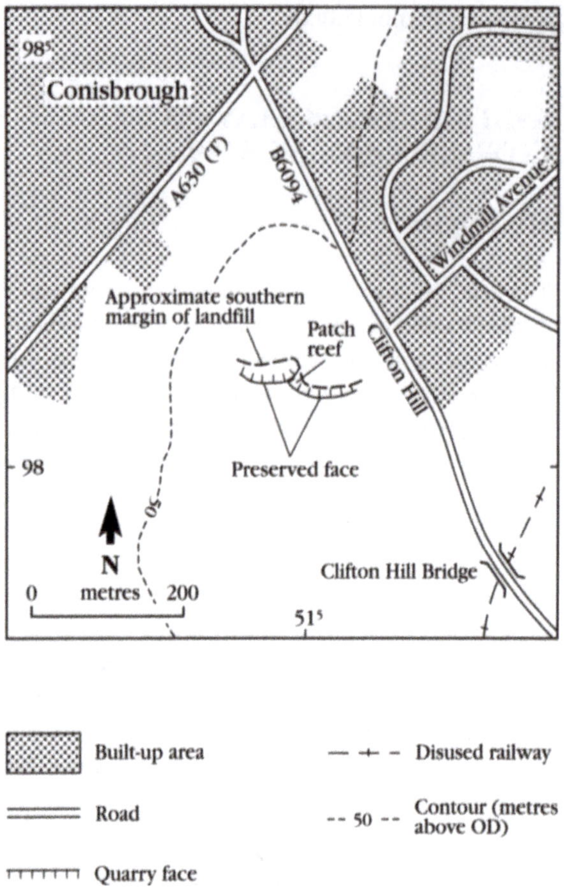

Figure 4.24 Ashfield Brick-clay Pit, Conisbrough, and its environs, showing the location of the main features of geological interest.

	Thickness (m)
Soil and dolomite brash	0.2-0.4
Cadeby Formation, Wetherby Member (open shelf facies)	
Dolomite, cream and buff, saccharoidal (fine sand-grade), porous, in fairly even beds 0.20–0.70 m thick. Passes sharply laterally into a bryozoan boundstone patch-reef about 30 m across, with some arching of overlying strata	*c.* 4.00
Dolomite, cream and buff, saccharoidal (fine sand-grade), unevenly thin- and medium-bedded (locally merging to thick-bedded)	0.60
Dolomite, buff, saccharoidal (fine sand-grade), in one to three beds, with 0–06 m of grey and dark red mottled clayey mudstone filling hollows at the top and a 0.07–0.12 m basal group of thin wavy-bedded dolomites with several laminae of red dolomitic clay	0.45
Dolomite, cream-buff, saccharoidal (fine sand-grade), in four beds 0.12–0.55 m thick; bivalves are abundant at several levels including *c.* 0.28 and 0.75 m above base. Red friable dolomite on bedding plane *c.* 0.70 m above base	1.25
Dolomite grainstone, buff, ooidal, a single bed, with traces of large ripples and cut-and-fill structures and with very abundant casts of bivalves (mainly *Bakevellia*). The 'Bakevellia Bed'	0.90
Dolomite, grey-buff, saccharoidal (very fine sand-grade), in unevenly flaggy beds, with two discontinuous brick-red clayey 0–0.15 mm layers 0.03 m apart at the top and other red layers in the upper part. Some irregular dark red patches. A few poor casts of bivalves,especially of *Bakevellia*	0.45
Dolomite, buff-cream, dense, saccharoidal (very fine sand-grade), in two 0.15–0.20 m beds, separated by a 0.02–0.05 m irregular layer of denser finely-mottled buff and purple-red dolomite. Both main beds contain bivalve casts and the lower also contains bryozoan casts. Probably an altered ooid grainstone	0.40
Dolomite, cream, porous, soft, saccharoidal (sand-grade), in one bed, with fairly abundant bivalve casts; brick-red patches near top. Scattered U-burrows about 13–16 mm diameter. Uneven sharp base on channelled erosion surface. Probably an altered ooid grainstone	*c.* 0.55
Dolomite, cream, saccharoidal (very fine sand-grade), thin-bedded and flasery	0.18
Dolomite, buff, saccharoidal (very fine sand-grade), one bed	0.28

Underlying strata were not visible in 1993 but are expected to be re-exposed late in 1994. The following section, based mainly on information given by Gilligan (1918), lies stratigraphically directly below that now visible and is likely to differ only in detail from that to be re-exposed.

	Thickness (m)
Cadeby Formation, Wetherby Member (Lower Marl facies)	
'Marl' (mudstone), red and grey, dolomitic	0.7
'Marl' (mudstone), kaolin-rich	0.4
Interbedded thin calcitic dolomite and red and grey 'marls'	1.3
'Marl' (mudstone), grey calcitic (0.05 m) on 'marl', red, finely bedded (0.05 m)	0.1
Dolomite, buff with thin 'marl' layers	0.4
'Marl' (mudstone), brown, slightly dolomitic and calcitic	0.2
'Marl' (mudstone), dark grey, slightly dolomitic and calcitic	0.2
'Marl' (mudstone), grey grading up to dull brown-red, calcitic, with scattered casts of *Schizodus*. Sharp base	0.2
Basal Permian deposits	
Breccio-conglomerate with persistent beds of fine-grained, weakly cemented sand, red, passing into pebbly sand where thin	0.4 - 1.7
----unconformity-----	
Carboniferous (Westphalian, Upper Coal Measures)	
Sandstone and silty shale, red at top	

Basal Permian deposits

Gilligan (1918) was moved to document this section by the unusual (to him) character of the basal Permian deposits and of the overlying basal beds of the Cadeby Formation (then known as the Lower Magnesian Limestone). The face measured by Gilligan has since been quarried away, but the basal Permian deposits were described as a friable fine sandy breccia or conglomerate with fairly persistent thin beds of sand. The deposits filled a hollow (?minor valley) on the old land surface (i.e. the unconformity), thinning westwards in 4.5 m from 1.7 m to 0.5 m and passing into sparingly pebbly sand-rock. Pebbles in the breccia and sand were generally less than 25 mm across and mainly comprised local Coal Measures (Westphalian) sandstone, hematite, chert and pink felspar. Polycrystalline quartz and chert were identified by Gilligan as the main constituents of the sand, the more resistant grains of which were 'exceedingly well rounded'; large flakes of muscovite were also recorded, and an extensive suite of heavy minerals was identified, the latter all potentially derived from Coal Measures rocks in the region. Gilligan noted that the sand in the deposits was in all respects except colour, like typical Basal Permian Sands, but it was left to Mitchell *et al.* (1947) to note that the colour was red. The face had retreated appreciably when Downie (1967) recorded a basal bed 0.75 m thick consisting of coarse, friable, gritty sand with layers of pebbles (mostly brown ironstone eroded from the local Coal Measures).

Cadeby Formation, Wetherby Member, ?Lower Marl facies

This unit, 3.5 m thick according to Gilligan (1918), 4.9 m thick according to Mitchell *et al.* (1947) and about 5.3 m thick according to Downie (1967), was described in detail by Gilligan (see tabulation) who included chemical analyses of each bed; these analyses showed that the unit mainly comprised red, brown and grey 'marls' (dolomitic and siliciclastic calcitic mudstone and argillaceous dolomite) with subordinate beds of calcitic dolomite. In addition to Gilligan's (1918) record of casts of *Schizodus* in the basal bed of this unit, Downie (1967) noted that the unit yielded unspecified marine microfossils. This unit is assigned to the Lower Marl facies on the basis of its bulk composition and clay mineral content but differs in several respects from its more extensive counterpart farther south.

Cadeby Formation, Wetherby Member, open shelf facies

This unit here comprises two main parts, a lower varied sub-unit mainly of shelly dolomite with several laminae and thin beds of red and grey dolomitic siliciclastic mudstone, and an upper sub-unit (4.6 m+) of relatively pure dolomite that, in its upper part, locally passes into a lenticular patch-reef at least 4 m thick.

Beds in the lower sub-unit range from thin and semi-nodular to thick and even, and several feature patchy purple-red hematite staining. At least two beds are of altered ooid grainstone but, as Mitchell *et al.* (1947) noted, other beds that superficially appear to be of porous saccharoidal dolomite also prove, on close inspection, to be of highly altered ooid grainstone. Casts of *Bakevellia* are common to abundant in some beds and reach rock-forming proportions in others; commonly they are associated with lesser numbers of other bivalves (*Liebea* and *Schizodus*). Casts of small fragments of the ramose bryozoan *Acanthocladia* are present in at least one bed.

Bedding in the upper sub-unit is less even than that in the lower sub-unit, but the rock-type is more uniform and less fossiliferous; most, if not all, is of highly altered dolomite ooid grainstone and most faunal remains comprise casts of *Bakevellia* that are abundant only at certain levels. The coarsely saccolithic boundstone reef into which the beds of this upper sub-unit sharply pass, however, is rich in the remains of frame-forming *Acanthocladia*. Arching of the youngest beds in the quarry against the flanks of the reef may be a compactional effect.

Interpretation

The sequence of Permian strata conserved at Ashfield Brick-clay Pit provides a window into the rarely-seen part of the sequence and an uncommon view of the Carboniferous–Permian unconformity. The latter, in this part of Yorkshire, is generally flat to very gently rolling, and bears a thin scattering of mainly small subangular resistant pebbles probably loosened from the old land surface by extreme temperature variations and chemical (mainly salt) weathering. It is, perhaps, not surprising that such pebbles should be concentrated in hollows like that here, swept there by occasional flash floods and typically high rates of run-off. Elsewhere, comparable hollows in the unconformity are extremely rare, perhaps the best known being those in the A1(M) road cutting (NZ 247128) at Cleasby, near Darlington. Breccias of local rock are, of course, relatively common in places such as Knaresborough, where the desert surface had a steep local relief (Fox-Strangways, 1874); amongst GCR sites, breccias of local debris are present around a minor sandstone hill at Newsome Bridge Quarry, near Wetherby. Such breccias differ considerably from the well-cemented rocks that are classed as Basal Permian Breccias; the latter form extensive sheets across much of the Cleveland High and comprise resistant multi-cyclic pebbles that probably accumulated as desert piedmont pavements.

Cadeby Formation, Wetherby Member, Lower Marl Facies

Rocks of this facies are uncommon in the Conisbrough area, being most widespread in the southern part of the outcrop and at depth farther east (see Smith, 1989, fig. 6 for distribution). The facies here is clearly a local variant, not connected with the main area of Lower Marl and one of a number of relatively small similar patches distributed unevenly along the outcrop from Sheffield northwards. The Lower Marl is an argillaceous facies of the Wetherby Member and it is not to be confused with the wholly older Marl Slate which does not crop out in the Yorkshire Province and is a deeper-water deposit formed under anoxic conditions.

The 'marly' rocks at Ashfield are assigned to the Lower Marl on the basis of their general stratigraphical equivalence with the argillaceous rocks in the main (southern) outcrop, their marine fauna and their overall lithological character. They differ, however, in their content of red and brown beds (uncommon in the main area), in the relative rarity of shelly fossils and in the high dolomite content of most of the carbonate rocks present. Their depositional environment is uncertain, but it may be speculated that they accumulated slowly in a low-energy inshore setting such as a shallow lagoon lying landward of a minor oolite shoal or barrier bar.

Cadeby Formation, Wetherby Member, normal facies

The lower sub-unit of this member at Ashfield is relatively normal for the area, though the number of dark red argillaceous layers and of red-stained patches is atypically high. The origin of this colour is not known, but it may have been derived from the strongly reddened Carboniferous clay-rocks immediately beneath and elsewhere in the area. Rocks composed mainly of ooids are widespread in this district at and near the base of the Wetherby Member, and the abundance of casts of a restricted range of bivalves is typical. The latter commonly reach rock-forming proportions (as here) in a 0.8–2.5 m bed near the base of the Magnesian Limestone sequence in many central outcrop areas; it seems likely (but cannot be proved) that this informally-named 'Bakevellia Bed' is the product of an unusually extensive single sheet coquina, and that shallow water and at least moderate energy are implied.

The upper sub-unit of the Wetherby Member at Ashfield is normal for the area. The altered ooid grainstones of which it is composed are almost entirely of dolomite and the patch-reef has a typical lithology and biota (Smith, 1981b) and is of typical size. Its base, not currently exposed, may extend below the 4 m bed at the top of the section, but no reefs have ever been reported in or below the 'Bakevellia Bed' and this reef is not likely to be an exception. As throughout the north to south belt of patch-reefs, formation on a broad, shallow, clear, open marine shelf with moderate energy is inferred (Smith, 1981b, 1989).

Conclusions

The site is an exposure of the basal Permian deposits overlain by the lowest basal part of the Wetherby Member of the Cadeby Formation. The actual unconformity is not exposed at the time of writing, but has been well-described by previous workers and is likely to be exposed by excavation planned for 1994. The basal deposits are red, and contain pebbles of derived Carboniferous ironstone and sandstone. The lower part of the Wetherby Member consists of multi-coloured ?lagoonal mudstones and argillaceous dolomites, overlain by open-shelf shelly dolomite containing abundant bivalve remains; dolomite at the top of the exposed sequence passes into a lenticular bryozoanal patch-reef.

NEW EDLINGTON BRICK-CLAY PIT (SK 530987)

This site was selected by Smith *et al.* (1986) as the type locality of the newly-defined Edlington Formation (formerly the Middle Permian Marls) and its acceptance as part of the GCR network followed. The site was subsequently covered but is included here because no other site has yet been identified as a suitable substitute.

Introduction

New Edlington Brick-clay Pit was cut into the 'Middle Permian Marls', which are here about 8–11 m thick and occupy a faulted north-east to south-west trough 800–900 m wide. The excavation was already large when the 1931 edition of the 1:10,560 Ordnance Survey map was surveyed, and brief notes on the section then visible were shown on the ensuing 1:10,560 geological map (Mitchell, 1932b). Details of the sequence in the eastern part of the workings were later given by Mitchell *et al.* (1947) and Downie (1967), and a sedimentologically updated sequence, combining sections in the west and east of the excavations, was given by Harwood *et al.* (1982) and Smith *et al.* (1986).

The sequence at New Edlington Brick-clay Pit

Before being covered, the Edlington Formation at its type locality at New Edlington Brick-clay Pit was seen to be about 8 m thick in the westernmost workings but thickened gradually to about 11 m in the eastern workings. In the west and north of the pit, it was worked beneath a thin cover of the Brotherton Formation and in the south the pit locally extended a few metres into the underlying Cadeby Formation. Strata exposed in 1982 in the west, at the type locality, were as shown below.

	Thickness (m)
Clay-loam, red-brown	0–0.30
Brotherton Formation	
Carbonate mudstone (mainly dolomite) and *Calcinema*-bivalve packstone, grey and buff, partly ripple-laminated, with semi-nodular very uneven bedding. Uneven smoothly rounded base, relief *c.* 0.1 m	*c.* 2.50
Sandstone, mainly brown-red and brown but pale khaki in lowest 0.08 m, very fine-grained to medium-grained, with scattered lenses of grey dolomite mudstone near top and two thin uneven beds of grey dolomite mudstone near base	*c.* 0.45
Edlington Formation	
Mudstone, purple-brown and red-brown, silty, blocky, with laminae of pale grey and pale grey-green silty sandstone	*c.* 0.30

Siltstone, sandy, grey-green, with thin beds of green medium-grained sandstone	*c.* 0.40
Mudstone, mottled dark red-brown and pale grey-green, blocky	*c.* 0.40+

Lower parts of the section, now wholly covered, comprised a few metres of poorly-exposed, dark red-brown, blocky mudstone with a number of thin beds of grey-green and red siltstone and fine-grained sandstone about 3–5 m below the top of the Edlington Formation. Some of these thin beds feature ripple lamination and others bear desiccation cracks (Figure 4.25) and hollow-faced casts of halite hoppers (Figure 4.26). Lower beds of the formation, especially the blocky mudstones, also contain a varied network of fibrous gypsum veins, the thickest of which are up to 0.10 m thick and are roughly concordant.

Faces in the east and south of the excavation revealed a similar sequence to that in the west with, additionally, about 3 m of the Cadeby Formation (Harwood *et al.*, 1982). This comprised pale-grey ooidal dolomite grainstone with a contemporaneously lithified bevelled and bored surface near the top, overlain by up to 1.2 m of monomict breccia derived from the underlying lithified grainstone (Harwood, 1986, fig. 3d).

Discussion and interpretation of the sequence

The short section now exposed at New Edlington is of interest in that the usual sharp simple contact between the Edlington and Brotherton formations is missing, its place being taken by a transition of interdigitating strata over about 0.45 m. Although it is rare to find evidence of substantial reworking of the top of the Edlington Formation during the succeeding marine transgression, such reworking may be implied here; if this is so, the contact between the two formations (i.e. the transgression surface) probably lies at the base of the 0.45 m sandstone bed which lies about 0.08–0.10 m below the lowest dolomite bed.

Special note

The choice by Smith *et al.* (1986) of New Edlington Brick-clay Pit as the type locality of the

Figure 4.25 Sandstone-filled desiccation cracks at the type locality of the Edlington Formation. Coin: 30 mm across. (Photo: D.B. Smith.)

Figure 4.26 Casts of halite crystals on the underside of an argillaceous siltstone bed at the type locality of the Edlington Formation. The large cast is about 10 mm across. (Photo: D.B. Smith.)

newly-defined Edlington Formation rested mainly on the rarity of exposures of these predominantly vale-forming recessive strata. Over the years these soft rocks have been widely worked for use in brick and tile manufacture but one by one the excavations have been flooded or otherwise filled, and the exposures lost. As a type locality, the pits at New Edlington had the advantage that both the base and the top of the formation were exposed and thus readily defined, but had the disadvantage that the sequence was both atypically thin and also lacked several of the main rock types commonly present (especially beds of ooid grainstone and anhydrite).

The loss of its type locality prompts nomination of an alternative, the main candidate amongst surface exposures probably being the River Ure Cliff site, near Ripon. This section, however, does not expose the full thickness of the formation nor its base, and is therefore not wholly satisfactory. A preferred option would probably be a well-documented borehole core such as that from, for example, the Barlow No 2, Camblesforth No 1, Synthetic Chemicals No 1 (Askern), Whitemoor or Wistow Wood boreholes; substantial numbers of core specimens from most of these have been retained by either British Coal or the British Geological Survey.

WOOD LEE COMMON, MALTBY (SK 532915)

Highlights

Wood Lee Common, Maltby (not shown in Figure 4.2), is the most accessible and amongst the best localities for the study of the structure and fabric of typical bryozoan patch-reefs in the Wetherby Member of the Cadeby Formation. The reef-rock is completely dolomitized and comprises an untidy assemblage of sack-like masses of dense bryozoan-rich rock with associated bivalve, gastropod and brachiopod fossils.

Introduction

The late Permian marine patch-reefs at Wood Lee Common form scattered natural upstanding tor-like crags in south-westwards-sloping scrubby grassland on the south-western fringe of Maltby;

most of the 'tors' are only a few metres high and less than 30 m across. No other strata are exposed and so there is no information on the nature of contacts or on reef dimensions; it is not even clear whether there is one reef or several. The outstanding feature of the exposures is that differential weathering has revealed that the reefs comprise a large number of stacked sack-like bodies up to 2.5 m across, each with a complex framework of straggling bryozoans.

The craggy exposures on Wood Lee Common were mentioned by Sedgwick (1829) and illustrated photographically by Eden *et al.* (1957, pl. 4) and Smith (1981b, figs 6, 14 and 17); the detailed ecology and make-up of reefs like those at Wood Lee Common was discussed fully by Smith (1981b).

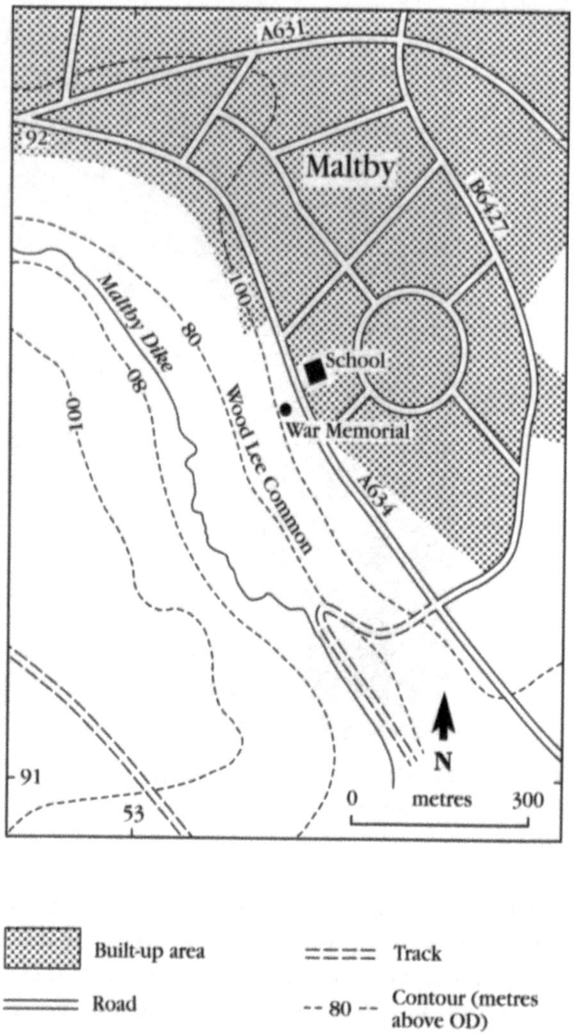

Figure 4.27 Wood Lee Common GCR site, Maltby, South Yorkshire. Most of the reef 'tors' are in the central and northern parts of the designated area.

Description

Wood Lee Common lies on the south-west side of the A634 Maltby–Blyth road and is shown in detail in Figure 4.27. Rock exposures cover only a small proportion of the common and are unevenly scattered both in geographical position and at different levels on the slope. All exposures are of rock in the lower and middle part of the Wetherby Member of the Cadeby Formation, the base of which trends NNW to SSE across the middle of the slope. All the reef-rock is of dolomite.

Examination of the reef exposures reveals little of the shape or size of the reef body or bodies, but shows that the 'tors' are almost wholly composed of dense masses ('saccoliths') of bryozoan boundstone (framestone) piled apparently haphazardly beside and on top of each other. Most of the saccoliths are roughly horizontally elongated, locally giving the rock a crudely thick-bedded aspect (Figure 4.28). They range from less than 1 m across to up to 2.5 m across and 1 m thick. Many of the saccoliths are in tight mutual contact, forming a coarse mosaic, but others are partly or wholly separated by irregular pockets and lenses of fine-grained shelly detritus; shell remains in the detritus are mainly of small bivalves (*Bakevellia, Liebea, Permophorus, Schizodus*) and small gastropods, but also locally include fragments of ramose bryozoans (probably mainly *Acanthocladia*) and of the small pedunculate brachiopod *Dielasma*.

Close inspection of the rock face shows that the saccoliths comprise 5–?25% of a twig-like framework of branching *Acanthocladia* colonies (with some possible *Thamniscus*) spread unevenly throughout a dense, fine-grained dolomite matrix. Thin sections (Smith, 1981b, figs 14 and 17) reveal that the matrix is of patchily turbid dolomite microspar and dolomicrite and that the rock has undergone a complex history of diagenesis and cavity-fill. Early cementation is suggested by a general lack of crushing of the skeletal remains, and this may have been initiated by the formation of fibrous isopachous fringes (0.05–0.25 mm thick) that coat and line most of the bryozoan frame elements and also many other organic remains.

An additional feature of interest in the reefy 'tors' of Wood Lee Common is the presence of well-developed honeycomb weathering on some faces, and the more restricted occurrence of narrow linenfold-like vertical dissolutional fluting.

Figure 4.28 Reef 'tor' in the central part of Wood Lee Common GCR site, showing the characteristic subdivision into 'saccoliths'. Hammer (centre-right): 0.33 m. (Photo: D.B. Smith.)

Interpretation

Patch-reefs in the Wetherby Member of the Cadeby Formation in Yorkshire are featured in five GCR sites and display different features in each; those at Wood Lee Common are special in that, in addition to being freely and readily accessible, they display *par excellence* the saccolithic structure that typifies the mainly bryozoan reefs in the lower part of the member (Smith, 1981b). Three of the four other reef GCR sites, Newsome Bridge Quarry, South Elmsall Quarry and Ashfield Brick-clay Pit, may have bryozoan saccolithic cores but, if so, this structure has subsequently almost been obliterated by diagenesis; reefs at the fifth site, Cadeby Quarry, have a rather different structure and biota from the other three. Additional places where a pronounced saccolithic reef structure in bryozoan reefs may be seen include an old quarry (SE 488176) on the northern fringes of Wentbridge and the many exposures in Hooton Pagnell village (SE 4808) where reef/grainstone contacts and relationships are also well exposed (Smith, 1981b, fig. 4); others were noted by Edwards *et al.* (1947) at Aberford (SE 4337) and Boston Spa (SE 4245). Upstanding crags of reef limestone, not unlike those at Maltby, have been reported at Minney Moor (SK 519989), Conisbrough by Mitchell (1932a) and near South Anston (SK 525838), east of Sheffield, by Eden *et al.* (1957).

The distribution and general characteristics of patch-reefs in the Wetherby Member have been investigated by Smith (1974b, 1981b, 1989) and are summarized in the account on South Elmsall Quarry. They lie at all levels in the Wetherby Member between the top of the Bakevellia Bed (commonly 1–3 m above the base) and the Hampole Discontinuity, and range from scattered to abundant in an 8–12 km wide belt that coincides roughly with the outcrop between Brearton (SE 322610, near Harrogate) and Barlborough (SK 4777, near Sheffield). Most simple reefs are a few metres thick and 10–25 m across (although some exceed 100 m), but closely-spaced reefs locally merged to form complexes more than 20 m thick and 120 m across.

Although Mitchell (1932a) was the first to apply the word 'reef' to unbedded or 'brecciated' bryozoan rock in the Wetherby Member, it is clear that Kirkby (1861, p. 315) recognized that the 'polyzoan beds' probably formed part of a sessile, organic community built up by and around ramose cryptostome bryozoans such as *Acanthocladia*. By their growth, the bryozoans gave rise to a variety of minor sub-environments that were occupied by, and sheltered, a more varied range of invertebrates than inhabited the surrounding more uniform grainstones; such forms include encrusting foraminifera and pedunculate small brachiopods. The roles played by the various organisms in the life and construction of the reefs were discussed by Smith (1981b), together with a preliminary analysis of reef diagenesis. It was concluded, partly from the contributary evidence of the surrounding grainstones, that the reefs were formed entirely subaqueously on an open marine shelf under a few metres of water of slightly above-normal salinity.

Because no margins are exposed, it is not possible to determine whether one large reef is present at Wood Lee Common or several smaller ones. If only one reef is present, however, it would be at least 150 m across and more than 20 m thick, which is very large for a single reef; it seems more likely therefore, and taking the several separate and linked reefs at Hooton Pagnell as a guide, that a number of reefs is present rather than one large one.

Future research

The main features of the patch-reefs in the Wetherby Member of the Cadeby Formation are now reasonably well documented and understood, but there remains much scope for detailed research on both the ecology and diagenesis of the reefs; the Maltby reefs are particularly well suited for this purpose.

Conclusions

This site is one of the best localities for the study of bryozoan patch-reefs in the Cadeby Formation. Such reefs are seen in several localities in Yorkshire, notably Cadeby, Newsome Bridge and South Elmsall Quarries. The exposures at Wood Lee Common allow easy access, and as natural outcrops, have undergone differential weathering which has highlighted the internal structure of the reef. The exposures reveal that the reefs are mainly composed of dense masses of bryozoan-rich rock known as 'saccoliths', which are elongate structures that are piled one on another so as to impart a bedded appearance to the rock. The saccoliths are separated by irregular patches of shelly debris.

Although the number of reefs and their relationship to surrounding strata is not known at site, this locality is significant in that its well-developed weathering allows details of the reef to be studied which cannot readily be seen in the fresher quarry faces of most other patch-reef sites.

References

Abbott, G. (1903) Report on excursion to Southwick, Fulwell and Roker. *Proceedings of the Geologists' Association*, **18**, 322-4.

Abbott, G. (1907) Concretions. *South-eastern Naturalist*, 1-7.

Abbott, G. (1914) Discoid limestones which simulate organic characters. A case of inorganic evolution. *The Pioneer*, 1-8.

Al-Rekabi, Y. (1982) Petrography, porosity and geochemistry of the Upper Magnesian Limestone of NE England. Unpublished Ph.D. Thesis, University of Dundee.

Aplin, G. (1985) Diagenesis of the Zechstein main reef complex, NE England. Unpublished Ph.D. Thesis, University of Nottingham.

Arthurton, R.S. and Hemingway, J.E. (1972) The St Bees Evaporites - a carbonate-evaporite formation of Upper Permian age in West Cumberland, England. *Proceedings of the Yorkshire Geological Society*, **38**, 565-92.

Arthurton, R.S., Burgess, I.C. and Holliday, D.W. (1978) Permian and Triassic, in *The Geology of the Lake District* (ed. F. Moseley), Yorkshire Geological Society, Occasional Publication No. 3, pp. 189-206.

Aveline, W.T., Dakyns, J.R. and Fox-Strangways, C. (1874) One-inch Geological Sheet 62 (Harrogate). Geological Survey of England and Wales.

Binney, E.W. (1855) On the Permian beds of the north-west of England. *Proceedings of the Manchester Literary and Philosophical Society (Series 2)*, **12**, 209-69.

Braithwaite, C.J.R. (1988) Calcitization and compaction in the Upper Permian Concretionary Limestone and Seaham formations of north-east England. *Proceedings of the Yorkshire Geological Society*, **47**, 33-45.

Browell, E.J.J. and Kirkby, J.W. (1866) On the chemical composition of various beds of the Magnesian Limestone and associated Permian rocks of Durham. *Transactions of the Natural History Society of Northumberland, Durham and Newcastle upon Tyne*, **1**, 204-30.

Burton, R.C. (1911) On the occurrence of beds of the Yellow Sands and marl in the Magnesian Limestone of Durham. *Geological Magazine*, **8**, 299-306.

Cameron, A.G. (1881) Subsidences over the Permian boundary between Hartlepool and Ripon. *Proceedings of the Yorkshire Geological and Polytechnic Society*, **7**, 342-51.

Card, G.W. (1892) On the flexibility of rocks; with special reference to the Flexible Limestone of Durham. *Geological Magazine*, **9**, 117-24.

Clapham, R.C. (1863) Analyses and description of Magnesian Limestone from the Trow Rocks. *Transactions of the Tyneside Naturalists' Field Club*, **5**, 122-4.

Clark, D.N. (1980) The diagenesis of Zechstein carbonate sediments, in *The Zechstein Basin with Emphasis on Carbonate Sequences* (eds H. Füchtbauer and T.M. Peryt), Elsevier, Amsterdam, pp. 167-203.

Clark, D.N. (1984) The Zechstein in NW Europe, in *Carbonate Geology* (course notes), Open University, Milton Keynes, pp. 150-71.

Clarke, R.F.A. (1965) British Permian saccate and monosulcate miospores. *Palaeontology*, **8**, 322-54.

Colter, V.S. and Reed, G.E. (1980) Zechstein 2 Fordon Evaporites of the Atwick No. 1 borehole, surrounding areas of NE England and the adjacent southern North Sea, in *The Zechstein Basin with Emphasis on Carbonate Sequences* (eds H. Füchtbauer and T.M. Peryt). Contributions to Sedimentology, **9**, Elsevier, Amsterdam, pp. 115-29.

Cooper, A.H. (1986) The subsidence hazard and foundering of strata caused by the dissolution of Permian gypsum in the Ripon and Bedale areas, North Yorkshire, in *The English Zechstein and Related Topics* (eds G.M. Harwood and D.B. Smith), Geological Society of London, Special Publication No. 22, pp. 127-39.

Cooper, A.H. (1987a) *The Permian Rocks of the Thirsk District; Geological Notes and Local Details of 1:50,000 Sheet 52*. British Geological Survey, Keyworth.

Cooper, A.H. (1987b) *The Permian Rocks of the Harrogate District; Geological Notes and Local Details of 1:50,000 Sheet 62*. British Geological Survey, Keyworth.

Cooper, A.H. (1988) Subsidence resulting from the dissolution of Permian gypsum in the Ripon area: its relevance to mining and water abstraction, in *Engineering Geology of Underground Movements* (eds F.G. Bell, M.G. Culshaw, J.C. Cripps and M. Lovell). Geological Society Engineering Geology Special Publication No. 5, pp. 387-90.

Donovan, S.K., Hollingworth, N.T.J. and Veltcamp, C.J. (1986) The British Permian crinoid *Cyathocrinites ramosus* (Schlotheim). *Palaeontology*, **29**, 809-25.

Downie, C. (1967) Conisborough, in *Geological Excursions in the Sheffield Region and the Peak District National Park* (eds R. Neves and C. Downie). University of Sheffield, pp. 145-9.

Eastwood, T., Dixon, E.E.L., Hollingworth, S.E. and Smith, B. (1931) *The Geology of the Whitehaven and Workington District*. Memoir of the Geological Survey of Great Britain, Sheet 28.

Ebburn, J. (1981) Geology of the Morecambe Bay gas field, in *Petroleum Geology of the Continental Shelf of North-west Europe* (eds L.V. Illing and G.D. Hobson), Heyden, London, pp. 485-93.

Eden, R.A., Stevenson, I.P. and Edwards, W. (1957) *Geology of the Country around Sheffield*. Memoir of the Geological Survey of Great Britain, Sheet 100.

Edwards, W., Wray, D.A. and Mitchell, G.H. (1940) *Geology of the Country around Wakefield*. Memoir of the Geological Survey of Great Britain, Sheet 78.

Edwards, W., Mitchell, G.H. and Whitehead, T.H. (1950) *Geology of the Country North and East of Leeds*. Memoir of the Geological Survey of Great Britain, Sheet 70.

Eriksson, K.A. (1977) Tidal flat and subtidal sedimentation in the 2250 MY Malmani Dolomite, Transvaal, South Africa. *Sedimentary Geology*, **18**, 223-44.

Evans, A.L., Fitch, F.J. and Miller, J.A. (1973) Potassium-argon age determinations on some British Tertiary igneous rocks. *Journal of the Geological Society of London*, **129**, 419-43.

Forbes, B.G. (1958) Folded gypsum of Ripon Parks, Yorkshire. *Proceedings of the Yorkshire Geological Society*, **31**, 351-8.

Fowler, A. (1943) On fluorite and other minerals in Lower Permian rocks of south Durham. *Geological Magazine*, **80**, 41-51.

Fowler, A. (1957) Minerals in the Permian and Trias of north-east England. *Proceedings of the Geologists' Association*, **67**, 251-65.

Fox-Strangways, C. (1874) *The Geology of the Country North and East of Harrogate*, Memoir of the Geological Survey of Great Britain, Sheet 62.

Fox-Strangways, C. (1908) *The Geology of the Country North and East of Harrogate*, 2nd edn, Memoir of the Geological Survey of Great Britain, Sheet 62.

Fox-Strangways, C., Cameron, A.G. and Barrow, G. (1885) *The Geology of the Country around Northallerton and Thirsk*, Memoir of the Geological Survey of Great Britain, Sheet 52.

Francis, E.A. (1964) 1:10,560 Geological map sheet NZ 34 NW. Geological Survey of Great Britain, London.

Füchtbauer, H. (1968) Carbonate sedimentation and subsidence in the Zechstein Basin (northern Germany), in *Recent Developments in Carbonate Sedimentology in Central Europe* (eds G. Müller and G.M. Friedman), Springer-Verlag, Berlin, pp. 196-204.

Fuzezy, L.M. (1970) Petrology of the Lower Magnesian Limestone in the neighbourhood of Selby, Yorkshire. Unpublished Ph.D. Thesis, University of Cambridge.

Fuzezy, L.M. (1980) Origin of nodular limestones, calcium sulphates and dolomites in the Lower Magnesian Limestone in the neighbourhood of Selby, Yorkshire, England, in *The Zechstein Basin with Emphasis on Carbonate*

Sequences (eds H. Füchtbauer and T.M. Peryt), Contributions to Sedimentology, **9**, Elsevier, Amsterdam, pp. 35-44.

Garwood, E.J. (1891) On the origin and mode of formation of the concretions in the Magnesian Limestone of Durham. *Geological Magazine*, **8**, 433-40 + 1 plate.

Geinitz, H.B. (1861) *Dyas, oder die Zechstein formation und das Rotliegend*, Leipzig, 130 pp.

Gilligan, A. (1918) The Lower Permian at Ashfield Brick and Tile Works, Conisbrough. *Proceedings of the Yorkshire Geological Society*, **19**, 289-97.

Glennie, K.W. (1984) Early Permian-Rotliegend, in *Introduction to the Petroleum Geology of the North Sea* (ed. K.W. Glennie), Blackwell Scientific Publications, Oxford, pp. 41-61.

Glennie, K.W. and Buller, A. (1983) The Permian Weisslegend of NW Europe: the partial deformation of aeolian sand caused by the Zechstein transgression. *Sedimentary Geology*, **35**, 43-81.

Goodall, I.G. (1987) Sedimentology and diagenesis of the Edlington Formation (Upper Permian) of Teesside. Unpublished Ph.D. Thesis, University of Reading.

Goodchild, J.W. (1893) Observations on the New Red Series of Cumberland and Westmorland, with especial reference to classification. *Transactions of the Cumberland and Westmorland Association*, **17**, 1-24.

Harwood, G.M. (1981) Controls of mineralization in the Cadeby Formation (Lower Magnesian Limestone). Unpublished Ph.D. Thesis, Open University, Milton Keynes.

Harwood, G.M. (1986) The diagenetic history of Cadeby Formation carbonate rocks (EZ1 Ca), Upper Permian, eastern England, in *The English Zechstein and Related Topics* (eds G.M. Harwood and D.B. Smith), Geological Society of London, Special Publication No. 22, pp. 75-86.

Harwood, G.M. (1989) *Contrasting Nearshore Sedimentation in Zechstein 1 Carbonate Rocks in Yorkshire*. British Sedimentological Research Group, Field Excursions Guide, Leeds, pp. 6.11-6.20.

Harwood, G.M. and Smith, D.B. (eds) (1986) *The English Zechstein and Related Topics*. Geological Society of London, Special Publication No. 22, 244 pp.

Harwood, G.M. and Smith, F.W. (1986) Mineralization in Upper Permian carbonates at outcrop in eastern England, in *The English Zechstein and Related Topics* (eds G.M. Harwood and D.B. Smith), Geological Society of London, Special Publication No. 22, pp. 103-11.

Hickling, G. (1906) On footprints from the Permian of Mansfield. *Quarterly Journal of the Geological Society of London*, **62**, 125-31.

Hickling, G. and Holmes, A. (1931) The brecciation of the Permian rocks, in *Contributions to the Geology of Northumberland and Durham, Proceedings of the Geologists' Association*, **42**, 252-5.

Hirst, D.M. and Smith, F.W. (1974) Controls of barite mineralization in the Lower Magnesian Limestone of the Ferryhill area, County Durham. *Transactions of the Institution of Mining and Metallurgy, Section B*, **83**, 49-55.

Holliday, D.W. (1993) Geophysical log signatures in the Eden Shales (Permo-Triassic) of Cumbria and their regional significance. *Proceedings of the Yorkshire Geological Society*, **49**, 345-54.

Hollingworth, N.T.J. (1987) Palaeoecology of the Upper Permian Zechstein Cycle 1 reef of NE England. Unpublished Ph.D. Thesis, University of Durham.

Hollingworth, N.T.J. and Barker, M.J. (1991) Gastropods from the Upper Permian Zechstein (Cycle 1) reef of north-east England. *Proceedings of the Yorkshire Geological Society*, **48**, 347-65.

Hollingworth, N.T.J. and Pettigrew, T.H. (1988) *Zechstein Reef Fossils and their Palaeoecology*. Palaeontological Association Field Guide to Fossils No. 3, University Printing House, Oxford, 75 pp.

Hollingworth, N.T.J. and Tucker, M.E. (1987) The Upper Permian (Zechstein) Tunstall Reef of north-east England: palaeoecology and early diagenesis, in *The Zechstein Facies in Europe* (ed. T.M. Peryt), Springer-Verlag, Berlin, pp. 25-51.

Hollingworth, S.E. (1942) Correlation of gypsum-anhydrite deposits and the associated strata in the north of England. *Proceedings of the Geologists' Association*, **53**, 141-51.

Holmes, A. (1931) Concretionary and oolitic structures of the Permian rocks, in *Contributions to the Geology of Northumberland and Durham, Proceedings of the Geologists' Association*, **42**, 255-8.

Holtedahl, O. (1921) On the occurrence of structures like Walcott's Algonkian algae in the Permian of England. *American Journal of Science*, **5**, 195-206.

Howse, R. (1848) A catalogue of the fossils of the

Permian System of the counties of Northumberland and Durham. *Transactions of the Tyneside Naturalists' Field Club*, **1**, 219-64.

Howse, R. (1858) Notes on the Permian System of the counties of Northumberland and Durham. *Annals and Magazine of Natural History*, **19**, 304-12.

Howse, R. (1864) On the glaciation of the counties of Durham and Northumberland. *Transactions of the North of England Institute of Mining and Mechanical Engineers*, **13**, 169-85.

Howse, R. (1891) Note on the discovery in 1836-7 of a fossil fish (*Acrolepis kirkbyi n.* sp.) in the Upper Division of the Magnesian Limestone of Marsden. *Transactions of the Natural History Society of Northumberland and Durham*, **12**, 171-2.

Howse, R. and Kirkby, J.W. (1863) *A Synopsis of the Geology of Durham and Part of Northumberland*, Tyneside Naturalists' Field Club, Newcastle upon Tyne, 33 pp.

Hudson, R.G.S., Edwards, W., Tonks, L. and Versey, H.C. (1938) Summer field meeting to the Harrogate district, July 1938. *Proceedings of the Geologists' Association*, **49**, 353-72.

Jackson, D.I. and Mulholland, P. (1993) Tectonic and stratigraphic aspects of the East Irish Sea Basin and adjacent areas: contrasts in their post-Carboniferous structural styles, in *Petroleum Geology of North West Europe: Proceedings of the 4th Conference* (ed. J.R. Parker). Geological Society, London, Vol. 2, pp. 791-808.

Jackson, D.I., Mulholland, P., Jones, S.M. and Warrington, G. (1987) The geological framework of the East Irish Sea Basin, in *Petroleum Geology of North West Europe* (eds J. Brooks and K.W. Glennie), Graham and Trotman, London, pp. 191-203.

James, A.N., Cooper, A.H. and Holliday, D.W. (1981) Solution of the gypsum cliff (Permian, Middle Marl) by the River Ure at Ripon Parks, North Yorkshire. *Proceedings of the Yorkshire Geological Society*, **43**, 433-50.

Jones, K. (1969) Mineralogy and geochemistry of the Lower and Middle Magnesian Limestone of County Durham. Unpublished Ph.D. Thesis, University of Durham.

Jones, K. and Hirst, D.M. (1972) The distribution of barium, lead and zinc in the Lower and Middle Magnesian Limestone of County Durham, Great Britain. *Chemical Geology*, **10**, 223-36.

Kaldi, J. (1980) Aspects of the sedimentology of the Lower Magnesian Limestone (Permian) of eastern England. Unpublished Ph.D. Thesis, University of Cambridge.

Kaldi, J. (1986a) Sedimentology of sandwaves in an oolite shoal complex in the Cadeby Formation (Upper Permian) of eastern England, in *The English Zechstein and Related Topics* (eds G.M. Harwood and D.B. Smith), Geological Society of London, Special Publication No. 22, pp. 62-74.

Kaldi, J. (1986b) Diagenesis of nearshore carbonate rocks in the Sprotbrough Member of the Cadeby (Magnesian Limestone) Formation (Upper Permian) of eastern England, in *The English Zechstein and Related Topics* (eds G.M. Harwood and D.B. Smith), Geological Society of London, Special Publication No. 22, pp. 87-102.

Kendall, P.F. and Wroot, H.E. (1924) *Geology of Yorkshire*, Vol. 1, Vienna, 660pp.

King, W. (1848) *A Catalogue of the Organic Remains of the Permian Rocks of Northumberland and Durham*, Newcastle upon Tyne.

King, W. (1850) *A Monograph of the Permian Fossils of England*, Palaeontographical Society, London, 258 pp.

Kirkby, J.W. (1857) On some Permian fossils from Durham. *Quarterly Journal of the Geological Society of London*, **8**, 213-8 and plate 7.

Kirkby, J.W. (1858) On Permian Entomostraca from the Shell-Limestone of Durham. *Transactions of the Tyneside Naturalists' Field Club*, **4**, 122-71.

Kirkby, J.W. (1859) On the Permian Chitonidae. *Quarterly Journal of the Geological Society of London*, **15**, 607-26.

Kirkby, J.W. (1860) On the occurrence of 'sand-pipes' in the Magnesian Limestone of Durham. *The Geologist*, **3**, 293-8, 329-36.

Kirkby, J.W. (1861) On the Permian rocks of South Yorkshire; and on their palaeontological relations. *Quarterly Journal of the Geological Society of London*, **7**, 287-325.

Kirkby, J.W. (1863) Fossil fish in Magnesian Limestone at Fulwell Hill. *Transactions of the Tyneside Naturalists' Field Club*, **5**, 248.

Kirkby, J.W. (1864) On some remains of fish and plants from the 'Upper Limestone' of the Permian series of Durham. *Quarterly Journal of the Geological Society of London*, **20**, 345-58 (reprinted 1867 in the *Transactions of the Natural History Society of Northumberland, Durham and Newcastle upon Tyne*, **1**, 64-83).

Kirkby, J.W. (1867) On the fossils of the Marl Slate and Lower Magnesian Limestone. *Transactions of the Natural History Society of Northumbria*, **1**, 184-200.

Kirkby, J.W. (1870) Notes on the 'Geology' of Messrs Baker and Tate's New Flora of Northumberland and Durham. *Transactions of the Natural History Society of Northumberland, Durham and Newcastle upon Tyne*, **3**, 357-60.

Kitson, D.C. (1982) Stratigraphical relationships, morphology and diagenesis of the Hesleden Dene algal biostrome. Unpublished M.Sc. Thesis, University of Reading.

Land, D.H. and Smith, D.B. (1981) 1:10,560 Geological Map Sheet NZ 36 NE. Geological Survey of Great Britain, London.

Lebour, G.A. (1884) On the breccia-gashes of the Durham coast and some recent earth-shakes at Sunderland. *Transactions of the North of England Institute of Mining and Mechanical Engineers*, **33**, 165-77.

Lebour, G.A. (1902) The Marl Slate and Yellow Sands of Northumberland and Durham. *Transactions of the Institution of Mining Engineers*, **24**, 370-91.

Lee, M.R. (1990) The sedimentology and diagenesis of the Raisby Formation (EZ1 carbonate), northern England. Unpublished Ph.D. Thesis, University of Newcastle upon Tyne.

Lee, M.R. (1993) Formation and diagenesis of slope limestones within the Upper Permian (Zechstein) Raisby Formation, north-east England. *Proceedings of the Yorkshire Geological Society*, **49**, 215-27.

Lee, M.R. and Harwood, G.M. (1989) Dolomite calcitization and cement zonation related to uplift of the Raisby Formation (Zechstein carbonate), north-east England. *Sedimentary Geology*, **65**, 285-305.

Logan, A. (1962) A revision of the palaeontology of the Permian limestones of County Durham. Unpublished Ph.D. Thesis, University of Newcastle upon Tyne.

Logan, A. (1967) *The Permian Bivalvia of Northern England*. Monograph of the Palaeontographical Society, London, 72 pp + 12 plates.

Macchi, L. (1990) *A Field Guide to the Continental Permo-Triassic Rocks of Cumbria and North-west Cheshire*, Liverpool Geological Society, 88 pp.

Magraw, D. (1975) Permian [beds] of the offshore and coastal region of Durham and SE Northumberland. *Journal of the Geological Society of London*, **131**, 397-414.

Magraw, D., Clarke, A.M. and Smith, D.B. (1963) The stratigraphy and structure of the south-east Durham coalfield. *Proceedings of the Yorkshire Geological Society*, **34**, 153-208 + 2 plates.

Marley, J. (1892) On the Cleveland and south Durham salt industry. *Transactions of the Northern England Institute of Mining Engineers*, **39**, 91-125.

Meyer, H.O.A. (1965) Revision of the stratigraphy of the Permian evaporites and associated strata in north-western England. *Proceedings of the Yorkshire Geological Society*, **35**, 71-89.

Mitchell, G.H. (1932a) Notes on the Permian rocks of the Doncaster district. *Proceedings of the Yorkshire Geological Society*, **22**, 133-41.

Mitchell, G.H. (1932b) 1:10,560 Geological Map Sheet Yorkshire 284 SE. Geological Survey of Great Britain, London.

Mitchell, G.H., Stephens, J.V., Bromehead, C.E.N. and Wray, D.A. (1947) *Geology of the Country around Barnsley,* Memoir of the Geological Survey of Great Britain, Sheet 87.

Monty, Cl.L.V. (1973) Precambrian background and Phanerozoic history of stromatolitic communities; an overview. *Annales de la Société Geologique de Belgique*, **96**, 584-624.

Moss, M. (1986) The geochemistry and environmental evolution of the Hampole Beds at the type area of the Cadeby Formation (Lower Magnesian Limestone). *Mercian Geologist*, **10**, 115-25.

Murchison, R.I. and Harkness, R. (1864) On the Permian rocks of the north-west of England, and their extension into Scotland. *Quarterly Journal of the Geological Society of London*, **20**, 144-65.

Mussett, A.E., Dagley, P. and Skelhorn, R.R. (1988) Time and duration of activity in the British Tertiary Igneous Province, in *Early Tertiary Volcanism and the Opening of the NE Atlantic* (eds A.C. Morton and L.M. Parson), Geological Society of London, Special Publication, No. 39, 337-48.

Pattison, J. (1969) Some Permian foraminifera from north-western England. *Geological Magazine*, **106**, 197-205.

Pattison, J. (1970) A review of the marine fossils from the Upper Permian rocks of Northern Ireland and north-west England. *Bulletin of the Geological Survey of Great Britain*, No. **32**, 123-65.

Pattison, J. (1974) (Summary of Upper Permian faunas), in *A Correlation of the Permian Rocks in the British Isles* (D.B. Smith, R.G.W. Brunstrom, P.I. Manning, S. Simpson and F.W. Shotton), Geological Society of London, Special Report No. 5, pp. 11-12.

Pattison, J. (1977) Catalogue of the type, figured and cited specimens in the King Collection of Permian fossils. *Bulletin of the Geological Survey of Great Britain*, No. **62**, 33-44.

Pattison, J. (1978) Permian communities, in *Ecology of Fossils* (ed. W.S. McKerrow), Duckworth, London, pp. 187-93.

Pattison, J. (1981) Permian, in *Stratigraphical Atlas of Fossil Foraminifera,* 2nd edn (eds D.G. Jenkins and J.W. Murray), Ellis Harwood, Chichester, pp. 70-7.

Pattison, J., Smith, D.B. and Warrington, G. (1973) A review of late Permian and early Triassic biostratigraphy in the British Isles, in *The Permian and Triassic Systems and their Mutual Boundary* (eds A. Logan and L.V. Hills), Canadian Society of Petroleum Geologists, Memoir No. 2, Calgary, pp. 220-60.

Peryt, T.M. and Peryt, D. (1975) Association of sessile tubular foraminifera and cyanophytic algae. *Geological Magazine*, **112**, 612-14.

Peryt, T.M. and Piatkowski, T.S. (1976) Osady caliche w wapieniu cechsztynskim zachodniej czésci syneklizny perybaltyciej. *Kwartalnik Geologiczny*, **20**, 525-37.

Pettigrew, T.H. (1980) Geology, in *The Magnesian Limestone of Durham County* (ed. T.C. Dunn), Durham County Conservation Trust, pp. 4-26.

Pettigrew, T.H., Athersuch, J., Keen, M. and Wilkinson, I. (eds) (in press) *Biostratigraphical Atlas of British Ostracods*, Chapman and Hall, London.

Powell, J.H., Cooper, A.H. and Benfield, A.C. (1992) *Geology of the Country around Thirsk,* Memoir of the British Geological Survey, Sheet 52.

Pryor, W.A. (1971) Petrology of the Permian Yellow Sands of north-eastern England and their North Sea Basin equivalents. *Sedimentary Geology*, **6**, 221-54.

Richter-Bernburg, G. (1952) Excursion zu den Dolomitkalk und Gipsvorkommen des Sudwestharzes. *Sonderdruch aus der Zeitschrift der Deutschen Geologischen Gesellschaft*, **103**, 428-30.

Richter-Bernburg, G. (1982) Stratogenese des Zechsteinkalkes am Westharz. *Zeitschrift der Deutschen Geologischen Gesellschaft*, **133**, 381-401.

Robinson, J.E. (1978) Permian, in *A Stratigraphical Index of British Ostracoda* (eds R.H. Bate and J.E. Robinson), Seel House Press, Liverpool, pp. 47-96.

Schweitzer, H.-J. (1986) The land flora of the English and German Zechstein sequences, in *The English Zechstein and Related Topics* (eds G.M. Harwood and D.B. Smith), Geological Society of London, Special Publication No. 22, pp. 31-54.

Sedgwick, A. (1829) On the geological relations and internal structure of the Magnesian Limestone, and the lower portions of the New Red Sandstone in their range through Nottinghamshire, Derbyshire, Yorkshire and Durham, to the southern extremity of Northumberland. *Transactions of the Geological Society of London*, **3**, 37-124.

Sedgwick, A. (1836) On the New Red Sandstone Series in the basin of the Eden, and north-western coasts of Cumberland and Lancashire. *Transactions of the Geological Society of London*, **4**, 383-407.

Shearman, D.J. (1971) Discussion on paper by F.W. Beales and E.P. Onasick. *Transactions of the Institution of Mining and Metallurgy*, **80**, B50-2.

Smith, B. (1924) On the west Cumberland Brockram and its associated rocks. *Geological Magazine*, **61**, 289-308.

Smith, D.B. (1958) Some observations on the Magnesian Limestone reefs of north-eastern Durham. *Bulletin of the Geological Survey of Great Britain*, No. **15**, 71-84.

Smith, D.B. (1962) 1:10,560 Geological Map Sheet NZ 44 NW. Institute of Geological Sciences.

Smith, D.B. (1964) 1:10,560 Geological Map Sheet NZ 43 NE. Institute of Geological Sciences.

Smith, D.B. (1968) The Hampole Beds - a significant marker in the Lower Magnesian Limestone of Yorkshire, Derbyshire and Nottinghamshire. *Proceedings of the Yorkshire Geological Society*, **36**, 463-77.

Smith, D.B. (1969a) 1:10,560 Geological Map Sheet NZ 35 NE. Institute of Geological Sciences.

Smith, D.B. (1969b) Report on excursion to the Upper Permian rocks of Yorkshire. *Proceedings of the Yorkshire Geological Society*, **37**, 175-8.

Smith, D.B. (1970a) Permian and Trias, in *The Geology of Durham County* (compiler G.A.L.

Johnson), *Transactions of the Natural History Society of Northumberland, Durham and Newcastle upon Tyne*, **41**, 66-91.

Smith, D.B. (1970b) The palaeogeography of the British Zechstein, in *Third Symposium on Salt* (eds J.L. Rau and L.F. Dellwig), Northern Ohio Geological Society, Cleveland, pp. 20-3.

Smith, D.B. (1970c) Submarine slumping and sliding in the Lower Magnesian Limestone of Northumberland and Durham. *Proceedings of the Yorkshire Geological Society*, **38**, 1-36.

Smith, D.B. (1971a) *The Stratigraphy of the Upper Magnesian Limestone: a revision based on the Institute's Seaham Borehole*, Report 71/3, Institute of Geological Sciences, 12 pp.

Smith, D.B. (1971b) 1:10,560 Geological Map Sheet NZ 35 SE. Institute of Geological Sciences.

Smith, D.B. (1972) Foundered strata, collapse-breccias and subsidence features of the English Zechstein, in *Geology of Saline Deposits* (ed. G. Richter-Bernburg), UNESCO (Earth Sciences 7), pp. 255-69.

Smith, D.B. (1973a) The Permian in north-east Durham, in *The Durham Area* (compiler G.A.L. Johnson), Geologists' Association Guide No. 15, pp. 15-21.

Smith, D.B. (1973b) Discussion on paper by V.C. Kelley entitled 'Geometry and correlation along Permian Capitan Escarpment, New Mexico and Texas'. *Bulletin of the American Association of Petroleum Geologists*, **57**, 940-3.

Smith, D.B. (1974a) The stratigraphy and sedimentology of Permian rocks at outcrop in North Yorkshire. *Journal of Earth Sciences* (Leeds), **8**, 365-86.

Smith, D.B. (1974b) Permian, in *The Geology and Mineral Resources of Yorkshire* (eds D.H. Rayner and J.E. Hemingway), Yorkshire Geological Society, Occasional Publication No. 2, Leeds, pp. 115-44.

Smith, D.B. (1975a) 1:10,560 Geological Map Sheet NZ 46 SW. Institute of Geological Sciences.

Smith, D.B. (1975b) 1:10,560 Geological Map Sheet NZ 36 SE. Institute of Geological Sciences.

Smith, D.B. (1976) The Permian sabkha sequence at Quarry Moor, Ripon, Yorkshire. *Proceedings of the Yorkshire Geological Society*, **40**, 639-52.

Smith, D.B. (1979) Rapid marine transgressions of the Upper Permian Zechstein Sea. *Journal of the Geological Society of London*, **136**, 155-6.

Smith, D.B. (1980a) The evolution of the English Zechstein Basin, in *The Zechstein Basin with Emphasis on Carbonate Sequences* (eds H. Füchtbauer and T.M. Peryt). Contributions to Sedimentology, 9, Elsevier, Amsterdam, pp. 7-34.

Smith, D.B. (1980b) Permian and Triassic rocks, in *The Geology of North-east England* (ed. D.A. Robson), Special Publication of the Natural History Society of Northumbria, pp. 36-48.

Smith, D.B. (1980c) The shelf-edge reef of the Middle Magnesian Limestone (English Zechstein Cycle 1) of north-east England - summary, in *The Zechstein Basin with Emphasis on Carbonate Sequences* (eds H. Füchtbauer and T.M. Peryt), Contributions to Sedimentology, 9, Elsevier, Amsterdam, pp. 3-5.

Smith, D.B. (1981a) The Magnesian Limestone (Upper Permian) reef complex of north-eastern England, in *European Fossil Reef Models* (ed. D.F. Toomey), Special Publication No. 30, Society of Economic Palaeontologists and Mineralogists, pp. 161-86.

Smith, D.B. (1981b) Bryozoan-algal patch-reefs in the Upper Permian Lower Magnesian Limestone of Yorkshire, north-east England, in *European Fossil Reef Models* (ed. D.F. Toomey), Special Publication No. 30, Society of Economic Paleontologists and Mineralogists, pp. 187-202.

Smith, D.B. (1981c) The Quaternary geology of the Sunderland district, in *The Quaternary in Britain* (eds J.W. Neale and J. Flenley), Pergamon Press, Oxford, pp. 146-67.

Smith, D.B. (1981d) Account of field excursion to the Permian of Tyne and Wear. *Proceedings of the Yorkshire Geological Society*, **43**, 467-70.

Smith, D.B. (1984, January) Blackhall Rocks: a personal view. *Bulletin of the Durham County Conservation Trust*, 21-6.

Smith, D.B. (1985a) Gravitational movements in Zechstein carbonate rocks in north-east England, in *The Role of Evaporites in Hydrocarbon Exploration* (compiler J.C.M. Taylor), JAPEC course notes No. 39, London, F1-F12.

Smith, D.B. (1985b) Zechstein reefs and associated facies, in *The Role of Evaporites in Hydrocarbon Exploration* (compiler J.C.M. Taylor), JAPEC course notes No. 39, London, E1-E23.

Smith, D.B. (1986) The Trow Point Bed - a deposit of Upper Permian marine oncoids, peloids and columnar stromatolites in the Zechstein of NE England, in *The English Zechstein and Related Topics* (eds G.M. Harwood and D.B. Smith), Geological Society of London, Special Publication No. 22, pp. 113-25.

Smith, D.B. (1989) The late Permian palaeogeography of north-east England. *Proceedings of the Yorkshire Geological Society*, **47**, 285-312.

Smith, D.B. (1992) Permian, in *Geology of England and Wales* (eds P.McL.D. Duff and A.J. Smith), The Geological Society, London, pp. 275-305.

Smith, D.B. (1994) *The Geology of the Sunderland District*, Memoir of the British Geological Survey, Sheet 21.

Smith, D.B. (in press) Discussion on Paper by Dr D.W. Holliday entitled 'Geophysical log signatures in the Eden Shales (Permo-Triassic) of Cumbria and their regional significance'. *Proceedings of the Yorkshire Geological Society*, **50.**

Smith, D.B. and Francis, E.A. (1967) *The Geology of the Country between Durham and West Hartlepool*, Memoir of the Geological Survey of Great Britain, Sheet 27.

Smith, D.B., Brunstrom, R.G.W., Manning, P.I., Simpson, S. and Shotton, F.W. (1974) *A Correlation of the Permian Rocks in the British Isles.* Geological Society of London, Special Report No. 5, 45 pp.

Smith, D.B., Harwood, G.M., Pattison, J. and Pettigrew, T.H. (1986) A revised nomenclature for Upper Permian strata in eastern England, in *The English Zechstein and Related Topics* (eds G.M. Harwood and D.B. Smith), Geological Society of London, Special Publication No. 22, pp. 9-17.

Smith, D.B. and Taylor, J.C.M. (1989) A 'north-west passage' to the southern Zechstein Basin of the UK North Sea. *Proceedings of the Yorkshire Geological Society* **47**, 313-20.

Smith, D.B. and Taylor, J.C.M. (1992) Permian, in *Atlas of Palaeogeography and Lithofacies* (eds J.C.W. Cope, J.K. Ingham and P.F. Rawson), Geological Society Memoir 13, London, pp. 87-96.

Southwood, D.A. (1985) The taxonomy and palaeoecology of bryozoa from the Upper Permian Zechstein reef of NE England. Unpublished Ph.D. Thesis, University of Durham.

Stoneley, H.M.M. (1958) *The Upper Permian Flora of England*, Bulletin of the British Museum (Natural History), Geology 3, 295-337.

Swift, A. (1986) The conodont *Merrillina divergens* (Bender and Stoppel) from the Upper Permian of England, in *The English Zechstein and Related Topics* (eds G.M. Harwood and D.B. Smith), Geological Society of London. Special Publication No. 22, pp. 55-62.

Swift, A. and Aldridge, R.J. (1982) Conodonts from Upper Permian strata of Nottinghamshire and North Yorkshire. *Palaeontology*, **25**, 845-56.

Tarr, W.A. (1933) Origin of the concretionary structures of the Magnesian Limestone at Sunderland, England. *Journal of Geology*, **41**, 268-87.

Taylor, B.J. (1961) The stratigraphy of exploratory boreholes in the west Cumberland coalfield. *Bulletin of the Geological Survey of Great Britain*, No. **17**, 1-74.

Taylor, J.C.M. (1984) Late Permian-Zechstein, in *Introduction to the Petroleum Geology of the North Sea* (ed K.W. Glennie), Blackwell Scientific Publications, Oxford, pp. 61-83 (2nd edn, 1986, 87-111; 3rd edn 1990, 153-90).

Taylor, J.C.M. and Fong, G. (1969) Correlation of Upper Permian strata in east Yorkshire and Durham. *Nature, Physical Science*, **224**, 173-5.

Taylor, J.C.M. and Colter, V.S. (1975) Zechstein of the English sector of the Southern North Sea Basin, in *Petroleum and the Continental Shelf of North-west Europe. Volume 1, Geology* (ed. A.W. Woodland), Applied Science Publishers Ltd, Barking: Institute of Petroleum, Great Britain, pp. 249-63.

Taylor, P.D. (1980) *Stomatopora voightiana* (King, 1850): a cyclostome bryozoan from the Permian of County Durham. *Proceedings of the Yorkshire Geological Society*, **42**, 621-6.

Trechmann, C.T. (1913) On a mass of anhydrite in the Magnesian Limestone at Hartlepool, and on the Permian of south-eastern Durham. *Quarterly Journal of the Geological Society of London*, **69**, 184-218.

Trechmann, C.T. (1914) On the lithology and composition of Durham Magnesian Limestones. *Quarterly Journal of the Geological Society of London*, **70**, 232-65 + 2 plates.

Trechmann, C.T. (1925) The Permian formation in Durham. *Proceedings of the Geologists' Association*, **36**, 135-45.

Trechmann, C.T. (1931) The Permian, in *Contributions to the Geology of*

Northumberland and Durham. Proceedings of the Geologists' Association, **42**, 246-52.

Trechmann, C.T. (1945) On some new Permian fossils from the Magnesian Limestone near Sunderland. *Quarterly Journal of the Geological Society of London*, **100**, 333-54 + 1 plate.

Trechmann, C.T. (1954) Thrusting and other movements in the Durham Permian. *Geological Magazine*, **91**, 193-208.

Tucker, M.E. and Hollingworth, N.T.J. (1986) The Upper Permian Reef Complex (EZ1) of North East England: Diagenesis in a marine to estuarine setting, in *Reef Diagenesis* (eds J.H. Schroeder and B.H. Purser), Springer-Verlag, Berlin, Heidelberg, pp. 270-90.

Tute, J.S. (1868a) The geology of the country near Ripon. *Proceedings of the Yorkshire Geological and Polytechnic Society*, **4**, 555-65.

Tute, J.S. (1868b) On certain natural pits in the neighbourhood of Ripon. *Geological Magazine*, **5**, 178-9.

Tute, J.S. (1870) On certain natural pits in the neighbourhood of Ripon. *Proceedings of the Yorkshire Geological and Polytechnic Society*, **5**, 2-7.

Tute, J.S. (1884) On the sequence of Permian rocks near Ripon. *Proceedings of the Yorkshire Geological and Polytechnic Society*, **12**, 218-20 + 1 plate.

Winch, N.J. (1817) Observations on the geology of Northumberland and Durham. *Transactions of the Geological Society of London*, **4**, 1-101.

Woolacott, D. (1897) *Geology of north-east Durham*, Hills and Co., Sunderland, 31 pp.

Woolacott, D. (1900a) On the boulder clay, raised beaches and associated phenomena in the east of Durham. *Proceedings of the University of Durham Philosophical Society*, **1**, 247-58.

Woolacott, D. (1900b) On a portion of a raised beach on the Fulwell Hills, near Sunderland. *Transactions of the Natural History Society of Northumberland, Durham and Newcastle upon Tyne*, **8**, 165-71.

Woolacott, D. (1903) Explanation of the Claxheugh section. *Transactions of the Natural History Society of Northumberland, Durham and Newcastle upon Tyne*, **14**, 211-21.

Woolacott, D. (1905) The landslip at Claxheugh, County Durham, September 1905. *Transactions of the Natural History Society of Northumberland, Durham and Newcastle upon Tyne*, **1**, 434-6 + 2 plates.

Woolacott, D. (1909) A case of thrust and crush-brecciation in the Magnesian Limestone of County Durham. *Proceedings of the University of Durham Philosophical Society, Memoir No. 1*, 16 pp. + 2 plates.

Woolacott, D. (1912) The stratigraphy and tectonics of the Permian of Durham (northern area). *Proceedings of the University of Durham Philosophical Society*, **4**, 241-331.

Woolacott, D. (1918) On sections in the Lower Permian rocks at Claxheugh and Down Hill, County Durham. *Transactions of the Natural History Society of Northumberland, Durham and Newcastle upon Tyne*, **5**, 155-62 + 3 plates.

Woolacott, D. (1919a) The Magnesian Limestone of Durham. *Geological Magazine*, **6**, 452-65, 485-98.

Woolacott, D. (1919b) Borings at Cotefield Close and Sheraton, Durham. *Geological Magazine*, **6**, 163-70.

Versey, H.C. (1925) The beds underlying the Magnesian Limestone in Yorkshire. *Proceedings of the Yorkshire Geological Society*, **20**, 200-14.

Glossary

Aeolian: produced by, or borne by, the wind.

Alabastrine: **gypsum** of a very fine-grained massive nature, generally white in mass but may be tinted.

Allochthonous: refers to rock formed elsewhere and transported to place where now found.

Anhydrite: anhydrous calcium sulphate ($CaSO_4$).

Anoxic: lacking in oxygen.

Aphanitic: a rock in which the individual grains or crystals cannot be seen by the naked eye.

Arborescent: tree-like.

Authigenic: a mineral formed in place in a sediment or rock either by replacing an earlier mineral or by displacive growth.

Autochthonous: refers to rock formed in place where now found.

Azurite: copper carbonate ($Cu_3(CO_3)_2(OH)_2$).

Backreef or back-reef: the environment lying landward of a linear reef, especially a barrier reef; can include the landward margin of a linear reef.

Bafflestone: a term used in a refinement of the Dunham system of limestone classification to denote a rock in which a sparse population of **sessile benthic** organisms caused grains to be deposited by functioning as baffles and thereby reducing current velocity.

Barite (barytes): barium sulphate ($BaSO_4$).

Benthic: refers to the flora and fauna of the sea floor.

Bindstone: a term used in a refinement of the Dunham system of limestone classification to denote a rock (commonly laminated) in which the constituent grains were held together by encrusting organisms such as **cyanophytes**.

Bioclasts: whole or fragmented organic remains, generally transported, in a sediment or rock.

Biota: faunal and floral assemblage of a bed or other stratigraphical unit.

Botryoidal: a term used to describe a smoothly mammilar accretionary surface, commonly on the free side of an encrusting mineral, facing a cavity.

Boundstone: a term used in the Dunham system of limestone classification to denote a rock in which the primary grains or constituents were bound together during formation or deposition (e.g. as in an organic reef).

Brash: a litter of broken pieces of rock, commonly in thin soil on rock.

Breccia: a rock composed of angular fragments, generally of varied sizes, produced in a wide range of ways.

Brockram: a term used in Cumbria for a sedimentary **breccia** of Permo-Triassic age; commonly red or purple.

Calcarenite: limestone formed mainly of calcium carbonate fragments of sand size.

Calcirudite: limestone formed mainly of calcium carbonate fragments of gravel size.

Calcite: calcium carbonate ($CaCO_3$).

Celestite (celestine): strontium sulphate ($SrSO_4$).

Chalcedony: a cryptocrystalline variety of silica (SiO_2), consisting essentially of fibrous or ultrafine quartz, some opal, together with water trapped in its structure.

Chalcocite: copper sulphide (Cu_2S).

Chert: cryptocrystalline silica (SiO_2) which may be of organic or inorganic origin, occurring as layers or nodules in sedimentary rocks (mainly limestones).

Chronostratigraphy: system of dividing up the geological column into convenient portions of time, leading to age classification of rocks according to hierarchal groupings of Systems, Series, Stages and Sub-stages.

Concretion: a hard, subspherical, discoidal or irregular mass or aggregate of mineral matter, generally formed by orderly and localized concentration from aqueous solution in the pores of a sedimentary rock.

Coquina: as **calcirudite**, but with most fragments being **bioclasts**.

Cyanophytic: related to microbes, especially blue-green algae, and the part they play in the creation of some laminated carbonate rocks.

Décollement (plane of): a surface separating rigid rock (below) from overlying, more plastic strata that have been detached and folded.

Dedolomite: a rock that previously has been composed of **dolomite** but which is now limestone.

Diachronous: a term used to describe a continuous rock body that is of different age in different places.

Diagenesis: the mainly physiochemical processes affecting sediments and sedimentary rocks between and including burial and re-emergence, but excluding metamorphism according to some authors.

Discontinuity: a break within a rock sequence indicating a cessation of deposition at the time of formation.

Dolomicrite: a **dolomite** rock composed of mud- to silt-size particles or crystals of dolomite.

Dolomite: (a) a mineral, carbonate of calcium and magnesium ($CaMg(CO_3)_2$) or (b) a rock composed mainly of the mineral dolomite; dolomite-rock.

Evaporite: a sedimentary rock composed mainly of minerals produced by chemical precipitation from a saline solution that became concentrated by evaporation of the solvent.

Fasciculate: as in a bundle of parallel rods.

Fenestral fabric: a texture characterized by very abundant primary or penecontemporaneous unsupported elongate cavities in a sediment or rock, generally carbonate; it may be open or filled with secondarily introduced sediment or minerals, commonly calcite or anhydrite.

Flaser: a sedimentary structure consisting of silt lenticles that are commonly aligned and usually cross-bedded.

Flowstone: a variety of **travertine** that coats existing surfaces (including the walls of caves and fissures) with laminar fine-grained deposits (generally calcium carbonate) precipitated from solution by trickling or slow-flowing mineral-rich water.

Foundering: the subsidence or collapse of strata overlying a sediment or rock that is undergoing dissolution.

Framestone: a term used in a refinement of the Dunham system of limestone classification to denote a variety of **boundstone** in which sessile skeletal organisms such as bryozoans construct a rigid or semi-rigid grain-trapping open framework.

Galena: lead sulphide (PbS).

Geode: a roughly equidimensional cavity up to a few centimetres across, in a rock; commonly lined with **botryoidal** deposits and/or inward-projecting crystals. Also called a vugh or vug.

Grainstone: a term used in the Dunham system of limestone classification to denote a carbonate rock composed of sand-sized grains in mutual contact and with no carbonate mud matrix.

Grapestone: a carbonate rock composed of grape-like clusters of silt-sized carbonate grains or crystals.

Gypsum: hydrated calcium sulphate ($CaSO_4.2H_2O$)

Halite: crystalline sodium chloride (rock-salt) (NaCl).

Infauna: the assemblage of fossil remains of organisms that lived below the sea floor, especially in sediments but also including some boring organisms.

Kaolinite: a clay mineral ($A1_4Si_4O_{10}(OH)_8$) of the kaolin group.

Lamellar drapes: thin layers of sediment, commonly laminated, that conform to substrate irregularities such as ripple marks.

Liesegang banding, rings: roughly concentric secondary rings or fronts caused in a sediment or rock by the rhythmic precipitation of pigmented minerals (commonly iron oxides) by groundwater.

Lithostratigraphy: the description, definition and naming of rock units. Units are named according to their perceived rank in a formal hierarchy, namely Supergroup, Group, Formation, Member and Bed.

Malachite: copper carbonate ($Cu_2CO_3(OH)_2$).

Mammilar: as **botryoidal**.

Marl: a loosely-used term properly applied to a calcareous clay but widely misapplied in geology to describe a thick-bedded claystone or mudstone, whether calcareous or not.

Micrite: a limestone composed of microcrystalline calcite.

Microspar: a mosaic of crystals of any mineral in the 4–50 micron range; commonly applied to calcite and dolomite in the context used here.

Monomict: refers to a **breccia**/conglomerate composed of clasts of a single rock type, generally locally derived and accumulated.

Mucilage: a layer or mass of organic matter, commonly coating the shells of marine organisms and some grains such as **ooids**.

Muscovite: the commonest form of white mica; a silicate of aluminium and potassium, with hydroxyl and fluorine ($KAl_2(AlSi_3)O_{10}(OH,F)_2$).

Mylonite: a roughly laminated finely fragmental rock created at the mutual contact of two rock-masses that have been moved forcefully against each other.

Olistolith: a large coherent mass of rock that has been transported down a submarine slope by gravity sliding, and which forms part of a body of rock ('olistostrome') composed of similar masses in a varied fragmental matrix.

Oncoid or oncolith: a pisoid or pisolith of algal origin (= a subspherical algal **stromatolite**).

Ooid or oolith: a subspherical grain of sand-size, with or without a nucleus and with at least two concentric layers of roughly uniform thickness. Generally used to describe calcium carbonate grains but can be composed of other minerals.

Packstone: a term used in the Dunham system of limestone classification to denote a rock in which constituent grains in point-contact have mud-size carbonate grains in the interstices.

Palaeosol: a fossil soil.

Patch-reef: an isolated body of **autochthonous** reef-rock, generally 10–50 m across and 3–10 m thick in the sense used in this book.

Pellicle: a thin resistant coating on a grain of any size.

Pelloid: a sand-sized to granule-sized grain of finely crystalline carbonate of any origin, including pellets and **ooids** (or ooliths).

Peritidal: within or close to the tidal range; slightly broader than 'intertidal'.

Pinnate: leaf-like, with a central stalk.

Polyzoan: bryozoan.

Proximal turbidite: an obsolescent term used to describe a rock comprising an accumulation of coarse debris near the upslope limit of a submarine slump or slide. Now being replaced by 'debris flow'.

Pycnocline: a plane or thin transitional zone separating a dense lower layer in a density-stratified water body, from a less dense upper layer.

Pyrite: crystalline iron sulphide (FeS_2).

Ramose: a term used for a fossil bryozoan or other sessile benthic organism with thin twiggy branches. Dendritic.

Recessive: forming a step-back or cleft in a cliff profile.

Reef crest: the junction between the basinward side of a reef flat and the top of the basinward reef slope (or reef wall, reef face).

Reef slope: the basinward slope (wall, face) of a shelf-edge or barrier reef.

Regression, marine: withdrawal of the sea from a large area of land.

Reticulate: having a net-like, equidimensional structure produced by rod-like frame elements crossing at right-angles and outlining square spaces or interstices.

Sabkha: a broad, very gently-sloping arid alluvial plain, generally understood to border a tropical or sub-tropical sea or lake and to have a high water table.

Saccharoidal: sugar-like, used to describe a carbonate rock formed of calcite or dolomite crystals of sand size.

Saccolith: a sack-shaped and sack-sized mass in a reef, thought to be a single colony of frame-building organisms such as bryozoans.

Scalenohedron: a crystal shape, essentially a twinned form of rhombohedra, especially in calcite, in which the twin plane is the basal pinacoid 0001.

Sessile: attached, applied to an organism that remains in one place during adult life.

Siliciclastic: a sediment or sedimentary rock comprising a high proportion of silica-rich grains or clasts.

Slickensides: parallel striations or scratches on the faces of a movement plane.

Speleothem: (= dripstone). A secondary mineral deposit, generally of calcium carbonate, formed in caves by deposition from saturated groundwater.

Sphalerite: zinc sulphide (ZnS).

Stellate: an aggregate of crystals in a star-like arrangement.

Stromatolite: a variously shaped (commonly domal) laminated, generally calcareous sedimentary structure, now mainly formed in a shallow-water, tropical environment under the influence of a mat or assemblage of sediment-binding blue-green algae (**cyanophytes**).

Stylolite: an irregular interpenetrant suture-like boundary, mainly in carbonate rocks, which is caused by pressure-dissolution; can lie at any angle relative to the bedding.

Sucrosic: a granular or crystalline texture resembling that of sugar.

Talus: an accumulation of rock litter at the foot of a slope, generally with a wide size-range (up to several metres) and ungraded; commonly used to denote debris shed from the high part of a reef slope and transported basinward by gravity ('reef talus', 'talus apron').

Transgression, marine: the invasion of a large area of land by the sea.

Travertine: see **flowstone**. Use of term broadened by some to include deposits of silica or other mineral formed in a similar manner.

Trepostome: an organism belonging to an extinct order of bryozoan.

Vor-riff, vorreef: an accumulation of debris near the basinward margin of a reef.

Wackestone: a term used in the Dunham system of limestone classification to denote a rock mainly of carbonate mud that contains more than 10%, but less than 50% of coarser clasts.

Index

Page numbers in **bold** type refer to figures and page numbers in *italic* type refer to tables.

Index

Index

Index

Index